AF876085

GCSE OCR
Mathematics

GCSE Maths is a big subject, but that's OK — this is a big book. Better still, it's packed with absolutely everything you'll need for the OCR Higher Level exams.

For starters... clear study notes, helpful examples, brilliant practice questions and realistic exam papers. Plus, we've included a free Online Edition of the whole book, digital quizzes and video solutions from CGP's Maths experts. Crikey.

And since we're the UK's No.1, you can be 100% sure this is the best book around.

Unlock your Digital Extras

Just scan a QR code in the book to get your digital extras. Or go to **cgpbooks.co.uk/extras** and enter this code to unlock your Online Edition!

2802 5040 9065 1762

Online Edition

By the way, this code only works for one person. If somebody else has used this book before you, they might have already claimed the code.

Complete
Revision & Practice

Includes Free Online Edition, Video Solutions & Digital Quizzes!

Contents

Throughout this book you'll see **grade stamps** like these: 9 GRADE 5 GRADE 9 GRADE 5 GRADE
You can use these to focus your revision on easier or harder work.
But remember — to get a top grade you have to know **everything**, not just the hardest topics.

Section Five — Geometry and Measures

Section Six — Pythagoras and Trigonometry

Section Seven — Probability and Statistics

This book includes digital extras — just scan the QR code to access them! You'll find **video solutions**, loads of **online quizzes** and **answers** to the Exam Practice Questions.

P.S. You can also find these extras at **cgpbooks.co.uk/mountain**

Online Extras

Published by CGP

From original material by Richard Parsons.
Updated by: Sean McParland and Maddie Wright

Contributor: Alastair Duncombe

With thanks to Jake McGuffie for the proofreading.
With thanks to Alice Dent for the copyright research.

ISBN: 978 1 83774 115 1

Printed by Elanders Ltd, Newcastle upon Tyne.
Clipart from Corel®

Types of Number and BODMAS

Welcome to the GCSE that never goes out of fashion — Maths. To start you off, here are some handy definitions of different types of number, and a bit about what order to do things in.

Integers:

GRADE 3

You need to make sure you know the meaning of this word — it'll come up all the time in GCSE Maths. An integer is another name for a whole number — either a positive or negative number, or zero.

Examples

Integers: –365, 0, 1, 17, 989, 1 234 567 890

Not integers: 0.5, $\frac{2}{3}$, $\sqrt{7}$, $13\frac{3}{4}$, –1000.1, 66.66, π

All Numbers are Either **Rational** or **Irrational**

GRADE 5

Rational numbers can be written as fractions. Most numbers you deal with are rational.

Rational numbers come in 3 different forms:

1) Integers e.g. $4\ (=\frac{4}{1})$, $-5\ (=\frac{-5}{1})$, $-12\ (=\frac{-12}{1})$
2) Fractions p/q, where p and q are (non-zero) integers, e.g. $\frac{1}{4}$, $\frac{-1}{2}$, $\frac{7}{4}$
3) Terminating or recurring decimals e.g. $0.125\ (=\frac{1}{8})$, $0.33333333...\ (=\frac{1}{3})$, $0.143143143...\ (=\frac{143}{999})$

- Irrational numbers are messy. They can't be written as fractions — they're never-ending, non-repeating decimals.
- Square roots of positive integers are either integers or irrational (e.g. $\sqrt{2}$ is irrational, but $\sqrt{4} = 2$ isn't).
- Surds (see p.29) are numbers or expressions containing irrational roots. π is also irrational.

BODMAS

Brackets, Other, Division, Multiplication, Addition, Subtraction

GRADE 3

BODMAS tells you the ORDER in which these operations should be done:
Work out Brackets first, then Other things like squaring, then Divide / Multiply groups of numbers before Adding or Subtracting them.

You can use BODMAS when it's not clear what to do next, or if there's more than one thing you could do.

EXAMPLE **Find the reciprocal of $\sqrt{4+6\times(12-2)}$.**

$\sqrt{4+6\times(12-2)} = \sqrt{4+6\times10}$ — It's not obvious what to do inside the square root — so use BODMAS. Brackets first...

$= \sqrt{4+60}$ — ... then multiply...

$= \sqrt{64}$ — ... then add.

$= 8$ — Take the square root

The reciprocal of 8 is $\frac{1}{8}$. — Finally, take the reciprocal (the reciprocal of a number is just 1 ÷ the number).

An integer can be positive, negative or zero

It's really important to check your working on BODMAS questions. Try this question and see how you do.

Q1 Without using a calculator, find the value of 3 + 22 × 3 – 14. [2 marks] GRADE 3

Multiples, Factors and Prime Factors

You need to know what multiples, factors, primes and prime factors are — and how to find them.

Multiples and Factors

The MULTIPLES of a number are just its times table.

EXAMPLE **Find the first 8 multiples of 13.**

You just need to find the first 8 numbers in the 13 times table:
13 26 39 52 65 78 91 104

The FACTORS of a number are all the numbers that divide into it.

There's a method that guarantees you'll find them all:

1) Start off with 1 × the number itself, then try 2 ×, then 3 × and so on, listing the pairs in rows.
2) Try each one in turn. Cross out the row if it doesn't divide exactly.
3) Eventually, when you get a number repeated, stop.
4) The numbers in the rows you haven't crossed out make up the list of factors.

EXAMPLE **Find all the factors of 24.**

Increasing by 1 each time

1 × 24
2 × 12
3 × 8
4 × 6
~~5 × —~~
6 × 4

So the factors of 24 are:
1, 2, 3, 4, 6, 8, 12, 24

Prime Numbers:

2 3 5 7 11 13 17 19 23 29 31 37 41 43...

A prime number is a number which doesn't divide by anything, apart from itself and 1 — i.e. its only factors are itself and 1. (The only exception is 1, which is NOT a prime number.)

Finding **Prime Factors** — The **Factor Tree**

Any number can be broken down into a string of prime factors all multiplied together — this is called 'prime factor decomposition' or 'prime factorisation'.

EXAMPLE **Express 420 as a product of prime factors.**

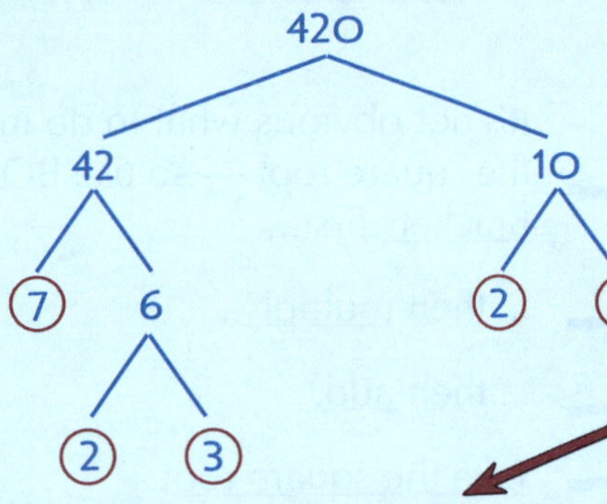

So $420 = 2 \times 2 \times 3 \times 5 \times 7$
$= 2^2 \times 3 \times 5 \times 7$

To write a number as a product of its prime factors, use the Factor Tree method:

1) Start with the number at the top, and split it into factors as shown.
2) Every time you get a prime, ring it.
3) Keep going until you can't go further (i.e. you're just left with primes), then write the primes out in order. If there's more than one of the same factor, you can write them as powers.

No matter which numbers you choose at each step, you'll find that the prime factorisation is exactly the same. Each number has a unique set of prime factors.

Factors and multiples are easy marks

Learn the Factor Tree method for finding prime numbers inside out, then try this Practice Question...

Q1 Express 990 as a product of its prime factors. [2 marks]

Q1 Video Solution

LCM and HCF

*You'll need to know about **multiples** and **factors** from the previous page before you have a go at this one...*

LCM — 'Least Common Multiple'

The SMALLEST number that will DIVIDE BY ALL the numbers in question.

If you're given two numbers and asked to find their LCM, just LIST the MULTIPLES of BOTH numbers and find the SMALLEST one that's in BOTH lists.

So, to find the LCM of 12 and 15, list their multiples (multiples of 12 = 12, 24, 36, 48, 60, 72, ... and multiples of 15 = 15, 30, 45, 60, 75, ...) and find the smallest one that's in both lists — so LCM = 60.

However, if you already know the prime factors of the numbers, you can use this method instead:

1) List all the PRIME FACTORS that appear in EITHER number.
2) If a factor appears MORE THAN ONCE in one of the numbers, list it THAT MANY TIMES.
3) MULTIPLY these together to give the LCM.

EXAMPLE **$18 = 2 \times 3^2$ and $30 = 2 \times 3 \times 5$. Find the LCM of 18 and 30.**

$18 = 2 \times 3 \times 3$ $\quad$ $30 = 2 \times 3 \times 5$

So the prime factors that appear in either number are: 2, 3, 3, 5 — List 3 twice as it appears twice in 18.

$LCM = 2 \times 3 \times 3 \times 5 = 90$

HCF — 'Highest Common Factor'

The BIGGEST number that will DIVIDE INTO ALL the numbers in question.

If you're given two numbers and asked to find their HCF, just LIST the FACTORS of BOTH numbers and find the BIGGEST one that's in BOTH lists.

Take care listing the factors — make sure you use the proper method (as shown on the previous page).

So, to find the HCF of 36 and 54, list their factors (factors of 36 = 1, 2, 3, 4, 6, 9, 12, 18 and 36 and factors of 54 = 1, 2, 3, 6, 9, 18, 27 and 54) and find the biggest one that's in both lists — so HCF = 18.

Again, there's a different method you can use if you already know the prime factors of the numbers:

1) List all the PRIME FACTORS that appear in BOTH numbers.
2) MULTIPLY these together to find the HCF.

EXAMPLE **$180 = 2^2 \times 3^2 \times 5$ and $84 = 2^2 \times 3 \times 7$. Use this to find the HCF of 180 and 84.**

$180 = (2) \times (2) \times (3) \times 3 \times 5$ $\quad$ $84 = (2) \times (2) \times (3) \times 7$

2, 2 and 3 are prime factors of both numbers, so $HCF = 2 \times 2 \times 3 = 12$

Don't be put off by the fancy names

Method 1 is much simpler in both cases, but make sure you learn Method 2 as well — just in case the exam question specifically tells you to use the prime factors or the numbers are really big.

Q1 Video Solution

Q1 a) Find the lowest common multiple (LCM) of 9 and 12. [2 marks]
b) Given that $28 = 2^2 \times 7$ and $8 = 2^3$, find the LCM of 28 and 8. [2 marks]

Q2 a) Find the highest common factor (HCF) of 36 and 84.
b) Given that $150 = 2 \times 3 \times 5^2$ and $60 = 2^2 \times 3 \times 5$, find the HCF of 150 and 60. [3 marks]

Warm-Up and Worked Exam Questions

This stuff is pretty straightforward, but that doesn't mean you can get away without learning the facts and practising the questions. You should have learnt the facts already — try these and we'll see.

Warm-Up Questions

1) Choose from the numbers 1, 2.3, 3.2312, 10, –4, 7, $\sqrt{2}$, –5.1, $\frac{2}{7}$, 6π:
   Which numbers are a) integers? b) irrational?
2) Find all the factors of 40.
3) Write down the multiples of 17 between 20 and 70.
4) Explain why 231 is not a prime number.
5) Write 40 as a product of its prime factors.
6) a) Find the lowest common multiple (LCM) of 6 and 14.
   b) Find the highest common factor (HCF) of 18 and 42.

Worked Exam Questions

Take a look at these worked exam questions. They're not too hard, but they should give you a good idea of what to write. Make the most of the handy hints now — they won't be there in the exam.

1 Use your calculator to work out $\sqrt{\dfrac{12.71 + 137.936}{\cos 50° \times 13.2^2}}$ (Grade 4)
Give your answer to 2 decimal places.

Use BODMAS.

$$\sqrt{\frac{12.71 + 137.936}{\cos 50° \times 13.2^2}} = \sqrt{\frac{150.646}{0.642787609... \times 174.24}}$$

$$= \sqrt{1.34506182...}$$

$$= 1.1597680...$$

1.16
[2 marks]

2 Express: (Grade 4)

a) 210 as a product of its prime factors.

Draw a factor tree...

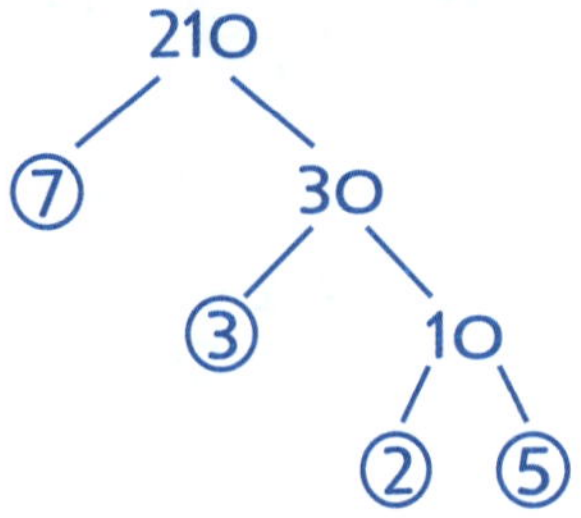

$210 = 2 \times 3 \times 5 \times 7$
[2 marks]

b) 105^2 as a product of its prime factors.

Draw a factor tree...

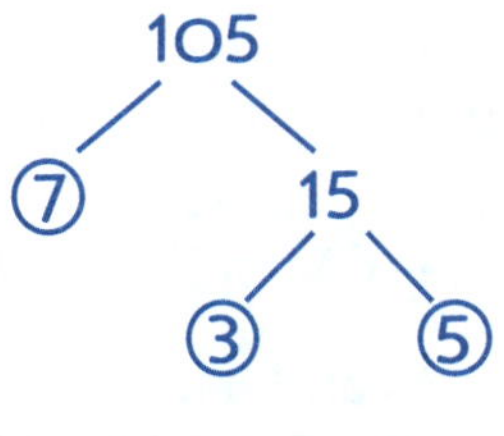

You could also write this as $105^2 = 3^2 \times 5^2 \times 7^2$

$105^2 = 3 \times 3 \times 5 \times 5 \times 7 \times 7$
[2 marks]

Exam Questions

3 $P = 3^7 \times 11^2$ and $Q = 3^4 \times 7^3 \times 11$. (Grade 4)

Write as the product of prime factors:

a) the LCM of P and Q,

....
[1 mark]

b) the HCF of P and Q.

....
[1 mark]

4 x and y are integers and $0 < x < y$.
Write down two sets of values for x and y such that $6 = \sqrt{3x + 2y}$. (Grade 5)

$x =$ .... , $y =$ ....

or $x =$ .... , $y =$ ....
[2 marks]

5 A number, x, is a common multiple of 6 and 7, and a common factor of 252 and 420. Given that $50 < x < 150$, find the value of x. (Grade 6)

$x =$ ....
[4 marks]

6 A and B are different prime numbers. Find the LCM of A and B. (Grade 6)

....
[2 marks]

Fractions

*These pages show you how to cope with fraction calculations without your **calculator**.*

1) Cancelling down

GRADE 3

To cancel down or simplify a fraction, divide top and bottom by the same number till they won't go further:

EXAMPLE **Simplify $\frac{18}{24}$.**

Cancel down in a series of easy steps — keep going till the top and bottom don't have any common factors.

$$\frac{18}{24} = \frac{6}{8} = \frac{3}{4}$$

(÷3 then ÷2, top and bottom)

The number on the top of the fraction is the numerator, and the number on the bottom is the denominator.

2) Mixed numbers

GRADE 3

Mixed numbers are things like $3\frac{1}{3}$, with an integer part and a fraction part. Improper fractions are ones where the top number is larger than the bottom number. You need to be able to convert between the two.

EXAMPLES **1. Write $4\frac{2}{3}$ as an improper fraction.**

1) Think of the mixed number as an addition:

$$4\frac{2}{3} = 4 + \frac{2}{3}$$

2) Turn the integer part into a fraction:

$$4 + \frac{2}{3} = \frac{12}{3} + \frac{2}{3} = \frac{12+2}{3} = \frac{14}{3}$$

2. Write $\frac{31}{4}$ as a mixed number.

Divide the top number by the bottom.

1) The answer gives the whole number part.

2) The remainder goes on top of the fraction.

$31 \div 4 = 7$ remainder 3 so $\frac{31}{4} = 7\frac{3}{4}$

3) Multiplying

GRADE 3

Multiply top and bottom separately. It usually helps to cancel down first if you can.

EXAMPLE **Find $\frac{8}{15} \times \frac{5}{12}$.**

Cancel down by dividing top and bottom by any common factors you find in either fraction:

8 and 12 both divide by 4

$$\frac{8}{15} \times \frac{5}{12} = \frac{2}{15} \times \frac{5}{3}$$

15 and 5 both divide by 5

Now multiply the top and bottom numbers separately:

$$= \frac{2}{3} \times \frac{1}{3} = \frac{2 \times 1}{3 \times 3} = \frac{2}{9}$$

Fractions

Here are some more tricks for dealing with fractions.

4) Dividing

GRADE 3

Turn the 2nd fraction UPSIDE DOWN and then multiply:

EXAMPLE **Find $2\frac{1}{3} \div 3\frac{1}{2}$.**

Rewrite the mixed numbers as fractions: $2\frac{1}{3} \div 3\frac{1}{2} = \frac{7}{3} \div \frac{7}{2}$

Turn $\frac{7}{2}$ upside down and multiply: $= \frac{7}{3} \times \frac{2}{7}$

Simplify by cancelling the 7s: $= \frac{1}{3} \times \frac{2}{1} = \frac{2}{3}$

When you're multiplying or dividing with mixed numbers, always turn them into improper fractions first.

5) Common denominators

GRADE 4

This comes in handy for ordering fractions by size, and for adding or subtracting fractions.

You need to find a number that all the denominators divide into — this will be your common denominator. The simplest way is to find the lowest common multiple of the denominators:

EXAMPLE **Put these fractions in ascending order of size:**

$$\frac{8}{3} \quad \frac{5}{4} \quad \frac{12}{5}$$

The LCM of 3, 4 and 5 is 60, so make 60 the common denominator:

$$\frac{8}{3} = \frac{160}{60} \ (\times 20) \qquad \frac{5}{4} = \frac{75}{60} \ (\times 15) \qquad \frac{12}{5} = \frac{144}{60} \ (\times 12)$$

So the correct order is $\frac{75}{60}, \frac{144}{60}, \frac{160}{60}$ i.e. $\frac{5}{4}, \frac{12}{5}, \frac{8}{3}$

Don't forget to use the original fractions in the final answer.

Don't be put off by mixed numbers

You can easily turn mixed numbers into 'normal' fractions — this is particularly important when you are multiplying or dividing. Finding a common denominator will often come in handy when you are adding and subtracting fractions — if you need a reminder on how to find the LCM, flick back to p.3.

Fractions

6) Adding, subtracting — sort the denominators first

GRADE 4

If you're adding or subtracting mixed numbers, it usually helps to convert them to improper fractions first.

1) Make sure the denominators are the same (see previous page).

2) Add (or subtract) the top lines (numerators) only.

EXAMPLE **Calculate $2\frac{1}{5} - 1\frac{1}{2}$.**

Rewrite the mixed numbers as fractions: $2\frac{1}{5} - 1\frac{1}{2} = \frac{11}{5} - \frac{3}{2}$

Find a common denominator: $= \frac{22}{10} - \frac{15}{10}$

Combine the top lines: $= \frac{22-15}{10} = \frac{7}{10}$

People usually find adding and subtracting fractions harder than multiplying and dividing — but it's actually pretty easy as long as you make sure the denominators are the same.

7) Fractions of something

GRADE 3

Multiply the 'something' by the TOP of the fraction, and divide it by the BOTTOM.

EXAMPLE **What is $\frac{9}{20}$ of £360?**

Start by dividing by 20 — that's easiest: $\frac{9}{20}$ of £360 = (£360 ÷ 20) × 9
= £18 × 9 = £162

It doesn't matter which order you do those two steps in — just start with whatever's easiest.

8) Expressing as a Fraction

GRADE 3

EXAMPLE **Write 180 as a fraction of 80.**

Just write the first number over the second and cancel down. $\frac{180}{80} = \frac{9}{4}$

You have to learn to handle fractions in these 8 situations

When you think you've learnt all this, try all of these Exam Practice Questions without a calculator.

Q1 Calculate: a) $\frac{3}{8} \times 1\frac{5}{12}$ [3 marks] b) $1\frac{7}{9} \div 2\frac{2}{3}$ [3 marks]

c) $4\frac{1}{9} + 2\frac{2}{27}$ [3 marks] d) $5\frac{2}{3} - 9\frac{1}{4}$ [3 marks]

GRADE 3

Q2 Dean has made 30 sandwiches. $\frac{7}{15}$ of the sandwiches he has made are vegetarian, and $\frac{3}{7}$ of the vegetarian sandwiches are cheese sandwiches. How many cheese sandwiches has he made? [2 marks]

GRADE 3

Q2 Video Solution

Fractions, Decimals and Percentages

The one word that describes all these three is PROPORTION. Fractions, decimals and percentages are simply three different ways of expressing a proportion of something — and it's pretty important you should see them as closely related and completely interchangeable with each other.

These tables show the really common conversions which you should know straight off without having to work them out:

Fraction	Decimal	Percentage
$\frac{1}{2}$	0.5	50%
$\frac{1}{4}$	0.25	25%
$\frac{3}{4}$	0.75	75%
$\frac{1}{3}$	0.333333...	$33\frac{1}{3}\%$
$\frac{2}{3}$	0.666666...	$66\frac{2}{3}\%$
$\frac{1}{10}$	0.1	10%
$\frac{2}{10}$	0.2	20%

Fraction	Decimal	Percentage
$\frac{1}{5}$	0.2	20%
$\frac{2}{5}$	0.4	40%
$\frac{1}{8}$	0.125	12.5%
$\frac{3}{8}$	0.375	37.5%
$\frac{5}{2}$	2.5	250%
$\frac{7}{2}$	3.5	350%
$\frac{9}{4}$	2.25	225%

The more of those conversions you learn, the better — but for those that you don't know, you must also learn how to convert between the three types. These are the methods:

Fraction —Divide→ **Decimal** —× by 100→ **Percentage**

E.g. $\frac{7}{20}$ is 7÷20 = 0.35 e.g. 0.35 × 100 = 35%

Fraction ←The awkward one— **Decimal** ←÷ by 100— **Percentage**

Converting decimals to fractions is awkward, because it's different for different types of decimal. There are two different methods you need to learn:

1) Terminating decimals to fractions — this is fairly easy. The digits after the decimal point go on the top, and a power of 10 on the bottom — with the same number of zeros as there were decimal places.

$0.6 = \frac{6}{10}$ $\quad 0.3 = \frac{3}{10}$ $\quad 0.7 = \frac{7}{10}$ etc.

$0.12 = \frac{12}{100}$ $\quad 0.78 = \frac{78}{100}$ $\quad 0.05 = \frac{5}{100}$ etc.

$0.345 = \frac{345}{1000}$ $\quad 0.908 = \frac{908}{1000}$ $\quad 0.024 = \frac{24}{1000}$ etc.

These can often be cancelled down — see p.6.

2) Recurring decimals to fractions — this is trickier. See next page...

Fractions, decimals and percentages are interchangeable

Learn the top tables and the 4 conversion processes. Then have a go at the Practice Questions below...

Q1 Turn the following decimals into fractions and reduce them to their simplest form.
a) 0.4 b) 0.02 c) 0.77 d) 0.555 e) 5.6 [5 marks]

Q2 Which is greater: a) 57% or $\frac{5}{9}$, b) 0.2 or $\frac{6}{25}$, c) $\frac{7}{8}$ or 90%? [3 marks]

Fractions and Recurring Decimals

Recurring *and* ***terminating*** *decimals can always be written as fractions.*

Recurring or Terminating...

1) Recurring decimals have a pattern of numbers which repeats forever, e.g. $\frac{1}{3}$ is the decimal 0.333333...
   Note, it doesn't have to be a single digit that repeats. You could have, for instance: 0.143143143...
2) The repeating part is usually marked with dots or a bar on top of the number. If there's one dot, then only one digit is repeated. If there are two dots, then everything from the first dot to the second dot is the repeating bit. E.g. $0.2\dot{5} = 0.2555555...$, $0.\dot{2}\dot{5} = 0.25252525...$, $0.\dot{2}5\dot{5} = 0.255255255...$
3) Terminating decimals are finite (they come to an end), e.g. $\frac{1}{20}$ is the decimal 0.05.

The denominator (bottom number) of a fraction in its simplest form tells you if it converts to a recurring or terminating decimal. Fractions where the denominator has prime factors of only 2 or 5 will give terminating decimals. All other fractions will give recurring decimals.

	Only prime factors: 2 and 5				Also other prime factors			
FRACTION	$\frac{1}{5}$	$\frac{1}{125}$	$\frac{1}{2}$	$\frac{1}{20}$	$\frac{1}{7}$	$\frac{1}{35}$	$\frac{1}{3}$	$\frac{1}{6}$
EQUIVALENT DECIMAL	0.2	0.008	0.5	0.05	$0.\dot{1}4285\dot{7}$	$0.0\dot{2}8571\dot{4}$	$0.\dot{3}$	$0.1\dot{6}$
	Terminating decimals				Recurring decimals			

For prime factors, see p.2.

Converting terminating decimals into fractions was covered on the previous page.
Converting recurring decimals is quite a bit harder — but you'll be OK once you've learnt the method...

Recurring Decimals into Fractions

1) Basic Ones

Turning a recurring decimal into a fraction uses a really clever trick. Just watch this...

EXAMPLE **Write $0.\dot{2}3\dot{4}$ as a fraction.**

1) Name your decimal — I've called it *r*. — Let $r = 0.\dot{2}3\dot{4}$
2) Multiply *r* by a power of ten to move it past the decimal point by one full repeated lump — here that's 1000. → $1000r = 234.\dot{2}3\dot{4}$
3) Now you can subtract to get rid of the decimal part:

$$\begin{aligned} 1000r &= 234.\dot{2}3\dot{4} \\ -\quad r &= 0.\dot{2}3\dot{4} \\ \hline 999r &= 234 \end{aligned}$$

4) Then just divide to leave *r*, and cancel if possible: $r = \frac{234}{999} = \frac{26}{111}$

The 'Just Learning the Result' Method:

1) For converting recurring decimals to fractions, you could just learn the result that the fraction always has the repeating unit on the top and the same number of nines on the bottom...
2) BUT this only works if the repeating bit starts straight after the decimal point (see the next page for an example where it doesn't).
3) AND some exam questions will ask you to 'show that' or 'prove' that a fraction and a recurring decimal are equivalent — and that means you have to use the proper method.

Fractions and Recurring Decimals

2) The Trickier Type

If the recurring bit doesn't come right after the decimal point, things are slightly trickier — but only slightly.

EXAMPLE **Write $0.1\dot{6}$ as a fraction.**

1) Name your decimal. — Let $r = 0.1\dot{6}$
2) Multiply r by a power of ten to move the non-repeating part past the decimal point. — $10r = 1.\dot{6}$
3) Now multiply again to move one full repeated lump past the decimal point. — $100r = 16.\dot{6}$
4) Subtract to get rid of the decimal part:

$$\begin{array}{rl} 100r = & 16.\dot{6} \\ -\quad 10r = & 1.\dot{6} \\ \hline 90r = & 15 \end{array}$$

5) Divide to leave r, and cancel if possible: $r = \frac{15}{90} = \frac{1}{6}$

Fractions into Recurring Decimals

You might find this cropping up in your exam too — and if they're being really unpleasant, they'll stick it in a non-calculator paper.

EXAMPLE **Write $\frac{8}{33}$ as a recurring decimal.**

There are two ways you can do this:

1. Find an equivalent fraction with all nines on the bottom. The number on the top will tell you the recurring part.

$$\frac{8}{33} \overset{\times 3}{=} \frac{24}{99}$$

Watch out — the number of nines on the bottom tells you the number of digits in the recurring part. E.g. $\frac{24}{99} = 0.\dot{2}\dot{4}$, but $\frac{24}{999} = 0.\dot{0}2\dot{4}$

$$\frac{24}{99} = 0.\dot{2}\dot{4}$$

2. Remember, $\frac{8}{33}$ means $8 \div 33$, so you could just do the division: (This is OK if you're allowed your calculator, but a bit tricky if not... you can use short or long division if you're feeling bold, but I recommend sticking with method 1 instead.)

$$33\overline{)8.0^{14}0^{8}0^{14}0^{8}00} \quad = 0.2\,4\,2\,4...$$

$$\frac{8}{33} = 0.\dot{2}\dot{4}$$

You could be asked to convert any recurring decimal into a fraction

So you need to learn how to turn basic and tricky recurring decimals into fractions. You should practise turning fractions into recurring decimals too. And as if by magic, here are some questions...

Q1 Express $0.\dot{1}2\dot{6}$ as a fraction in its simplest form. [2 marks] (Grade 7)

Q2 Show that $0.0\dot{7} = \frac{7}{99}$ [2 marks] (Grade 7)

Q3 Without using a calculator, convert $\frac{5}{111}$ to a recurring decimal. [2 marks] (Grade 7)

Q1 Video Solution

Warm-Up and Worked Exam Questions

These warm-up questions will help to check that you've learnt the basics from the last few pages — if you're struggling with any of them, go and look back over that page before you go any further.

Warm-Up Questions

1) Work these out, then simplify your answers:
   a) $\frac{2}{5} \times \frac{2}{3}$ b) $\frac{2}{5} \div \frac{2}{3}$ c) $\frac{2}{5} + \frac{2}{3}$ d) $\frac{2}{3} - \frac{2}{5}$
2) What percentage is the same as $\frac{2}{5}$?
3) What percentage is the same as $\frac{2}{3}$?
4) a) What fraction is the same as 0.44?
   b) What fraction is the same as 0.444444...?
   c) What fraction is the same as 0.45454545...?
5) a) What decimal is the same as $\frac{7}{10}$? b) What decimal is the same as $\frac{7}{9}$?

Worked Exam Questions

Make sure you understand what's going on in these questions before trying the next page for yourself.

1 Francis owns all the shares of his company. (Grade 4)
He sells $\frac{2}{15}$ of the shares to Dayana and $\frac{5}{12}$ of the shares to Mei.

What fraction of the shares does Francis still own?
Give your answer in its simplest form.

$$1 - \frac{2}{15} - \frac{5}{12} = 1 - \frac{8}{60} - \frac{25}{60}$$

Write over a common denominator

$$= \frac{27}{60}$$

$$= \frac{9}{20}$$

Remember to give your answer in its simplest form.

$\frac{9}{20}$ ....
[3 marks]

2 Write $0.\dot{2}\dot{6}$ in the form $\frac{a}{b}$. Simplify your answer as far as possible. (Grade 6)

Let $r = 0.\dot{2}\dot{6}$, so $100r = 26.\dot{2}\dot{6}$

Multiply by a power of ten to get one full repeated chunk on the left hand side of the decimal point.

$100r - r = 26.\dot{2}\dot{6} - 0.\dot{2}\dot{6}$

$99r = 26$

$r = \frac{26}{99}$

$\frac{26}{99}$ ....
[2 marks]

Exam Questions

3 Look at shapes X, Y and Z below.

$\frac{2}{5}$ of shape X is shaded and $\frac{6}{7}$ of shape Y is shaded.

What fraction of shape Z is shaded?

X Y Z

....
[3 marks]

4 If $a = \frac{3}{4}$ and $b = 2\frac{1}{2}$, find the value of $\frac{1}{a} + \frac{1}{b}$.

....
[3 marks]

5 A factory buys 25 tonnes of flour. $17\frac{1}{2}$ tonnes of the flour is used to make scones.

$\frac{1}{5}$ of the scones are cheese scones.

What fraction of the total amount of flour is used to make cheese scones?

....
[2 marks]

6 Write $\frac{7}{33}$ as a recurring decimal.

....
[2 marks]

7 Show that $0.5\dot{9}\dot{0} = \frac{13}{22}$

Hint: start by trying to get only the non-repeating part before the decimal point.

[3 marks]

Rounding Numbers

*There are **two different ways** of specifying **where** a number should be **rounded**. They are: 'Decimal Places' and 'Significant Figures'.*

Decimal Places (d.p.)

To round to a given number of decimal places:

1) Identify the position of the 'last digit' from the number of decimal places.
2) Then look at the next digit to the right — called the decider.
3) If the decider is 5 or more, then round up the last digit.
   If the decider is 4 or less, then leave the last digit as it is.
4) There must be no more digits after the last digit (not even zeros).

'Last digit' = last one in the rounded version, not the original number.

EXAMPLE

What is 7.45839 to 2 decimal places?

7.45839 = 7.46 ← The LAST DIGIT rounds UP because the DECIDER is 5 or more.

LAST DIGIT (5) DECIDER (8)

If you have to round up a 9 (to 10), replace the 9 with 0, and carry 1 to the left. Remember to keep enough zeros to fill the right number of decimal places — so to 2 d.p. 45.699 would be rounded to 45.70, and 64.996 would be rounded to 65.00.

65 has the same value as 65.00, but 65 isn't expressed to 2 d.p. so it would be marked wrong.

Significant Figures (s.f.)

The method for significant figures is identical to that for decimal places except that locating the last digit is more difficult — it wouldn't be so bad, but for the zeros...

1) The 1st significant figure of any number is simply the first digit which isn't a zero.

2) The 2nd, 3rd, 4th, etc. significant figures follow on immediately after the 1st, regardless of being zeros or not zeros.

0.002309 — SIG. FIGS: 1st (2), 2nd (3), 3rd (0), 4th (9)

2.03070 — SIG. FIGS: 1st (2), 2nd (0), 3rd (3), 4th (0)

(If we're rounding to say, 3 s.f., then the LAST DIGIT is simply the 3rd sig. fig.)

3) After rounding the last digit, end zeros must be filled in up to, but not beyond, the decimal point.

No extra zeros must ever be put in after the decimal point.

EXAMPLES

	to 3 s.f.	to 2 s.f.	to 1 s.f.
1) 54.7651	54.8	55	50
2) 0.0045902	0.00459	0.0046	0.005
3) 30895.4	30900	31000	30000

Estimating

*'**Estimate**' doesn't mean 'take a wild guess', it means 'look at the numbers, make them a bit easier, then do the calculation'. Your answer won't be as **accurate** as the real thing but it's easier on your brain.*

Estimating **Calculations**

GRADE 4

It's time to put your rounding skills to use and do some estimating.

EXAMPLE **Estimate the value of $\frac{127.8 + 41.9}{56.5 \times 3.2}$, showing all your working.**

1) Round all the numbers to easier ones — 1 or 2 s.f. usually does the trick.
2) You can round again to make later steps easier if you need to.

$$\frac{127.8 + 41.9}{56.5 \times 3.2} \approx \frac{130 + 40}{60 \times 3} = \frac{170}{180} \approx 1$$

EXAMPLE **A cylindrical glass has a height of 18 cm and a radius of 3 cm.**
a) Find an estimate in cm³ for the volume of the glass.

The formula for the volume of a cylinder is $V = \pi r^2 h$ (see p.130).

Round the numbers to 1 s.f.:
π = 3.14159... = 3 (1 s.f.), height = 20 cm (1 s.f.) and radius = 3 cm (1 s.f.).

Now just put the numbers into the formula:
$V = \pi r^2 h \approx 3 \times 3^2 \times 20 = 3 \times 9 \times 20 \approx 540 \text{ cm}^3$

≈ means 'approximately equal to'.

b) Use your answer to part a) to estimate the number of glasses that could be filled from a 2.5 litre bottle of lemonade.

2.5 litres = 2500 cm³
2500 ÷ 540 ≈ 2500 ÷ 500 = 5 glasses

The number of glasses must be an integer.

Estimating **Square Roots**

GRADE 5

Estimating square roots can be a bit tricky, but there are only 2 steps:

1) Find two square numbers, one either side of the number you're given.
2) Decide which number it's closest to, and make a sensible estimate of the digit after the decimal point.

EXAMPLE **Estimate the value of $\sqrt{87}$ to 1 d.p.**

87 is between 81 (= 9²) and 100 (= 10²).

It's closer to 81, so its square root will be closer to 9 than 10: $\sqrt{87} \approx 9.3$

(the actual value of $\sqrt{87}$ is 9.32737..., so this is a reasonable estimate).

Round the numbers first, then do the calculation like normal

If you're asked to estimate something in the exam, make sure you show all your steps (including what each number is rounded to) to prove that you didn't just use a calculator. Hassle, but it'll pay off.

Q1 Estimate the value of: a) $\frac{4.23 \times 11.8}{7.7}$ [2 marks] b) $\sqrt{136}$ [2 marks] (GRADE 4)

Q2 The volume of a sphere is given by the formula $V = \frac{4}{3}\pi r^3$.
a) Use this formula to estimate the volume of a sphere of radius 9 cm. [2 marks]
b) Will your estimate be bigger or smaller than the actual value? [1 mark] (GRADE 5)

Q2 Video Solution

Bounds

*Finding **upper and lower bounds** is pretty easy, but using them in **calculations** is a bit trickier.*

Upper and Lower Bounds

Grade 5

When a measurement is ROUNDED to a given UNIT, the actual measurement can be anything up to HALF A UNIT bigger or smaller.

EXAMPLE **The mass of a cake is given as 2.4 kg to the nearest 0.1 kg. Find the interval within which m, the actual mass of the cake, lies.**

lower bound = 2.4 – 0.05 = 2.35 kg
upper bound = 2.4 + 0.05 = 2.45 kg

So the interval is 2.35 kg ≤ m < 2.45 kg ← This is known as the error interval.

The actual value is greater than or equal to the lower bound but strictly less than the upper bound. The actual mass of the cake could be exactly 2.35 kg, but if it was exactly 2.45 kg it would round up to 2.5 kg instead.

When a measurement is TRUNCATED to a given UNIT, the actual measurement can be up to A WHOLE UNIT bigger but no smaller.

You truncate a number by chopping off decimal places, so if the mass of the cake was 2.4 truncated to 1 d.p. the interval would be 2.4 kg ≤ x < 2.5 kg.

If the mass was 2.49999, it would still be truncated to 2.4.

Maximum and Minimum Values for Calculations

Grade 7

When a calculation is done using rounded values there will be a DISCREPANCY between the CALCULATED VALUE and the ACTUAL VALUE:

EXAMPLE **A pinboard is measured as being 0.89 m wide and 1.23 m long, to the nearest cm.**
a) Calculate the minimum and maximum possible values for the area of the pinboard.

1) Find the bounds for the width and length:

0.885 m ≤ width < 0.895 m
1.225 m ≤ length < 1.235 m

2) Find the minimum area by multiplying the lower bounds, and the maximum by multiplying the upper bounds:

minimum possible area = 0.885 × 1.225
= 1.084125 m^2

maximum possible area = 0.895 × 1.235
= 1.105325 m^2

b) Use your answers to part a) to give the area of the pinboard to an appropriate degree of accuracy.

The area of the pinboard lies in the interval 1.084125 m^2 ≤ a < 1.105325 m^2. Both the upper bound and the lower bound round to 1.1 m^2 to 1 d.p. so the area of the pinboard is 1.1 m^2 to 1 d.p.

EXAMPLE **To 1 d.p., a = 5.3 and b = 4.2. What are the maximum and minimum values of a ÷ b?**

First find the bounds for a and b. → 5.25 ≤ a < 5.35, 4.15 ≤ b < 4.25

Now the tricky bit... The bigger the number you divide by, the smaller the answer, so:

max(a ÷ b) = max(a) ÷ min(b)
and min(a ÷ b) = min(a) ÷ max(b)

max. value of a ÷ b = 5.35 ÷ 4.15
= 1.289 (to 3 d.p.)

min. value of a ÷ b = 5.25 ÷ 4.25
= 1.235 (to 3 d.p.)

Bounds tell you the possible values of rounded numbers

Be careful with bounds if the quantity has to be a whole number. E.g. the maximum value of the bound 145 ≤ x < 155 is 154 for a number of people but 154.99999... for the height of a person.

Q1 Nia runs 200 m (to the nearest m) in a time of 32.2 seconds (to the nearest 0.1 second). Using bounds, find her speed in m/s to an appropriate degree of accuracy. [5 marks] Grade 7

Q1 Video Solution

Warm-Up and Worked Exam Questions

Without a good warm-up you're likely to strain a brain cell or two. So take the time to run through these simple questions and get the basic facts straight before plunging into the exam questions.

Warm-Up Questions

1) Round these numbers to the level of accuracy indicated:
   a) 40.218 to 2 d.p. b) 39.888 to 3 s.f. c) 27.91 to 2 s.f.
2) By rounding to 1 significant figure, estimate the answer to $\frac{94 \times 1.9}{0.328 + 0.201}$.
3) A distance is given as 14 km, to the nearest km.
   Find the upper and lower bounds for the distance.
4) $r = 6.3$ and $s = 2.9$, both to 1 d.p. Find the maximum and minimum possible values for:
   a) $r + s$ b) $r - s$
   c) $r \times s$ d) $r \div s$

Worked Exam Questions

With the answers written in, it's very easy to skim these worked examples and think you've understood. But that's not going to help you, so take the time to make sure you've really understood them.

1 Work out an estimate for $\sqrt{\frac{2321}{19.673 \times 3.81}}$ (Grade 4)

Show all of your working.

$$\sqrt{\frac{2321}{19.673 \times 3.81}} \approx \sqrt{\frac{2000}{20 \times 4}}$$
$$= \sqrt{\frac{100}{4}} = \sqrt{25}$$
$$= 5$$

Round each number to an easier one before doing the calculation.

..........5..........
[3 marks]

2 The width of a rectangular piece of paper is 23.6 centimetres, correct to 1 decimal place.
The length of the paper is 54.1 centimetres, correct to 1 decimal place.

a) Write down the lower bound for the length of the paper. (Grade 5)

Lower bound for length = 54.1 cm – 0.05 cm
= 54.05 cm

..........54.05.......... cm
[1 mark]

b) Calculate the lower bound for the perimeter of the piece of paper. (Grade 6)

Lower bound for width = 23.6 cm – 0.05 cm
= 23.55 cm

Lower bound for perimeter = (2 × 54.05 cm) + (2 × 23.55 cm)
= 108.1 cm + 47.1 cm = 155.2 cm

..........155.2.......... cm
[2 marks]

Exam Questions

3 Look at the following calculation: $\dfrac{215.7 \times 44.8}{460}$ (Grade 4)

a) By rounding each number to 1 significant figure, give an estimate for $\dfrac{215.7 \times 44.8}{460}$.

....
[3 marks]

b) Will your answer to part a) be larger or smaller than the exact answer? Explain why.

....

....
[2 marks]

4 Here is a rectangle.
$x = 55$ mm to the nearest 5 mm.
$y = 30$ mm to the nearest 5 mm. (Grade 7)

Calculate the upper bound for the area of this rectangle.
Give your answer to 3 significant figures.

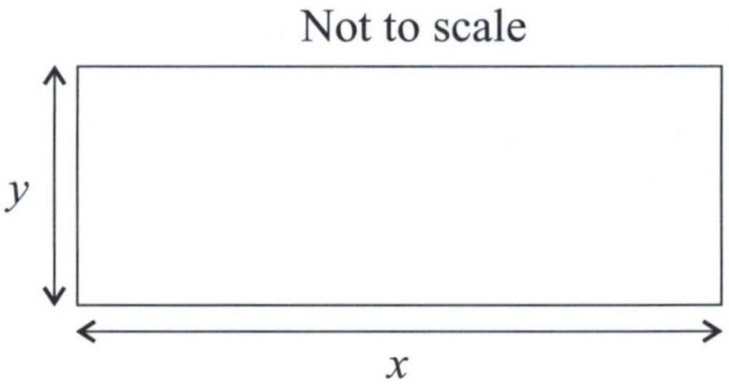

.... mm^2
[3 marks]

5 Given that $x = 2.2$ correct to 1 decimal place, find the interval that contains the value of $4x + 3$. Give your answer as an inequality. (Grade 7)

....
[4 marks]

6 A cuboid measures 0.94 m by 0.61 m by 0.21 m, each measured to the nearest cm.
Find the volume of the cuboid in m^3 to a suitable degree of accuracy. (Grade 7)

.... m^3
[4 marks]

Standard Form

Standard form is useful for writing ***VERY BIG*** *or* ***VERY SMALL*** *numbers in a more convenient way, e.g.*

56 000 000 000 would be 5.6×10^{10} in standard form.

0.000 000 003 45 would be 3.45×10^{-9} in standard form.

But ***ANY NUMBER*** *can be written in standard form and you need to know how to do it:*

What it Actually is:

A number written in standard form must always be in exactly this form:

$$A \times 10^n$$

This number (A) must always be between 1 and 10.
(The fancy way of saying this is $1 \le A < 10$)

This number (n) is just the number of places the decimal point moves.

Learn the THREE RULES:

1) **The front number must always be between 1 and 10.**
2) **The power of 10, n, is how far the decimal point moves.**
3) **n is positive for BIG numbers, n is negative for SMALL numbers.**
   (This is much better than rules based on which way the decimal point moves.)

Four Important Examples:

1 Express 35 600 in standard form.

1) Move the decimal point until 35 600 becomes 3.56 ($1 \le A < 10$)
2) The decimal point has moved 4 places so n = 4, giving: 10^4
3) 35 600 is a big number so n is +4, not –4

3 5 6 0 0.0
$= 3.56 \times 10^4$

2 Express 0.0000623 in standard form.

1) The decimal point must move 5 places to give 6.23 ($1 \le A < 10$).
   So the power of 10 is 5.
2) Since 0.0000623 is a small number it must be 10^{-5} not 10^{+5}

0.0 0 0 0 6 2 3
$= 6.23 \times 10^{-5}$

3 Express 4.95×10^{-3} as an ordinary number.

1) The power of 10 is negative, so it's a small number — the answer will be less than 1.
2) The power is –3, so the decimal point moves 3 places.

0 0 0 4.9 5 $\times 10^{-3}$
= 0.00495

4 What is 146.3 million in standard form?

Too many people get this type of question wrong. Just take your time and do it in two stages:

146.3 million = 146.3 × 1 000 000
= 146 300 000 — 1) Write the number out in full.
$= 1.463 \times 10^8$ — 2) Convert to standard form.

The two favourite wrong answers for this are:

146.3×10^6 — which is kind of right but it's not in standard form because 146.3 is not between 1 and 10

1.463×10^6 — this one is in standard form but it's not big enough

Standard Form

Calculations with Standard Form

Grade 5

These are really popular exam questions — you might be asked to add, subtract, multiply or divide using numbers in standard form without using a calculator.

Multiplying and Dividing — not too bad

1) Rearrange to put the front numbers and the powers of 10 together.
2) Multiply or divide the front numbers, and use the power rules (see p.25) to multiply or divide the powers of 10.
3) Make sure your answer is still in standard form.

EXAMPLES

1. Find $(2 \times 10^3) \times (6.75 \times 10^5)$ without using a calculator. Give your answer in standard form.

$(2 \times 10^3) \times (6.75 \times 10^5)$

$= (2 \times 6.75) \times (10^3 \times 10^5)$ — Multiply front numbers and powers separately

$= 13.5 \times 10^{3+5}$ — Add the powers (see p.25)

$= 13.5 \times 10^8$

$= 1.35 \times 10 \times 10^8$ — Not in standard form — convert it

$= 1.35 \times 10^9$

2. Calculate $240\,000 \div (4.8 \times 10^{10})$ without using a calculator. Give your answer in standard form.

$240\,000 \div (4.8 \times 10^{10})$ — Convert 240 000 to standard form

$= \frac{2.4 \times 10^5}{4.8 \times 10^{10}} = \frac{2.4}{4.8} \times \frac{10^5}{10^{10}}$ — Divide front numbers and powers separately

$= 0.5 \times 10^{5-10}$ — Subtract the powers (see p.25)

$= 0.5 \times 10^{-5}$

$= 5 \times 10^{-1} \times 10^{-5}$ — Not in standard form — convert it

$= 5 \times 10^{-6}$

Adding and Subtracting — a bit trickier

1) Make sure the powers of 10 are the same — you'll probably need to rewrite one of them.
2) Add or subtract the front numbers.
3) Convert the answer to standard form if necessary.

EXAMPLE

Calculate $(9.8 \times 10^4) + (6.6 \times 10^3)$ without using a calculator. Give your answer in standard form.

$(9.8 \times 10^4) + (6.6 \times 10^3)$

1) Rewrite one number so both powers of 10 are equal: $= (9.8 \times 10^4) + (0.66 \times 10^4)$
2) Now add the front numbers: $= (9.8 + 0.66) \times 10^4$
3) 10.46×10^4 isn't in standard form, so convert it: $= 10.46 \times 10^4 = 1.046 \times 10^5$

To put standard form numbers into your calculator, use the EXP or the $\times 10^x$ button.

E.g. enter 2.67×10^{15} by pressing 2.67 EXP 15 = or 2.67 $\times 10^x$ 15 = .

Remember, n tells you how far the decimal point moves

Make sure you understand all the examples on these pages. Then answer these Exam Practice Questions:

Q1 Express 0.854 million and 0.00018 in standard form. [2 marks] (Grade 4)

Q2 Work out the following without using a calculator. Give your answers in standard form.
a) $(3.2 \times 10^7) \div (1.6 \times 10^{-4})$ [2 marks] b) $(6.7 \times 10^{10}) + (5.8 \times 10^{11})$ [2 marks] (Grade 5)

Q3 Write $2^{25} \times 5^{27}$ in standard form. [3 marks] (Grade 7)

Q3 Video Solution

Warm-Up and Worked Exam Questions

I know that you'll be champing at the bit to get into the exam questions, but these basic warm-up questions are invaluable for getting the basic facts straight first.

Warm-Up Questions

1) The moon is about 240 000 miles away from Earth.
   Write this number in standard form.
2) The half-life of a chemical isotope is 0.0000027 seconds.
   Write this number in standard form.
3) An oxygen atom has a mass of 2.7×10^{-23} g. Write this as an ordinary number.
4) Work out $(4 \times 10^3) \times 30\,000$. Give your answer in standard form.
5) Find each of the following. Give your answers in standard form.
   a) $(3 \times 10^6) \times (8 \times 10^4)$
   b) $(8.4 \times 10^8) \div (4.2 \times 10^4)$
   c) $(7.65 \times 10^6) + (1.47 \times 10^5)$
   d) $(3.28 \times 10^{12}) - (7.12 \times 10^{10})$

Worked Exam Question

Instead of flicking past this worked exam question, go through it yourself — it'll help you to tackle the exam questions on the next page.

1 $A = 4.834 \times 10^9$, $B = 2.7 \times 10^5$, $C = 5.8 \times 10^3$ (Grade 4)

a) Express A as an ordinary number.

$A = 4.834 \times 10^9 = 4\,834\,000\,000$ — Since n = 9, the decimal point moves 9 places.

4 834 000 000

[1 mark]

b) Work out $B \times C$. Give your answer in standard form.

Multiply the front numbers together and use the power rules on the powers of 10:

$B \times C = (2.7 \times 10^5) \times (5.8 \times 10^3)$
$= (2.7 \times 5.8) \times (10^5 \times 10^3)$
$= 15.66 \times 10^8$
$= 1.566 \times 10^9$ — Make sure the final answer is in standard form.

1.566×10^9

[2 marks]

c) Put A, B and C in order from smallest to largest.

In this question, each number has a different power of 10, so you can order them just by looking at these.

C , B , A

[1 mark]

Exam Questions

2 Light travels at approximately 1.86×10^5 miles per second.
The distance from the Earth to the Sun is approximately 9.3×10^7 miles. (Grade 5)

How long will it take light to travel this distance?
Give your answer in standard form.

.... seconds
[2 marks]

3 A patient has been prescribed a dose of 4×10^{-4} grams of a certain drug to be given daily. (Grade 6)

a) The tablets that the hospital stocks each contain 8×10^{-5} grams of the drug.
How many tablets should the patient be given each day?

.... tablets
[3 marks]

b) The doctor increases the patient's daily dose of the drug by 6×10^{-5} grams.
What is the patient's new daily dose of the drug?

TIP: you need matching powers to be able to add two numbers together in standard form.

.... grams per day
[3 marks]

4 Express $\frac{3^2}{2^{122} \times 5^{120}}$ in standard form. (Grade 7)

You need to write the denominator in standard form.

....
[2 marks]

Revision Questions for Section One

That wraps up Section One — now find out how much you know.

- Try these questions and tick off each one when you get it right.
- When you're completely happy with the topic, tick that off too.

For even more practice, try the quiz for Section One — just scan this QR code!

Section One Quiz

Types of Number, Factors and Multiples (p1-3)

1) What are: a) integers b) rational numbers c) prime numbers?
2) Use BODMAS to answer the following questions: a) $7 + 8 \div 2$ b) $7 \div (5 + 9)$ c) $(2 - 5 \times 3)^2$
3) Buns are sold in packs of 6, cheese slices are sold in packs of 16 and hot dogs are sold in packs of 12. Noah wants to buy the same number of each item. What is the smallest number of packs of buns, cheese slices and hot dogs he can buy?
4) Find: a) the HCF of 42 and 28 b) the LCM of 8 and 10
5) a) Write 320 and 880 as products of their prime factors.
b) Use the prime factorisations to find the LCM and HCF of 320 and 880.

Fractions (p6-8)

You're not allowed to use a calculator for q6-18 and 23-26. Sorry.

6) How do you simplify a fraction?
7) a) Write $\frac{74}{9}$ as a mixed number b) Write $4\frac{5}{7}$ as an improper fraction
8) What are the rules for multiplying, dividing and adding/subtracting fractions?
9) Calculate: a) $\frac{2}{11} \times \frac{7}{9}$ b) $5\frac{1}{2} \div 1\frac{3}{4}$ c) $\frac{5}{8} - \frac{1}{6}$ d) $3\frac{3}{10} + 4\frac{1}{4}$
10) a) Find $\frac{7}{9}$ of 270 kg. b) Write 88 as a fraction of 56.
11) Which of $\frac{5}{8}$ and $\frac{7}{10}$ is closer in value to $\frac{3}{4}$?

Fractions, Decimals and Percentages (p9-11)

12) How do you convert: a) a fraction to a decimal? b) a terminating decimal to a fraction?
13) Write: a) 0.04 as: (i) a fraction (ii) a percentage b) 65% as: (i) a fraction (ii) a decimal
14) 25 litres of fruit punch is made up of 50% orange juice, $\frac{2}{5}$ lemonade and $\frac{1}{10}$ cranberry juice. How many litres of orange juice, lemonade and cranberry juice are there in the punch?
15) Show that $0.\dot{5}\dot{1} = \frac{17}{33}$

Rounding, Estimating and Bounds (p14-16)

16) Round 427.963 to: a) 2 d.p. b) 1 d.p. c) 2 s.f. d) 4 s.f.
17) Estimate the value of $(104.6 + 56.8) \div 8.4$
18) Estimate the value of $\sqrt{45}$ to 1 d.p.
19) How do you determine the upper and lower bounds of a rounded and truncated measurement?
20) The volume of water in a jug is given as 2.4 litres to the nearest 100 ml. Find the upper and lower bounds for the volume of the jug. Give your answer as an inequality.
21) A rectangle measures 15.6 m by 8.4 m, to the nearest 0.1 m. Find its maximum possible area.

Standard Form (p19-20)

22) What are the three rules for writing numbers in standard form?
23) Write these numbers in standard form: a) 970 000 b) 3 560 000 000 c) 0.00000275
24) Express 4.56×10^{-3} and 2.7×10^5 as ordinary numbers.
25) Calculate: a) $(3.2 \times 10^6) \div (1.6 \times 10^3)$ b) $(1.75 \times 10^{12}) + (9.89 \times 10^{11})$
Give your answers in standard form.
26) At the start of an experiment, there are 3.1×10^8 bacteria on a petri dish. The number of bacteria doubles every 10 minutes. How many bacteria will there be after 30 minutes?

Algebra Basics

Before you can really get your teeth into ***algebra****, there are some basics you need to get your head around.*

Negative Numbers (Grade 3)

Negative numbers crop up everywhere so you need to learn these rules for dealing with them:

+	+	makes	+
+	–	makes	–
–	+	makes	–
–	–	makes	+

Use these rules when:

1) Multiplying or dividing.
e.g. $-2 \times 3 = -6$, $-8 \div -2 = +4$, $-4p \times -2 = +8p$
2) Two signs are together.
e.g. $5 - -4 = 5 + 4 = 9$, $x + -y - -z = x - y + z$

Be careful when squaring or cubing. Squaring a negative number gives a positive number, e.g. $(-2)^2 = 4$ but cubing a negative number gives a negative number, e.g. $(-3)^3 = -27$.

Letters Multiplied Together (Grade 3)

Watch out for these combinations of letters in algebra that regularly catch people out:

1) abc means $a \times b \times c$. The ×'s are often left out to make it clearer.
2) gn^2 means $g \times n \times n$. Note that only the n is squared, not the g as well — e.g. πr^2 means $\pi \times r \times r$.
3) $(gn)^2$ means $g \times g \times n \times n$. The brackets mean that BOTH letters are squared.
4) $p(q - r)^3$ means $p \times (q - r) \times (q - r) \times (q - r)$. Only the brackets get cubed.
5) -3^2 is a bit ambiguous. It should either be written $(-3)^2 = 9$, or $-(3^2) = -9$ (you'd usually take -3^2 to be -9).

Terms (Grade 3)

Before you can do anything else with algebra, you must understand what a term is:

A TERM is a collection of numbers, letters and brackets, all multiplied/divided together

Terms are separated by + and – signs. Every term has a + or – attached to the front of it.

If there's no sign in front of the first term, it means there's an invisible + sign.

$4xy$ $+ 5x^2$ $- 2y$ $+ 6y^2$ $+ 4$

'xy' term — 'x^2' term — 'y' term — 'y^2' term — 'number' term

Simplifying or 'Collecting Like Terms' (Grade 3)

To simplify an algebraic expression, you combine 'like terms' — terms that have the same combination of letters (e.g. all the x terms, all the y terms, all the number terms etc.).

EXAMPLE **Simplify $2x - 4 + 5x + 6$**

Invisible + sign → $2x$; number terms: -4, $+6$; x-terms: $2x$, $+5x$

$2x$ -4 $+5x$ $+6$ $=$ $+2x$ $+5x$ -4 $+6$

$= 7x + 2 = 7x + 2$

1) Put bubbles round each term — be sure you capture the +/– sign in front of each.
2) Then you can move the bubbles into the best order so that like terms are together.
3) Combine like terms.

These simple bits and bobs are the key to all algebra

It's best to get this stuff in your head now or you'll be lost in a few pages' time. Try out this Practice Question.

Q1 A rectangle has sides measuring $5x$ cm and $(3y + 1)$ cm.
Find an expression for its perimeter. [2 marks] (Grade 3)

Powers and Roots

*Make sure you know what powers are. Then, learn the **ten special rules** for working with them — seven easy ones on this page and three trickier ones on the next page. They're not very exciting, but you need to learn them.*

Powers are a very useful shorthand:
$2 \times 2 \times 2 \times 2 \times 2 \times 2 \times 2 = 2^7$ ('two to the power 7')

The **Seven** Easy Rules:

GRADE 5 GRADE

1. When MULTIPLYING, you ADD THE POWERS.

   e.g. $3^6 \times 3^4 = 3^{6+4} = 3^{10}$, $a^2 \times a^7 = a^{2+7} = a^9$

Warning: Rules 1 & 2 don't work for things like $2^3 \times 3^7$, only for powers of the same number.

2. When DIVIDING, you SUBTRACT THE POWERS.

   e.g. $5^4 \div 5^2 = 5^{4-2} = 5^2$, $b^8 \div b^5 = b^{8-5} = b^3$

3. When RAISING one power to another, you MULTIPLY THEM.

   e.g. $(3^2)^4 = 3^{2\times4} = 3^8$, $(c^3)^6 = c^{3\times6} = c^{18}$

4. $x^1 = x$, ANYTHING to the POWER 1 is just ITSELF.

   e.g. $3^1 = 3$, $d \times d^3 = d^1 \times d^3 = d^{1+3} = d^4$

5. $x^0 = 1$, ANYTHING to the POWER 0 is just 1

   e.g. $5^0 = 1$, $67^0 = 1$, $e^0 = 1$

6. $1^x = 1$, 1 TO ANY POWER is STILL JUST 1.

   e.g. $1^{23} = 1$, $1^{89} = 1$, $1^2 = 1$

7. FRACTIONS — Apply the power to both TOP and BOTTOM.

   e.g. $\left(1\frac{3}{5}\right)^3 = \left(\frac{8}{5}\right)^3 = \frac{8^3}{5^3} = \frac{512}{125}$, $\left(\frac{u}{v}\right)^5 = \frac{u^5}{v^5}$

Remember, rules 1 & 2 only work for powers of the same number

If you can add, subtract and multiply, there's nothing on this page you can't do — as long as you learn the rules. Try copying them over and over until you can do it with your eyes closed.

Powers and Roots

The Three Tricky Rules:

Grade 7

8 NEGATIVE Powers — Turn it Upside Down

People have real difficulty remembering this — whenever you see a negative power you need to immediately think: "Aha, that means turn it the other way up and make the power positive".

e.g. $7^{-2} = \frac{1}{7^2} = \frac{1}{49}$ $\quad a^{-4} = \frac{1}{a^4}$ $\quad \left(\frac{3}{5}\right)^{-2} = \left(\frac{5}{3}\right)^{+2} = \frac{5^2}{3^2} = \frac{25}{9}$

9 FRACTIONAL POWERS

The power $\frac{1}{2}$ means Square Root,
The power $\frac{1}{3}$ means Cube Root,
The power $\frac{1}{4}$ means Fourth Root etc.

e.g. $25^{\frac{1}{2}} = \sqrt{25} = 5$
$64^{\frac{1}{3}} = \sqrt[3]{64} = 4$
$81^{\frac{1}{4}} = \sqrt[4]{81} = 3$
$z^{\frac{1}{5}} = \sqrt[5]{z}$

The one to really watch is when you get a negative fraction like $49^{-1/2}$ — people get mixed up and think that the minus is the square root, and forget to turn it upside down as well.

e.g. $49^{-\frac{1}{2}} = \frac{1}{\sqrt{49}} = \frac{1}{7}$

10 TWO-STAGE FRACTIONAL POWERS

With fractional powers like $64^{\frac{5}{6}}$ always split the fraction into a root and a power, and do them in that order: root first, then power: $(64)^{\frac{1}{6} \times 5} = (64^{\frac{1}{6}})^5 = (2)^5 = 32$.

EXAMPLE **Simplify $(3a^2b^4c)^3$**

Just deal with each bit separately:

$= (3)^3 \times (a^2)^3 \times (b^4)^3 \times (c)^3$

$= 27 \times a^{2\times3} \times b^{4\times3} \times c^3$

$= 27a^6b^{12}c^3$

You simplify algebraic fractions using the power rules (though you might not realise it).

So if you had to simplify e.g. $\frac{p^3q^6}{p^2q^3}$, you'd just cancel using the power rules to get $p^{3-2}q^{6-3} = pq^3$.

These three rules might be a bit trickier — but they are essential

Because these are things which people often get muddled, examiners love to sneak them into the exam — so scribble down these rules and learn them really well. Then in the exam, you'll have the last laugh.

Q1 Simplify: a) $e^4 \times e^7$ [1 mark] b) $f^9 \div f^5$ [1 mark]
c) $(g^6)^{\frac{1}{2}}$ [1 mark] d) $2h^5j^{-2} \times 3h^2j^4$ [2 marks] Grade 5

Q2 Evaluate without a calculator:
a) $625^{\frac{3}{4}}$ [2 marks] b) $25^{-\frac{1}{2}}$ [2 marks] c) $\left(\frac{27}{216}\right)^{-\frac{1}{3}}$ [2 marks] Grade 7

Q1 Video Solution

Multiplying Out Brackets

You often find brackets in algebraic expressions. The first thing you need to be able to do is to expand them (multiply them out).

Single Brackets

The main thing to remember when multiplying out brackets is that the thing outside the bracket multiplies each separate term inside the bracket.

EXAMPLE **Expand the following:**

a) $4a(3b - 2c)$

$= (4a \times 3b) + (4a \times -2c)$

$= 12ab - 8ac$

b) $-4(3p^2 - 7q^3)$

$= (-4 \times 3p^2) + (-4 \times -7q^3)$

$= -12p^2 + 28q^3$

Note: both signs have been reversed.

Double Brackets

Double brackets are trickier than single brackets — this time, you have to multiply everything in the first bracket by everything in the second bracket. You'll get 4 terms, and usually 2 of them will combine to leave 3 terms. There's a handy way to multiply out double brackets — it's called the FOIL method:

First — multiply the first term in each bracket together
Outside — multiply the outside terms (i.e. the first term in the first bracket by the second term in the second bracket)
Inside — multiply the inside terms (i.e. the second term in the first bracket by the first term in the second bracket)
Last — multiply the second term in each bracket together

EXAMPLE **Expand and simplify $(2p - 4)(3p + 1)$**

$(2p - 4)(3p + 1) = (2p \times 3p) + (2p \times 1) + (-4 \times 3p) + (-4 \times 1)$

$= 6p^2 + 2p - 12p - 4$

$= 6p^2 - 10p - 4$

The two p terms combine together.

Always write out SQUARED BRACKETS as TWO BRACKETS (to avoid mistakes), then multiply out as above.
So $(3x + 5)^2 = (3x + 5)(3x + 5) = 9x^2 + 15x + 15x + 25 = 9x^2 + 30x + 25$.
(DON'T make the mistake of thinking that $(3x + 5)^2 = 9x^2 + 25$ — this is wrong wrong wrong.)

Triple Brackets

It doesn't matter which pair of brackets you multiply together first.

1) For three brackets, just multiply two together as above, then multiply the result by the remaining bracket.
2) If you end up with three terms in one bracket, you won't be able to use FOIL. Instead, you can reduce it to a series of single bracket multiplications — like in the example below.

EXAMPLE **Expand and simplify $(x + 2)(x + 3)(2x - 1)$**

$(x + 2)(x + 3)(2x - 1) = (x + 2)(2x^2 + 5x - 3) = x(2x^2 + 5x - 3) + 2(2x^2 + 5x - 3)$

$= (2x^3 + 5x^2 - 3x) + (4x^2 + 10x - 6)$

$= 2x^3 + 9x^2 + 7x - 6$

Use the FOIL method to make sure you don't miss out any terms

You can expand cubed brackets by writing them out as three brackets and expanding as above.

Q1 Expand and simplify: a) $(y + 4)(y - 5)$ [2 marks] b) $(2p - 3)^2$ [2 marks]

Q2 Expand and simplify: a) $(2t + \sqrt{2})(t - 3\sqrt{2})$ [3 marks] b) $(x - 2)^3$ [3 marks]

Q2 Video Solution

Factorising

*Right, now you know how to expand brackets, it's time to put them back in. This is known as **factorising**.*

Factorising — Putting Brackets In

Grade 4

This is the exact reverse of multiplying out brackets. Here's the method to follow:

1) Take out the biggest number that goes into all the terms.
2) For each letter in turn, take out the highest power (e.g. x, x^2 etc.) that will go into EVERY term.
3) Open the bracket and fill in all the bits needed to reproduce each term.
4) Check your answer by multiplying out the bracket and making sure it matches the original expression.

EXAMPLES

1. Factorise $3x^2 + 6x$

Biggest number that'll divide into 3 and 6

Highest power of x that will go into both terms

$3x(x + 2)$

Check: $3x(x + 2) = 3x^2 + 6x$ ✓

2. Factorise $8x^2y + 2xy^2$

Biggest number that'll divide into 8 and 2

Highest powers of x and y that will go into both terms

$2xy(4x + y)$

Check: $2xy(4x + y) = 8x^2y + 2xy^2$ ✓

REMEMBER: The bits taken out and put at the front are the common factors. The bits inside the bracket are what's needed to get back to the original terms if you multiply the bracket out again.

D.O.T.S. — The Difference Of Two Squares

Grade 6

The 'difference of two squares' (D.O.T.S. for short) is where you have 'one thing squared' take away 'another thing squared'. There's a quick and easy way to factorise it — just use the rule below:

$$a^2 - b^2 = (a + b)(a - b)$$

EXAMPLE

Factorise:

a) $9p^2 - 16q^2$ **Answer:** $9p^2 - 16q^2 = (3p + 4q)(3p - 4q)$

Here you had to spot that 9 and 16 are square numbers.

b) $3x^2 - 75y^2$ **Answer:** $3x^2 - 75y^2 = 3(x^2 - 25y^2) = 3(x + 5y)(x - 5y)$

This time, you had to take out a factor of 3 first.

c) $x^2 - 5$ **Answer:** $x^2 - 5 = (x + \sqrt{5})(x - \sqrt{5})$

Although 5 isn't a square number, you can write it as $(\sqrt{5})^2$.

Watch out — the difference of two squares can creep into other algebra questions. A popular exam question is to put a difference of two squares on the top or bottom of a fraction and ask you to simplify it. There's more on algebraic fractions on p.45.

EXAMPLE

Simplify $\frac{x^2 - 36}{5x + 30}$

$$\frac{x^2 - 36}{5x + 30} = \frac{(x + 6)(x - 6)}{5(x + 6)} = \frac{x - 6}{5}$$

The numerator is a difference of two squares.

Factorise the denominator.

D.O.T.S. is straightforward as long as you recognise the pattern

Once you've seen one D.O.T.S. question, you've seen them all — they all follow the same basic pattern.

Q1 Factorise $6xy + 15y^2$ [2 marks] (Grade 5)

Q2 Factorise $x^2 - 16y^2$ [2 marks] (Grade 6)

Q3 Factorise $x^2 - 11$ [2 marks] (Grade 7)

Q4 Simplify $\frac{6x - 42}{x^2 - 49}$ [3 marks] (Grade 7)

Manipulating Surds

Surds *are expressions with* ***irrational square roots*** *in them (remember from p.1 that irrational numbers are ones which* ***can't*** *be written as* ***fractions****, such as most square roots, cube roots and π).*

Manipulating Surds — 6 Rules to Learn

There are 6 rules you need to learn for dealing with surds...

1. $\sqrt{a} \times \sqrt{b} = \sqrt{a \times b}$ e.g. $\sqrt{2} \times \sqrt{3} = \sqrt{2 \times 3} = \sqrt{6}$ — also $(\sqrt{b})^2 = \sqrt{b} \times \sqrt{b} = \sqrt{b \times b} = b$

2. $\dfrac{\sqrt{a}}{\sqrt{b}} = \sqrt{\dfrac{a}{b}}$ e.g. $\dfrac{\sqrt{8}}{\sqrt{2}} = \sqrt{\dfrac{8}{2}} = \sqrt{4} = 2$

3. $\sqrt{a} + \sqrt{b}$ — Do nothing — in other words it is definitely NOT $\sqrt{a+b}$

4. $(a + \sqrt{b})^2 = (a + \sqrt{b})(a + \sqrt{b}) = a^2 + 2a\sqrt{b} + b$ — NOT just $a^2 + (\sqrt{b})^2$ (see p.27)

5. $(a + \sqrt{b})(a - \sqrt{b}) = a^2 + a\sqrt{b} - a\sqrt{b} - (\sqrt{b})^2 = a^2 - b$ (see p.28).

6. $\dfrac{a}{\sqrt{b}} = \dfrac{a}{\sqrt{b}} \times \dfrac{\sqrt{b}}{\sqrt{b}} = \dfrac{a\sqrt{b}}{b}$

This is known as 'RATIONALISING the denominator' — it's where you get rid of the $\sqrt{}$ on the bottom of the fraction. For denominators of the form $a \pm \sqrt{b}$, multiply by the denominator but change the sign in front of the root (see example 4 below).

Use the Rules to Simplify Expressions

EXAMPLES

1. Write $2\sqrt{75} - \sqrt{48}$ in the form $a\sqrt{3}$, where a is an integer.

Write each surd in terms of $\sqrt{3}$:

$2\sqrt{75} = 2\sqrt{25 \times 3} = 2 \times \sqrt{25} \times \sqrt{3} = 10\sqrt{3}$

$\sqrt{48} = \sqrt{16 \times 3} = \sqrt{16} \times \sqrt{3} = 4\sqrt{3}$

Then do the sum (leaving your answer in terms of $\sqrt{3}$):

$2\sqrt{75} - \sqrt{48} = 10\sqrt{3} - 4\sqrt{3} = 6\sqrt{3}$

2. Fully simplify $\dfrac{12\sqrt{2}}{\sqrt{6}}$.

Multiply top and bottom by $\sqrt{6}$:

$\dfrac{12\sqrt{2}}{\sqrt{6}} = \dfrac{12\sqrt{2} \times \sqrt{6}}{\sqrt{6} \times \sqrt{6}} = \dfrac{12\sqrt{12}}{6} = 2\sqrt{12}$

This can be simplified even further:

$2\sqrt{12} = 2\sqrt{4}\sqrt{3} = 2 \times 2\sqrt{3} = 4\sqrt{3}$

3. A rectangle with length $4x$ cm and width x cm has an area of 32 cm². Find the exact value of x, giving your answer in its simplest form.

Area of rectangle = length × width = $4x \times x = 4x^2$

So $4x^2 = 32$

$x^2 = 8$

$x = \pm\sqrt{8}$

You can ignore the negative square root (see p.33) as length must be positive.

'Exact value' means you have to leave your answer in surd form, so get $\sqrt{8}$ into its simplest form:

$\sqrt{8} = \sqrt{4 \times 2} = \sqrt{4}\sqrt{2} = 2\sqrt{2}$ So $x = 2\sqrt{2}$

4. Write $\dfrac{3}{2+\sqrt{5}}$ in the form $a + b\sqrt{5}$, where a and b are integers.

To rationalise the denominator, multiply top and bottom by $2 - \sqrt{5}$:

$$\frac{3}{2+\sqrt{5}} = \frac{3(2-\sqrt{5})}{(2+\sqrt{5})(2-\sqrt{5})}$$

$$= \frac{6-3\sqrt{5}}{2^2 - 2\sqrt{5} + 2\sqrt{5} - (\sqrt{5})^2}$$

$$= \frac{6-3\sqrt{5}}{4-5} = \frac{6-3\sqrt{5}}{-1} = -6+3\sqrt{5}$$

Once you get used to them, surds are quite easy

Learn the 6 rules for manipulating surds, then give these Exam Practice Questions a go...

Q1 Simplify $\sqrt{180} + \sqrt{20} + (\sqrt{5})^3$ [3 marks]

Q2 Write $\dfrac{2}{2+\sqrt{3}}$ in the form $a + b\sqrt{3}$, where a and b are integers. [3 marks]

Q2 Video Solution

Warm-Up and Worked Exam Questions

Take a deep breath and go through these warm-up questions one by one. Then you'll be ready for the really exciting bit (well, slightly more exciting anyway) — the exam questions.

Warm-Up Questions

1) Simplify: a) $4a + c - 2a - 6c$ b) $3r^2 - 2r + 4r^2 - 1 - 3r$.
2) Evaluate: a) $4^5 \times 4^{-2}$ b) $\left(1\frac{2}{7}\right)^2$ c) $27^{\frac{2}{3}}$ d) $\left(\frac{2}{3}\right)^{-2}$
3) Multiply out:
   a) $4(2p + 7)$ b) $(4x - 2)(2x + 1)$ c) $a(5a - 3)$.
4) Factorise:
   a) $6p - 12q + 4$ b) $4cd^2 - 2cd + 10c^2d^3$.
5) Factorise $x^2 - 4y^2$.
6) Work out $\sqrt{5} \times \sqrt{6}$, leaving your answer as a surd.

Worked Exam Questions

Don't skip over these worked exam questions just because they already have the answers written in. Work through them yourself so you know what's going on, then have a go at the next page.

1 Factorise the following expressions fully.

a) $x^2 - 81$

This is a difference of two squares:

$x^2 - 81 = x^2 - 9^2$

$= (x + 9)(x - 9)$

$(x + 9)(x - 9)$ ....

[1 mark]

b) $9n^2 - 4m^2$

$9n^2 - 4m^2 = (3n)^2 - (2m)^2$ — Here you have to spot that 9 and 4 are square numbers.

$= (3n + 2m)(3n - 2m)$

$(3n + 2m)(3n - 2m)$ ....

[2 marks]

2 Write $2\sqrt{50} - (\sqrt{2})^3$ in the form $a\sqrt{b}$, where a and b are integers.

$2\sqrt{50} = 2\sqrt{25 \times 2} = 2 \times 5\sqrt{2}$

$= 10\sqrt{2}$

$(\sqrt{2})^3 = \sqrt{2} \times \sqrt{2} \times \sqrt{2} = (\sqrt{2})^2 \times \sqrt{2}$

$= 2\sqrt{2}$

So $2\sqrt{50} - (\sqrt{2})^3 = 10\sqrt{2} - 2\sqrt{2}$

$= 8\sqrt{2}$

$8\sqrt{2}$ ....

[2 marks]

Exam Questions

3 Peter is making a sculpture using different pieces of metal tubing.
He makes a tower by stacking 7 pieces that are $(f + g)$ cm tall, 9 pieces that are $(h - g)$ cm tall and 5 pieces that are $2h$ cm tall on top of each other.

Find a simplified expression for the height of the tower in terms of f, g and h.

.... cm
[2 marks]

4 Estimate the value of each of the following to 1 decimal place:

a) x, where $x = \sqrt{70}$

....
[2 marks]

b) y, where $3^y = 20$

....
[2 marks]

5 Expand and simplify $(x - 1)(2x + 3)(2x - 3)$.

....
[3 marks]

6 Fully factorise $x^3 - 25x$.

....
[3 marks]

7 Express $\frac{1+\sqrt{7}}{3-\sqrt{7}}$ in the form $a + b\sqrt{7}$, where a and b are integers.

Multiply by $3 + \sqrt{7}$ to rationalise the denominator.

....
[4 marks]

Solving Equations

The basic idea of ***solving equations*** *is very simple — keep rearranging until you end up with x = number. The two most common methods for* ***rearranging*** *equations are: 1) '****same to both sides****' and 2) do the* ***opposite*** *when you cross the '='. We'll use the 'same to both sides' method on these pages.*

Rearrange Until You Have x = Number (Grade 3)

The easiest ones to solve are where you just have a mixture of x's and numbers.

1) First, rearrange the equation so that all the x's are on one side and the numbers are on the other. Combine terms where you can.
2) Then divide both sides by the number multiplying x to find the value of x.

EXAMPLE **Solve $5x + 4 = 8x - 5$**

(+5) $5x + 4 + 5 = 8x - 5 + 5$ — This means 'add 5 to both sides'.

$5x + 9 = 8x$

(−5x) $5x + 9 - 5x = 8x - 5x$

$9 = 3x$ — Numbers on left, x's on right.

(÷3) $9 \div 3 = 3x \div 3$ — Divide by number multiplying x.

$3 = x$

Once you're happy with the method, you don't have to write everything out in full — your working might be:

$5x + 9 = 8x$

$9 = 3x$

$3 = x$

Multiply Out Brackets First (Grade 4)

If your equation has brackets in it...

1) Multiply them out before rearranging.
2) Solve it in the same way as above.

EXAMPLE **Solve $3(3x - 2) = 5x + 10$**

$9x - 6 = 5x + 10$

(−5x) $9x - 6 - 5x = 5x + 10 - 5x$

$4x - 6 = 10$

(+6) $4x - 6 + 6 = 10 + 6$

$4x = 16$

(÷4) $4x \div 4 = 16 \div 4$

$x = 4$

Get Rid of Fractions (Grade 5)

1) Fractions make everything more complicated — so you need to get rid of them before doing anything else (yep, even before multiplying out brackets).
2) To get rid of fractions, multiply every term of the equation by whatever's on the bottom of the fraction. If there are two fractions, you'll need to multiply by both denominators.

EXAMPLES

1. Solve $\frac{x+2}{4} = 4x - 7$

(×4) $\frac{4(x+2)}{4} = 4(4x) - 4(7)$ — Multiply every term by 4 to get rid of the fraction.

$x + 2 = 16x - 28$

$30 = 15x$ — And solve.

$2 = x$

2. Solve $\frac{3x+5}{2} = \frac{4x+10}{3}$ — Multiply everything by 2 then by 3.

(×2), (×3) $\frac{2 \times 3 \times (3x+5)}{2} = \frac{2 \times 3 \times (4x+10)}{3}$

$3(3x + 5) = 2(4x + 10)$ — And solve.

$9x + 15 = 8x + 20$

$x = 5$

Remember that you're trying to get x on its own

You can always check your answer by putting your value of x back into both sides of the original equation — you should get the same number on each side. Now, practise what you've learned on these questions:

Q1 Solve $2x + 5 = 17 - 4x$ [2 marks] (Grade 3)

Q2 Solve $4(y + 3) = 3y + 16$ [3 marks] (Grade 4)

Q3 Solve $\frac{3x+2}{5} = \frac{5x+6}{9}$ [3 marks] (Grade 5)

Solving Equations

Now you know the basics of solving equations, it's time to put it all together into a step-by-step method.

Solving Equations Using the **6-Step Method** (GRADE 5)

Here's the method to follow (just ignore any steps that don't apply to your equation):

1. Get rid of any fractions.
2. Multiply out any brackets.
3. Collect all the x-terms on one side and all number terms on the other.
4. Reduce it to the form '$Ax = B$' (by combining like terms).
5. Finally divide both sides by A to give '$x =$ ', and that's your answer.
6. If you had '$x^2 =$ ' instead, square root both sides to end up with '$x = \pm$ '.

EXAMPLE Solve $\frac{3x+4}{5} + \frac{4x-1}{3} = 14$

1. Get rid of any fractions. (×5), (×3) $\frac{5 \times 3 \times (3x+4)}{5} + \frac{5 \times 3 \times (4x-1)}{3} = 5 \times 3 \times 14$ (Multiply everything by 5 then by 3.)

   $3(3x + 4) + 5(4x - 1) = 210$

2. Multiply out any brackets. $9x + 12 + 20x - 5 = 210$

3. Collect all the x-terms on one side and all number terms on the other.

   (–12), (+5) $9x + 20x = 210 - 12 + 5$

4. Reduce it to the form '$Ax = B$' (by combining like terms).

   $29x = 203$

5. Finally divide both sides by A to give '$x =$ ', and that's your answer.

   (÷29) $x = 7$ (You're left with '$x =$ ' so you can ignore step 6.)

Dealing With **Squares** (GRADE 5)

If you're unlucky, you might get an x^2 in an equation. If this happens, you'll end up with '$x^2 = ...$' at step 5, and then step 6 is to take square roots. There's one very important thing to remember: whenever you take the square root of a number, the answer can be positive or negative (unless there's a reason it can't be –ve).

EXAMPLE **There are 75 tiles on a roof. Each row contains three times the number of tiles as each column. How many tiles are there in one column?**

Let the number of tiles in a column be x. Write an equation for the total number of tiles in terms of x.

$3x \times x = 75$

$3x^2 = 75$

(÷3) $x^2 = 25$

($\sqrt{\ }$) $x = \pm 5$

So there are 5 tiles in one column.

Ignore the negative square root — you can't have a negative number of tiles.

You always get a +ve and -ve version of the same number (your calculator only gives the positive answer). E.g. $5^2 = 5 \times 5 = 25$ but also $(-5)^2 = (-5) \times (-5) = 25$.

Learn the 6-step method for solving equations

You might not need all 6 steps to solve your equation — ignore any that you don't need and move on to the next step. Make sure you do them in the right order though, otherwise you'll get it wrong.

Q1 Solve $2x^2 + 8 = 80$ [2 marks] (GRADE 5)

Q2 Solve $\frac{3x-2}{2} - \frac{4x-5}{3} = 2$ [3 marks] (GRADE 5)

Q2 Video Solution

Rearranging Formulas

***Rearranging formulas** means making one letter the **subject**, e.g. getting '$y =$ ' from '$2x + z = 3(y + 2p)$' — you have to get the subject **on its own**.*

Rearrange Formulas with the Solving Equations Method

GRADE 5

Rearranging formulas is remarkably similar to solving equations. The method below is identical to the method for solving equations, except that I've added an extra step at the start.

1. Get rid of any square root signs by squaring both sides.
2. Get rid of any fractions.
3. Multiply out any brackets.
4. Collect all the subject terms on one side and all non-subject terms on the other.
5. Reduce it to the form '$Ax = B$' (by combining like terms). You might have to do some factorising here too.
6. Divide both sides by A to give '$x =$ '.
7. If you're left with '$x^2 =$ ', square root both sides to get '$x = \pm$ ' (don't forget the $\pm$).

x is the subject term here. A and B could be numbers or letters (or a mix of both).

What To Do If...

...the Subject Appears in a Fraction

GRADE 5

You won't always need to use all 7 steps in the method above — just ignore the ones that don't apply.

EXAMPLE **Make b the subject of the formula $a = \frac{5b + 3}{4}$.**

There aren't any square roots, so ignore step 1.

2. Get rid of any fractions. (by multiplying every term by 4, the denominator) (×4) $4a = \frac{4(5b + 3)}{4}$

$4a = 5b + 3$

There aren't any brackets so ignore step 3.

4. Collect all the subject terms on one side and all non-subject terms on the other. (remember that you're trying to make b the subject) (−3) $5b = 4a - 3$

5. It's now in the form $Ab = B$. (where $A = 5$ and $B = 4a - 3$)

6. Divide both sides by 5 to give '$b =$ '. (÷5) $b = \frac{4a - 3}{5}$

b isn't squared, so you don't need step 7.

The subject is the letter on its own

Remember that rearranging formulas is exactly the same as solving equations, except that instead of ending up with '$x =$ number' (e.g. $x = 3$), you'll end up with '$x =$ expression' (e.g. $x = 2y + 4$). Now try these:

Q1 Make q the subject of the formula $p = \frac{q}{7} + 2r$ [2 marks] GRADE 5

Q2 Make v the subject of the formula $a = \frac{v - u}{t}$ [2 marks] GRADE 5

Rearranging Formulas

Carrying straight on from the previous page, now it's time for what to do if...

...there's a Square or Square Root Involved

GRADE 5

If the subject appears as a square or in a square root, you'll have to use steps 1 and 7 (not necessarily both).

EXAMPLE **Make u the subject of the formula $v^2 = u^2 + 2as$.**

There aren't any square roots, fractions or brackets so ignore steps 1-3 (this is pretty easy so far).

❹ Collect all the subject terms on one side and all non-subject terms on the other.

$(-2as)$ $u^2 = v^2 - 2as$

❺ It's now in the form $Au^2 = B$ (where A = 1 and $B = v^2 - 2as$)

A = 1, which means it's already in the form '$u^2 =$ ', so ignore step 6.

❼ Square root both sides to get '$u = \pm$ '. $(\sqrt{\ })$ $u = \pm\sqrt{v^2 - 2as}$

This formula is a real-life kinematics formula — v = final velocity, u = initial velocity, a = acceleration and s = displacement. Other kinematics formulas include $v = u + at$ and $s = ut + \frac{1}{2}at^2$, where t is the time taken.

EXAMPLE **Make n the subject of the formula $2(m + 3) = \sqrt{n + 5}$.**

❶ Get rid of any square roots by squaring both sides.

$[2(m + 3)]^2 = (\sqrt{n + 5})^2$

$4(m^2 + 6m + 9) = n + 5$

$4m^2 + 24m + 36 = n + 5$

There aren't any fractions so ignore step 2.
The brackets were removed when squaring so ignore step 3.

❹ Collect all the subject terms on one side and all non-subject terms on the other.

(-5) $n = 4m^2 + 24m + 31$ This is in the form '$n =$ ' so you don't need to do steps 5-7.

...the Subject Appears Twice

GRADE 6

You'll have to do some factorising, usually in step 5.

EXAMPLE **Make p the subject of the formula $q = \frac{p + 1}{p - 1}$.**

There aren't any square roots so ignore step 1.

❷ Get rid of any fractions. $q(p - 1) = p + 1$

❸ Multiply out any brackets. $pq - q = p + 1$

❹ Collect all the subject terms on one side and all non-subject terms on the other.

$pq - p = q + 1$

❺ Combine like terms on each side of the equation. $p(q - 1) = q + 1$

This is where you factorise — p was in both terms on the LHS so it comes out as a common factor.

❻ Divide both sides by $(q - 1)$ to give '$p =$ '. $p = \frac{q + 1}{q - 1}$

(p isn't squared, so you don't need step 7.)

Remember — you square first and square root last

Try this Exam Practice Question to have a go at rearranging more complicated formulas...

Q1 Make y the subject of: a) $x = \frac{y^2}{4}$ [2 marks] GRADE 5 b) $x = \frac{y}{y - z}$ [4 marks] GRADE 6

Q1 Video Solution

Warm-Up and Worked Exam Questions

It's easy to think you've learnt everything in the section until you try the warm-up questions. Don't panic if there are bits you've forgotten. Just go back over them until they're firmly fixed in your brain.

Warm-Up Questions

1) Solve these equations to find the value of x:
   a) $8x - 5 = 19$ b) $3(2x + 7) = 3$ c) $4x - 9 = x + 6$.
2) What is the subject of these formulas?
   a) $p = \sqrt{\dfrac{ml^2}{h}}$ b) $t = px - y^3$.
3) Make n the subject of the formula $m = 5 - \dfrac{n}{4}$
4) Make z the subject of the formula $x = \dfrac{y + 2z}{3}$

Worked Exam Questions

Here are a couple of exam questions that I've done for you. You won't get any help for the questions on the next page though — so make the most of it whilst you can.

1 Solve the equation $\frac{5}{4}(2c - 1) = 3c - 2$ (Grade 5)

$$\frac{5}{4}(2c - 1) = 3c - 2$$

Get rid of the fraction... (×4) $5(2c - 1) = 4(3c - 2)$

...multiply out the brackets... $10c - 5 = 12c - 8$

...and solve. (−10c) $-5 = 2c - 8$

(+8) $3 = 2c$

(÷2) $1.5 = c$

$c =$ ......1.5......
[3 marks]

2 Using the formula $v^2 = u^2 + 2as$: (Grade 5)

a) Rearrange the formula to make s the subject.

Follow the 7-step method to get s on its own:

$v^2 = u^2 + 2as$

($-u^2$) $v^2 - u^2 = 2as$

($\div 2a$) $\dfrac{v^2 - u^2}{2a} = s$

$s = \dfrac{v^2 - u^2}{2a}$
....
[2 marks]

b) Find the displacement of an object that starts with an initial velocity of 4 m/s and ends with a final velocity of 6 m/s, with an acceleration of 5 m/s².

Substitute the values for v, u and a into the equation found in part a):

$v = 6$, $u = 4$ and $a = 5$

So $s = \dfrac{6^2 - 4^2}{2 \times 5} = \dfrac{36 - 16}{10}$

$= \dfrac{20}{10} = 2$

......2...... m
[2 marks]

Exam Questions

3 Poppy, Felix and Aleena sell 700 raffle tickets between them.
Poppy sells twice as many tickets as Felix, and Aleena sells 25 more tickets than Poppy.

How many tickets did each of them sell?

Poppy ....

Felix ....

Aleena ....

[5 marks]

4 Hassan thinks of two different positive integers.
Their product is 147, and one number is three times the other number.

What are the two numbers Hassan is thinking of?

....

[3 marks]

5 The relationship between a, b and y is given by the formula $a + y = \dfrac{b - y}{a}$.

a) Rearrange this formula to make y the subject.

....

[4 marks]

b) Find the value of y when $a = 3$ and $b = 6$.

$y =$ ....

[1 mark]

6 Rearrange the formula below to make n the subject.

$$x = \sqrt{\frac{(1 + n)}{(1 - n)}}$$

....

[5 marks]

Factorising Quadratics

There are several ways of solving a quadratic equation, as detailed on the following pages.

Factorising a Quadratic (Grade 5)

1) 'Factorising a quadratic' means 'putting it into 2 brackets'.
2) The standard format for quadratic equations is: $ax^2 + bx + c = 0$.
3) If $a = 1$, the quadratic is much easier to deal with. E.g. $x^2 + 3x + 2 = 0$
4) As well as factorising a quadratic, you might be asked to solve the equation. This just means finding the values of x that make each bracket 0 (see example below).

See next page for when 'a' is not 1.

Factorising Method when a = 1 (Grade 5)

1) ALWAYS rearrange into the STANDARD FORMAT: $x^2 + bx + c = 0$.
2) Write down the TWO BRACKETS with the x's in: $(x \quad)(x \quad) = 0$.
3) Then find 2 numbers that MULTIPLY to give 'c' (the end number) but also ADD/SUBTRACT to give 'b' (the coefficient of x).
4) Fill in the +/– signs and make sure they work out properly.
5) As an ESSENTIAL CHECK, expand the brackets to make sure they give the original equation.
6) Finally, SOLVE THE EQUATION by setting each bracket equal to 0.

Ignore any minus signs at this stage.

You only need to do step 6) if the question asks you to solve the equation — if it just tells you to factorise, you can stop at step 5).

EXAMPLE **Solve $x^2 - x = 12$.**

1) $x^2 - x - 12 = 0$ — 1) Rearrange into the standard format.

2) $(x \quad)(x \quad) = 0$ — 2) Write down the initial brackets.

3)

1 × 12	Add/subtract to give:	13 or 11
2 × 6	Add/subtract to give:	8 or 4
3 × 4	Add/subtract to give:	7 or (1)

3) Find the right pairs of numbers that multiply to give c (= 12), and add or subtract to give b (= 1) (remember, we're ignoring the +/– signs for now).

$(x \quad 3)(x \quad 4) = 0$ This is what we want.

4) $(x + 3)(x - 4) = 0$ — 4) Now fill in the +/– signs so that 3 and 4 add/subtract to give –1 (= b).

5) Check:
$(x + 3)(x - 4) = x^2 - 4x + 3x - 12$
$= x^2 - x - 12$ ✓

5) ESSENTIAL check — EXPAND the brackets to make sure they give the original expression.

But we're not finished yet — we've only factorised it, we still need to...

6) $(x + 3) = 0 \Rightarrow x = -3$
$(x - 4) = 0 \Rightarrow x = 4$

6) SOLVE THE EQUATION by setting each bracket equal to 0.

Factorising quadratics is not easy — but it is important

To help you work out which signs you need, look at c. If c is positive, the signs will be the same (both positive or both negative), but if c is negative the signs will be different (one positive and one negative).

Q1 Factorise $x^2 + 2x - 15$ [2 marks] (Grade 5)

Q2 Solve $x^2 - 9x + 20 = 0$ [3 marks] (Grade 5)

Factorising Quadratics

It gets a bit more complicated when 'a' isn't 1, but don't panic — just follow the method on this page.

When 'a' is Not 1

Grade 7

The basic method is still the same but it's a bit messier — the initial brackets are different as the first terms in each bracket have to multiply to give 'a'. This means finding the other numbers to go in the brackets is harder as there are more combinations to try. The best way to get to grips with it is to have a look at an example.

EXAMPLE **Solve $3x^2 + 7x - 6 = 0$.**

1) $3x^2 + 7x - 6 = 0$

2) $(3x \quad)(x \quad) = 0$

3) Number pairs: 1×6 and 2×3

$(3x \quad 1)(x \quad 6)$ multiplies to give $18x$ and $1x$ which add/subtract to give $17x$ or $19x$

$(3x \quad 6)(x \quad 1)$ multiplies to give $3x$ and $6x$ which add/subtract to give $9x$ or $3x$

$(3x \quad 3)(x \quad 2)$ multiplies to give $6x$ and $3x$ which add/subtract to give $9x$ or $3x$

$(3x \quad 2)(x \quad 3)$ multiplies to give $9x$ and $2x$ which add/subtract to give $11x$ or $7x$ ✓

$(3x \quad 2)(x \quad 3)$

4) $(3x - 2)(x + 3)$

5) $(3x - 2)(x + 3) = 3x^2 + 9x - 2x - 6$
$= 3x^2 + 7x - 6$ ✓

6) $(3x - 2) = 0 \Rightarrow x = \frac{2}{3}$
$(x + 3) = 0 \Rightarrow x = -3$

1) Rearrange into the standard format.

2) Write down the initial brackets — this time, one of the brackets will have a $3x$ in it.

3) The tricky part: first, find pairs of numbers that multiply to give c (= 6), ignoring the minus sign for now.

Then, try out the number pairs you just found in the brackets until you find one that gives $7x$. But remember, each pair of numbers has to be tried in 2 positions (as the brackets are different — one has $3x$ in it).

4) Now fill in the +/– signs so that 9 and 2 add/subtract to give +7 (= b).

5) ESSENTIAL check — EXPAND the brackets.

6) SOLVE THE EQUATION by setting each bracket equal to 0 (if a isn't 1, one of your answers will be a fraction).

EXAMPLE **Solve $2x^2 - 9x = 5$.**

1) Put in standard form: $2x^2 - 9x - 5 = 0$

2) Initial brackets: $(2x \quad)(x \quad) = 0$

3) Number pairs: 1×5

$(2x \quad 5)(x \quad 1)$ multiplies to give $2x$ and $5x$ which add/subtract to give $3x$ or $7x$

$(2x \quad 1)(x \quad 5)$ multiplies to give $1x$ and $10x$ which add/subtract to give $9x$ or $11x$ ✓

$(2x \quad 1)(x \quad 5)$

4) Put in the signs: $(2x + 1)(x - 5)$

5) Check:
$(2x + 1)(x - 5) = 2x^2 - 10x + x - 5$
$= 2x^2 - 9x - 5$ ✓

6) Solve:
$(2x + 1) = 0 \Rightarrow x = -\frac{1}{2}$
$(x - 5) = 0 \Rightarrow x = 5$

Factorising quadratics when a is not 1 is quite a bit harder

The problem is that it's a lot harder to work out the right combination of numbers to go in the brackets. Don't get stressed out, just take your time and work through the possibilities one at a time.

Q1 Factorise $2x^2 - 5x - 12$ [2 marks] (Grade 7)

Q2 Solve $3x^2 + 10x - 8 = 0$ [3 marks] (Grade 7)

Q3 Factorise $3x^2 + 32x + 20$ [2 marks] (Grade 7)

Q4 Solve $5x^2 - 13x = 6$ [3 marks] (Grade 7)

Q4 Video Solution

The Quadratic Formula

The solutions to ANY quadratic equation $ax^2 + bx + c = 0$ are given by this formula:

$$x = \frac{-b \pm \sqrt{b^2 - 4ac}}{2a}$$

Using the formula isn't that hard, but there are a few pitfalls — so ***take heed of these crucial details****:*

Quadratic Formula — Five Crucial Details

GRADE 7

1) Take it nice and slowly — always write it down in stages as you go.
2) WHENEVER YOU GET A MINUS SIGN, THE ALARM BELLS SHOULD ALWAYS RING!
3) Remember it's '2a' on the bottom line, not just 'a' — and you divide ALL of the top line by 2a.
4) The ± sign means you end up with two solutions (by replacing it in the final step with '+' and '–').
5) If you get a negative number inside your square root, go back and check your working. Some quadratics do have a negative value in the square root, but they won't come up at GCSE.

If either 'a' or 'c' is negative, the –4ac effectively becomes +4ac, so watch out. Also, be careful if b is negative, as –b will be positive.

EXAMPLE **Solve $3x^2 + 7x = 1$, giving your answers to 2 decimal places.**

$3x^2 + 7x - 1 = 0$

a = 3, b = 7, c = –1

$$x = \frac{-b \pm \sqrt{b^2 - 4ac}}{2a}$$

$$= \frac{-7 \pm \sqrt{7^2 - 4 \times 3 \times -1}}{2 \times 3}$$

$$= \frac{-7 \pm \sqrt{49 + 12}}{6}$$

$$= \frac{-7 \pm \sqrt{61}}{6}$$

$$= \frac{-7 + \sqrt{61}}{6} \text{ or } \frac{-7 - \sqrt{61}}{6}$$

$= 0.1350...$ or $-2.468...$

So to 2 d.p. the solutions are:
$x = 0.14$ or -2.47

1) First get it into the form $ax^2 + bx + c = 0$.
2) Then carefully identify a, b and c.
3) Put these values into the quadratic formula and write down each stage.
4) Finally, as a check put these values back into the original equation:
   E.g. for $x = 0.1350$: $3 \times 0.135^2 + 7 \times 0.135 = 0.999675$, which is 1, as near as...

Notice that you do two calculations at the final stage — one + and one –.

When to use the quadratic formula:

- If you have a quadratic that won't easily factorise.
- If the question mentions decimal places or significant figures.
- If the question asks for exact answers or surds (though this could be completing the square instead — see next page).

Looks nightmarish — but you'll soon be chanting it in your sleep

You might have to do some rearranging to get the quadratic you're given into the form $ax^2 + bx + c$.
In Q2 below, it doesn't even look like a quadratic until you start rearranging it and get rid of the fraction.

Q1 Solve $x^2 + 10x - 4 = 0$, giving your answers to 2 decimal places. [3 marks] GRADE 7

Q2 Find the exact solutions of $2x + \frac{3}{x-2} = -2$. [4 marks] GRADE 8

Completing the Square

There's just one more method to learn for solving quadratics — and it's a bit of a nasty one. It's called 'completing the square', and takes a bit to get your head round it.

Solving Quadratics by 'Completing the Square'

GRADE 8

To 'complete the square' you have to:

1) Write down a SQUARED bracket, and then... 2) Stick a number on the end to 'COMPLETE' it.

$$x^2 + 12x - 5 = (x+6)^2 - 41$$

The SQUARE... ...COMPLETED

It's not that bad if you learn all the steps — some of them aren't all that obvious.

1) **As always, REARRANGE THE QUADRATIC INTO THE STANDARD FORMAT: $ax^2 + bx + c$ (the rest of this method is for a = 1).**
2) **WRITE OUT THE INITIAL BRACKET: $(x + \frac{b}{2})^2$ — just divide the value of b by 2.**
3) **MULTIPLY OUT THE BRACKETS and COMPARE TO THE ORIGINAL to find what you need to add or subtract to complete the square.**
4) **Add or subtract the ADJUSTING NUMBER to make it MATCH THE ORIGINAL.**

If a isn't 1, you have to divide through by 'a' or take out a factor of 'a' at the start — see next page.

EXAMPLE

a) Express $x^2 + 8x + 5$ in the form $(x + m)^2 + n$.

1) It's in the standard format. — $x^2 + 8x + 5$
2) Write out the initial bracket. — $(x + 4)^2$
3) Multiply out the brackets and compare to the original. — $(x + 4)^2 = x^2 + 8x + 16$ (Original equation had +5 here...) — $(x + 4)^2 - 11 = x^2 + 8x + 16 - 11$ (...so you need −11)
4) Subtract adjusting number (11). — $= x^2 + 8x + 5$ ✓ — matches original now!

So the completed square is: $(x + 4)^2 - 11$.

b) Hence solve $x^2 + 8x + 5 = 0$, leaving your answers in surd form.

Now use the completed square to solve the equation. There are three more steps for this:

$(x + 4)^2 - 11 = 0$

1) Put the number on the other side (+11). — $(x + 4)^2 = 11$
2) Square root both sides (don't forget the ±!) ($\sqrt{\ }$). — $x + 4 = \pm\sqrt{11}$
3) Get x on its own (−4). — $x = -4 \pm \sqrt{11}$

So the two solutions (in surd form) are: $x = -4 + \sqrt{11}$ and $x = -4 - \sqrt{11}$

If you really don't like steps 3-4, just remember that the value you need to add or subtract is always $c - \left(\frac{b}{2}\right)^2$.

Make a SQUARE (bracket) and COMPLETE it (add or take away)

Completing the square basically means working out a squared bracket which is almost the same as your quadratic and then working out what has to be added or subtracted to make it the same as the original.

Q1 Write $x^2 - 12x + 23$ in the form $(x + p)^2 + q$. [3 marks] GRADE 8

Q2 Solve $x^2 + 10x + 7 = 0$, by first writing it in the form $(x + m)^2 + n = 0$. Give your answers as simplified surds. [5 marks] GRADE 8

Completing the Square

Completing the square can still be done when a isn't 1 — it just takes an extra step.

Completing the Square When 'a' Isn't 1

If 'a' isn't 1, completing the square is a bit trickier. You follow the same method as on the previous page, but you have to take out a factor of 'a' from the x^2 and x-terms before you start (which often means you end up with awkward fractions). This time, the number in the brackets is $\frac{b}{2a}$.

EXAMPLE **Write $2x^2 + 5x + 9$ in the form $a(x + m)^2 + n$.**

1) It's in the standard format. — $2x^2 + 5x + 9$
2) Take out a factor of 2. — $2(x^2 + \frac{5}{2}x) + 9$
3) Write out the initial bracket. — $2(x + \frac{5}{4})^2$
4) Multiply out the bracket and compare to the original. — $2(x + \frac{5}{4})^2 = 2x^2 + 5x + \frac{25}{8}$
5) Add on adjusting number ($\frac{47}{8}$). — $2(x + \frac{5}{4})^2 + \frac{47}{8} = 2x^2 + 5x + \frac{25}{8} + \frac{47}{8}$

$= 2x^2 + 5x + 9$ ✓ — matches original now!

Original equation had +9 here...

...so you need $9 - \frac{25}{8} = \frac{47}{8}$

So the completed square is: $2(x + \frac{5}{4})^2 + \frac{47}{8}$

The Completed Square Helps You Sketch the Graph

There's more about sketching quadratic graphs on p.72, but you can use the completed square to work out key details about the graph — like the turning point (maximum or minimum) and whether it crosses the x-axis.

1) For a positive quadratic (where the x^2 coefficient is positive), the adjusting number tells you the minimum y-value of the graph. If the completed square is $a(x + m)^2 + n$, this minimum y-value will occur when the brackets are equal to 0 (because the bit in brackets is squared, so is never negative) — i.e. when $x = -m$.
2) The solutions to the equation tell you where the graph crosses the x-axis. If the adjusting number is positive, the graph will never cross the x-axis as it will always be greater than 0 (this means that the quadratic has no real roots).

EXAMPLE **Sketch the graph of $y = 2x^2 + 5x + 9$.**

From above, completed square form is $2(x + \frac{5}{4})^2 + \frac{47}{8}$.
The minimum point occurs when the brackets are equal to 0 — this will happen when $x = -\frac{5}{4}$.
At this point, the graph takes its minimum value, which is the adjusting number ($\frac{47}{8}$).
The adjusting number is positive, so the graph will never cross the x-axis.
Find where the curve crosses the y-axis by substituting into the equation and mark this on your graph. $y = 0 + 0 + 9 = 9$

This is only a sketch, so label the points you know.

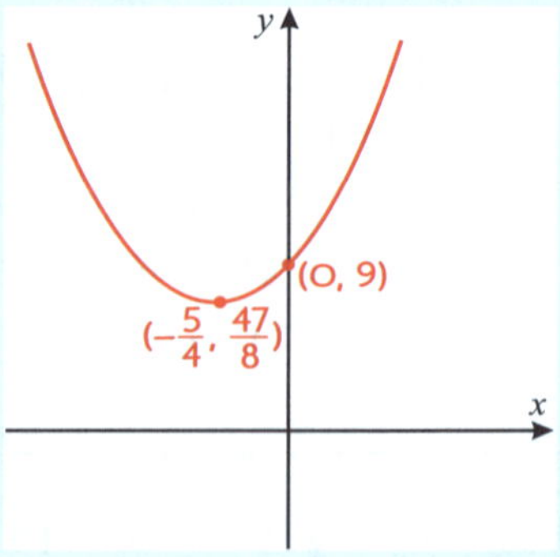

Take out a factor of 'a' from the x^2 and x-terms

After you've done that, you can complete the square like usual. Just don't forget to multiply out the brackets properly — you need to multiply by the number in front of them. It's tricky, but practice helps.

Q1 a) Write $2x^2 + 3x - 5$ in the form $a(x + b)^2 + c$. [4 marks]
b) Hence solve $2x^2 + 3x - 5 = 0$. [2 marks]
c) Use your answer to part a) to find the coordinates of the minimum point of the graph of $y = 2x^2 + 3x - 5$. [1 mark]

Q1 Video Solution

Warm-Up and Worked Exam Questions

This quadratic stuff isn't everyone's cup of tea. But once you get the knack of it, through lots of practice, you'll find a lot of the questions are really similar. Which is nice.

Warm-Up Questions

1) Factorise:
   a) $x^2 + 11x + 28$ b) $x^2 + 16x + 28$ c) $x^2 + 12x - 28$.
2) Solve by factorisation:
   a) $x^2 + 8x + 15 = 0$ b) $x^2 + 5x - 14 = 0$ c) $x^2 - 7x + 7 = -5$.
3) Factorise $4x^2 - 27x + 18$.
4) Solve $3x^2 + 20x = 7$.
5) Solve $3x^2 - 3x = 2$, giving your answers to 2 decimal places.
6) Express $x^2 + 8x - 5$ as a completed square, and hence solve $x^2 + 8x - 5 = 0$.
7) Complete the square for the expression $2x^2 + 16x + 39$.

Worked Exam Questions

Now, the exam questions — the good news is, if you've got the hang of the warm-up questions, you'll find that the exam questions are pretty much the same.

1 The expression $5x^2 - 19x + 18$ is an example of a quadratic expression. (Grade 7)

a) Fully factorise the expression $5x^2 - 19x + 18$.

Number pairs are 1×18, 2×9, 3×6.

$(5x \quad 9)(x \quad 2)$ multiplies to give $9x$ and $10x$ which add to give $19x$.

$5x^2 - 19x + 18 = (5x - 9)(x - 2)$

Be careful, you want $-19x$, so the signs are both $-$.

$(5x - 9)(x - 2)$

[2 marks]

b) Use your answer to part a) to factorise the expression $5(x - 1)^2 - 19(x - 1) + 18$.

$$5(x-1)^2 - 19(x-1) + 18 = (5(x-1) - 9)((x-1) - 2)$$
$$= (5x - 5 - 9)(x - 1 - 2)$$
$$= (5x - 14)(x - 3)$$

Replace x with $(x - 1)$ in the factorised expression from a).

$(5x - 14)(x - 3)$

[2 marks]

2 Solve the quadratic equation $x^2 + 5x + 3 = 0$, giving your answers to 2 decimal places. (Grade 7)

2 d.p. suggests you should use the quadratic formula:

a = 1, b = 5 and c = 3

$$x = \frac{-b \pm \sqrt{b^2 - 4ac}}{2a} = \frac{-5 \pm \sqrt{5^2 - 4 \times 1 \times 3}}{2 \times 1} = \frac{-5 \pm \sqrt{13}}{2} = -0.697... \text{ or } -4.302...$$

The answers are to 2 d.p. so you need to include the 0s on the end.

$x =$ ....−0.70.... or $x =$ ....−4.30....

[3 marks]

Exam Questions

3 Given that $x^2 + ax + b = (x + 2)^2 - 9$, work out the values of a and b.

a = ... and b = ...
[2 marks]

4 The shape on the right is made from a square and a triangle.
The sides of the square are $(x + 3)$ cm long and the height of the triangle is $(2x + 2)$ cm.
The area of the whole shape is 60 cm^2.

Find the value of x.

Don't forget — a length can't have a negative value.

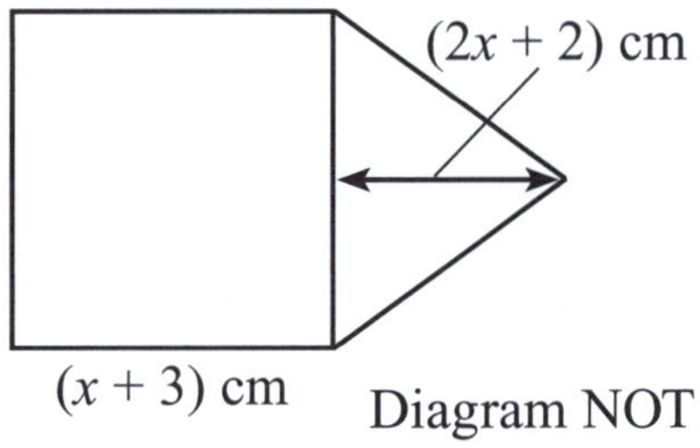

Diagram NOT drawn to scale

x = ...
[7 marks]

5 Solve the equation $3x^2 - 2x - 4 = 0$. Give your answers in simplified surd form.

x = ... or x = ...
[3 marks]

6 A curve has equation $y = 2x^2 - 8x + 19$.

a) Write the expression $2x^2 - 8x + 19$ in the form $a(x + b)^2 + c$.

....
[4 marks]

b) Find the coordinates of the minimum point of the graph.

....
[1 mark]

c) State if and where the graph of the equation crosses the x-axis.

Think about the minimum value of the graph.

....
[1 mark]

Algebraic Fractions

*Unfortunately, fractions aren't limited to numbers — you can get **algebraic fractions** too. Fortunately, everything you learnt about fractions on p.6-8 can be applied to algebraic fractions as well.*

Simplifying Algebraic Fractions

You can simplify algebraic fractions by cancelling terms on the top and bottom — just deal with each letter individually and cancel as much as you can. You might have to factorise first (see pages 28 and 38-39).

EXAMPLES

1. **Simplify** $\dfrac{21x^3y^2}{14xy^3}$

÷7 on the top and bottom

÷x on the top and bottom to leave x^2 on the top

÷y^2 on the top and bottom to leave y on the bottom

$$\frac{{}^{3}\cancel{21}\cancel{x^3}{}^{x^2}\cancel{y^2}}{{}_{2}\cancel{14}\cancel{x}\cancel{y^3}{}_{y}} = \frac{3x^2}{2y}$$

2. **Simplify** $\dfrac{x^2 - 16}{x^2 + 2x - 8}$

Factorise the top using D.O.T.S.

$$\frac{\cancel{(x+4)}(x-4)}{(x-2)\cancel{(x+4)}} = \frac{x-4}{x-2}$$

Factorise the quadratic on the bottom

Then cancel the common factor of $(x + 4)$

Multiplying/Dividing Algebraic Fractions

1) To multiply two fractions, just multiply tops and bottoms separately.
2) To divide, turn the second fraction upside down then multiply.

EXAMPLE **Simplify** $\dfrac{x^2 - 4}{x^2 + x - 12} \div \dfrac{2x + 4}{x^2 - 3x}$

Turn the second fraction upside down

Factorise and cancel

Multiply tops and bottoms

$$\frac{x^2-4}{x^2+x-12} \div \frac{2x+4}{x^2-3x} = \frac{x^2-4}{x^2+x-12} \times \frac{x^2-3x}{2x+4} = \frac{\cancel{(x+2)}(x-2)}{(x+4)\cancel{(x-3)}} \times \frac{x\cancel{(x-3)}}{2\cancel{(x+2)}} = \frac{x-2}{x+4} \times \frac{x}{2} = \frac{x(x-2)}{2(x+4)}$$

Adding/Subtracting Algebraic Fractions

For the common denominator, find something both denominators divide into.

Adding or subtracting is a bit more difficult:

1) Work out the common denominator (see p.7).
2) Multiply top and bottom of each fraction by whatever gives you the common denominator.
3) Add or subtract the numerators only.

Fractions		
$\dfrac{1}{x} + \dfrac{1}{3x}$	$\dfrac{1}{x+1} + \dfrac{1}{x-2}$	$\dfrac{1}{x} + \dfrac{1}{x(x+1)}$
$3x$	$(x + 1)(x - 2)$	$x(x + 1)$
Common denominator		

EXAMPLE **Write** $\dfrac{3}{(x+3)} + \dfrac{1}{(x-2)}$ **as a single fraction.**

1st fraction: × top & bottom by $(x - 2)$

2nd fraction: × top & bottom by $(x + 3)$

$$\frac{3}{(x+3)} + \frac{1}{(x-2)} = \frac{3(x-2)}{(x+3)(x-2)} + \frac{(x+3)}{(x+3)(x-2)}$$

Common denominator will be $(x + 3)(x - 2)$

Add the numerators

$$= \frac{3x-6}{(x+3)(x-2)} + \frac{x+3}{(x+3)(x-2)} = \frac{4x-3}{(x+3)(x-2)}$$

Put fractions over a common denominator

One more thing — never do this: $\dfrac{\cancel{x}}{\cancel{x}+y} = \dfrac{1}{y}$ ✗. It's WRONG and will lose you marks.

Q1 Simplify $\dfrac{x^4 - 4y^2}{x^3 - 2xy}$ [3 marks]

Q2 Simplify $\dfrac{x^2 - 3x - 10}{x^3 - 2x^2} \div \dfrac{x^2 - 25}{6x - 12}$ [6 marks]

Q3 Write $\dfrac{2}{x+5} + \dfrac{3}{x-2}$ as a single fraction in its simplest form. [3 marks]

Q3 Video Solution

Sequences

*You might be asked to "find an **expression** for the **nth term** of a sequence" — this is just a formula with n in, like 5n – 3. It gives you **every term in a sequence** when you put in different values for n.*

Finding the **nth Term** of a **Linear Sequence**

This method works for linear sequences — ones with a common difference (where the terms increase or decrease by the same amount each time). Linear sequences are also known as arithmetic sequences.

EXAMPLE **Find an expression for the *n*th term of the sequence that starts 5, 8, 11, 14, ...**

n:	1	2	3	4
term:	5	8	11	14
	+3	+3	+3	
3n:	3	6	9	12
	+2	+2	+2	+2
term:	5	8	11	14

The common difference is 3, so '3n' is in the formula.

You have to + 2 to get to the term.

So the expression for the nth term is 3n + 2

1) Find the common difference — this tells you what to multiply n by. So here, 3 gives '$3n$'.
2) Work out what to add or subtract. So for $n = 1$, '$3n$' is 3 so add 2 to get to the term (5).
3) Put both bits together. So you get $3n + 2$.

Always check your expression by putting the first few values of n back in, e.g. putting $n = 1$ into $3n + 2$ gives 5, $n = 2$ gives 8, etc. which is the original sequence you were given.

Finding the **nth Term** of a **Quadratic Sequence**

A quadratic sequence has an n^2 term — the difference between the terms changes as you go through the sequence, but the difference between the differences is the same each time.

EXAMPLE **Find an expression for the *n*th term of the sequence that starts 4, 15, 32, 55...**

n:	1	2	3	4
term:	4	15	32	55
	+11	+17	+23	
	+6	+6		

So the expression will contain a $3n^2$ term.

term:	4	15	32	55
$3n^2$:	3	12	27	48
term – $3n^2$:	1	3	5	7

The expression for this linear sequence is 2n – 1

So the expression for the nth term is $3n^2 + 2n - 1$

1) Find the difference between each pair of terms.
2) The difference is changing, so work out the difference between the differences.
3) Divide this value by 2 — this gives the coefficient of the n^2 term (here it's $6 \div 2 = 3$).
4) Subtract the n^2 term from each term in the sequence. This will give you a linear sequence.
5) Find the rule for the nth term of the linear sequence (see above) and add this on to the n^2 term.

Again, make sure you check your expression by putting the first few values of n back in — so $n = 1$ gives $3(1^2) + 2(1) - 1 = 4$, $n = 2$ gives $3(2^2) + 2(2) - 1 = 15$ and so on.

Look for the common difference

For a linear sequence, find the common difference between terms and multiply n by this. For a quadratic sequence, find the common difference between the differences, divide it by 2 and multiply n^2 by this.

Q1 Find an expression for the nth term of the linear sequence 2, 9, 16, 23, ... [2 marks]

Q2 A quadratic sequence starts 6, 10, 18, 30. Find an expression for the nth term. [4 marks]

Sequences

Now you know how to find the nth terms of linear and quadratic sequences, it's time to use your skills to solve problems involving sequences.

Deciding if a Term is in a Sequence

You might be given the nth term and asked if a certain value is in the sequence. The trick here is to set the expression equal to that value and solve to find n. If n is a whole number, the value is in the sequence.

EXAMPLE

The nth term of a sequence is given by $n^2 - 2$.

Have a look at p.33 for more on solving equations.

a) Find the 6th term in the sequence.

This is dead easy — just put $n = 6$ into the expression:

$6^2 - 2 = 36 - 2$
$= 34$

b) Is 45 a term in this sequence?

Set it equal to 45... $n^2 - 2 = 45$

$n^2 = 47$ ...and solve for n.

$n = \sqrt{47} = 6.8556...$

n is not a whole number, so 45 is not in the sequence.

Other Types of Sequence

You could be asked to continue a sequence that doesn't seem to be either linear or quadratic. These sequences usually involve doing something to the previous term(s) in order to find the next one.

EXAMPLE

Find the next two terms in each of the following sequences.

a) 0.2, 0.6, 1.8, 5.4, 16.2...

1) This is an example of a geometric progression — there is a common ratio (where you multiply or divide by the same number each time).

Common Ratio = 0.6 ÷ 0.2 = 3

2) So the next two terms are:

16.2 × 3 = 48.6 and 48.6 × 3 = 145.8

b) 1, 1, 2, 3, 5...

The rule for this sequence is 'add together the two previous terms', so the next two terms are:

This is known as the Fibonacci sequence.

3 + 5 = 8 and 5 + 8 = 13

You might sometimes see sequences like these written using u_1 for the first term, u_2 for the second, u_n for the nth term. Using this notation, the rule for part a) above would be written as $u_{n+1} = 3u_n$.

Using Sequences to Solve Problems

EXAMPLE

The nth term of a sequence is given by the expression $4n - 5$.
The sum of two consecutive terms is 186. Find the value of the two terms.

Call the two terms you're looking for n and $n + 1$.
Then their sum is:

$4n - 5 + 4(n + 1) - 5 = 4n - 5 + 4n + 4 - 5 = 8n - 6$

This is equal to 186, so solve the equation:

$8n - 6 = 186$
$8n = 192$
$n = 24$

So you need to find the 24th and 25th terms:

n = 24:
(4 × 24) – 5 = 96 – 5 = 91
n = 25:
(4 × 25) – 5 = 100 – 5 = 95

n is the number of the term in the sequence, n + 1 is the one after it

n represents the position of the term, so if it's the 5th term, $n = 5$. If you're checking if a term is in a sequence and you find that n isn't a whole number, it means the term isn't in the sequence.

Q1 A linear sequence has common difference of 8. Three consecutive terms in the sequence are added together to give a total of 126. Find the three terms. [4 marks]

Q1 Video Solution

Warm-Up and Worked Exam Questions

Have a go at these warm-up questions and check that you're comfortable with them before moving on to the exam questions. If you find anything a bit tricky, go back and read over it until you understand it.

Warm-Up Questions

1) Simplify $\frac{16ab^2c^2}{4bc^4}$

2) Simplify $\frac{2y^2 - xy}{4y^2 - x^2}$

3) a) Find the first 6 terms of the sequence whose nth term is $5n + 3$.
   b) Find the first 6 terms of the sequence whose nth term is $3n + 5$.

4) Find the nth term of the following sequences:
   a) 5, 10, 15, 20, 25, ... b) 7, 10, 13, 16, 19, ...

5) How many crosses are in the nth pattern?

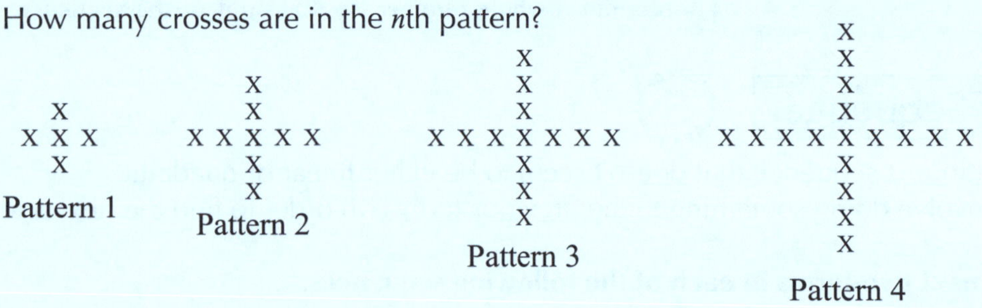

Worked Exam Questions

To ease you into the exam questions on the next page, I've done two for you (aren't I kind?). Have a look at these worked exam questions, and make sure you understand each step.

1 The first four terms in a sequence are $\sqrt{2}$, 2, $2\sqrt{2}$, 4... (Grade 5)

a) Find the next two terms in the sequence.

To get from one term to the next, multiply by $\sqrt{2}$.

The next term is $4\sqrt{2}$ and the next one is $4\sqrt{2} \times \sqrt{2} = 8$

$4\sqrt{2}$, 8 ....

[2 marks]

b) Circle the expression for the nth term of the sequence.

$\sqrt{2n}$ $n\sqrt{2}$ $(\sqrt{2})^n$ (circled) $n(\sqrt{2})^2$

[1 mark]

2 Write $\frac{2}{3} + \frac{m - 2n}{m + 3n}$ as a single fraction. (Grade 8)

$$\frac{2}{3} + \frac{m-2n}{m+3n} = \frac{2 \times (m+3n)}{3 \times (m+3n)} + \frac{3 \times (m-2n)}{3 \times (m+3n)}$$

$$= \frac{2m + 6n + 3m - 6n}{3(m+3n)}$$

$$= \frac{5m}{3(m+3n)}$$

Finding the common denominator is the tricky bit — you often just need to multiply the denominators together.

$\frac{5m}{3(m+3n)}$ ....

[3 marks]

Exam Questions

3 The term-to-term rule of a sequence is $u_{n+1} = \dfrac{-1}{2u_n}$.

a) If $u_1 = 2$, find the values of the next three terms in the sequence.

....
[2 marks]

b) Write down the value of u_{50}.

....
[1 mark]

4 The patterns below are made up of grey and white squares.

Pattern 1 Pattern 2 Pattern 3 Pattern 4

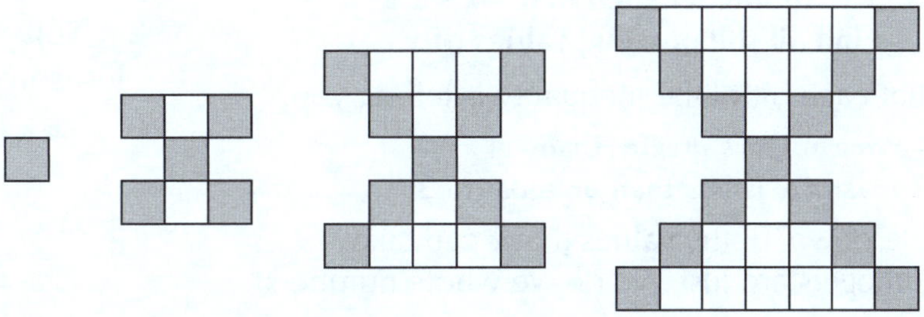

a) Find an expression for the number of **grey** squares in the *n*th pattern.

....
[2 marks]

b) Giles makes two consecutive patterns in the sequence.
He uses 414 grey squares in total. Which 2 patterns has he made?

....
[3 marks]

c) Find an expression for the **total** number of squares in the *n*th pattern.

....
[3 marks]

5 Simplify the calculation below as much as possible.

$$\frac{2a-8}{a^2-9} \div \frac{a^2-2a-8}{a^2+5a+6} \times (2a^2-a-15)$$

....
[5 marks]

Inequalities

*Inequalities aren't **half as difficult as they look**. Once you've learned the tricks involved, most of the algebra for them is **identical to ordinary equations** (have a look back at p.32-33 if you need a reminder).*

The **Inequality Symbols** (Grade 3)

$>$ means '<u>Greater than</u>'	$\geq$ means '<u>Greater than or equal to</u>'
$<$ means '<u>Less than</u>'	$\leq$ means '<u>Less than or equal to</u>'

Algebra with **Inequalities** (Grade 5)

The key thing about inequalities — solve them just like regular equations but WITH ONE BIG EXCEPTION:

Whenever you MULTIPLY OR DIVIDE by a NEGATIVE NUMBER, you must FLIP THE INEQUALITY SIGN.

EXAMPLES

1. x is an integer such that $-4 < x \leq 3$. Find all the possible values of x.

Work out what each bit of the inequality is telling you:

$-4 < x$ means 'x is greater than -4',
$x \leq 3$ means 'x is less than or equal to 3'.

Now just write down all the values that x can take. (Remember, integers are just +ve or –ve whole numbers)

$-3, -2, -1, 0, 1, 2, 3$

2. Solve $6x + 7 > x + 22$.

Just solve it like an equation:

$(-7)\quad 6x + 7 - 7 > x + 22 - 7$
$6x > x + 15$
$(-x)\quad 6x - x > x + 15 - x$
$5x > 15$
$(\div 5)\quad 5x \div 5 > 15 \div 5$
$x > 3$

3. Solve $-2 \leq \frac{x}{4} + 3 \leq 5$.

Don't be put off because there are two inequality signs — just do the same thing to each bit of the inequality:

$(-3)\quad -2 - 3 \leq \frac{x}{4} + 3 - 3 \leq 5 - 3$
$-5 \leq \frac{x}{4} \leq 2$
$(\times 4)\quad 4 \times -5 \leq \frac{4 \times x}{4} \leq 4 \times 2$
$-20 \leq x \leq 8$

4. Solve $9 - 2x > 15$.

Again, solve it like an equation:

$(-9)\quad 9 - 2x - 9 > 15 - 9$
$-2x > 6$
$(\div -2)\quad -2x \div -2 < 6 \div -2$
$x < -3$

The $>$ has turned into a $<$, because we divided by a negative number.

You Can Show Inequalities on **Number Lines** (Grade 4)

Drawing inequalities on a number line is dead easy — all you have to remember is that you use an open circle (○) for $>$ or $<$ and a coloured-in circle (●) for $\geq$ or $\leq$.

You might be asked to write inequalities in set notation (see p.169). This one would be $\{x : -4 < x \leq 3\}$.

EXAMPLE **Show the inequality $-4 < x \leq 3$ on a number line.**

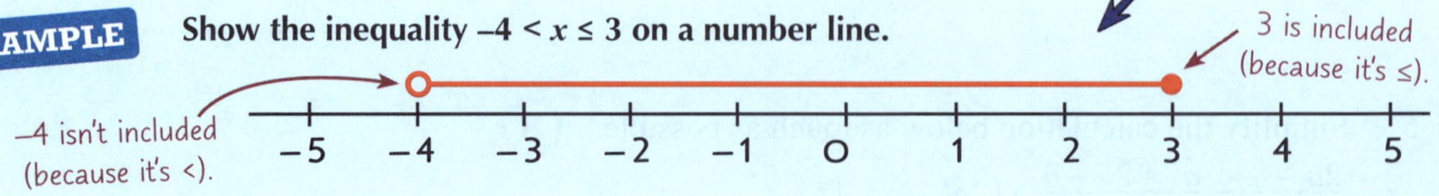

Treat inequalities like equations — but remember the exception

To check you've got the inequality sign right, substitute in a value for x and check the inequality's true.

Q1 Solve: a) $11x + 3 < 42 - 2x$ [2 marks] b) $6 - 4x \geq 18$ [2 marks] (Grade 5)

Q2 Solve the inequality $-8 \leq 5x + 2 \leq 22$ and represent the solution on a number line. [3 marks] (Grade 5)

Q2 Video Solution

Inequalities

Quadratic inequalities *are a bit tricky — you have to remember that there are* ***two solutions*** *(just like quadratic equations), so you might end up with a solution in* ***two separate bits****, or an* ***enclosed region****.*

Take Care with **Quadratic Inequalities**

GRADE 8

If $x^2 = 4$, then $x = +2$ or -2. So if $x^2 > 4$, $x > 2$ or $x < -2$ and if $x^2 < 4$, $-2 < x < 2$.

As a general rule:

If $x^2 > a^2$ then $x > a$ or $x < -a$
If $x^2 < a^2$ then $-a < x < a$

EXAMPLES

1. Solve the inequality $x^2 \leq 25$.

If $x^2 = 25$, then $x = \pm 5$.
As $x^2 \leq 25$, then $-5 \leq x \leq 5$

2. Solve the inequality $x^2 > 9$.

If $x^2 = 9$, then $x = \pm 3$.
As $x^2 > 9$, then $x < -3$ or $x > 3$

If you're confused by the '$x < -3$' bit, try putting some numbers in. E.g. $x = -4$ gives $x^2 = 16$, which is greater than 9, as required.

3. Solve the inequality $3x^2 \geq 48$.

(÷3) $\frac{3x^2}{3} \geq \frac{48}{3}$

$x^2 \geq 16$

$x \leq -4$ or $x \geq 4$

4. Solve the inequality $-2x^2 + 8 > 0$.

(−8) $-2x^2 + 8 - 8 > 0 - 8$

$-2x^2 > -8$

(÷ −2) $-2x^2 \div -2 < -8 \div -2$

$x^2 < 4$

$-2 < x < 2$

You're dividing by a negative number, so flip the sign.

Sketch the **Graph** to Help You

GRADE 9

Worst case scenario — you have to solve a quadratic inequality such as $-x^2 + 2x + 3 > 0$. Don't panic — you can use the graph of the quadratic to help (there's more on sketching quadratic graphs on p.72).

EXAMPLE **Solve the inequality $-x^2 + 2x + 3 > 0$.**

1) Start off by setting the quadratic equal to 0 and factorising:

$-x^2 + 2x + 3 = 0$

$x^2 - 2x - 3 = 0$

$(x - 3)(x + 1) = 0$

2) Now solve the equation to see where it crosses the x-axis:

$(x - 3)(x + 1) = 0$

$(x - 3) = 0$, so $x = 3$

$(x + 1) = 0$, so $x = -1$

3) Then sketch the graph — it'll cross the x-axis at -1 and 3, and because the x^2 term is negative, it'll be an n-shaped curve.

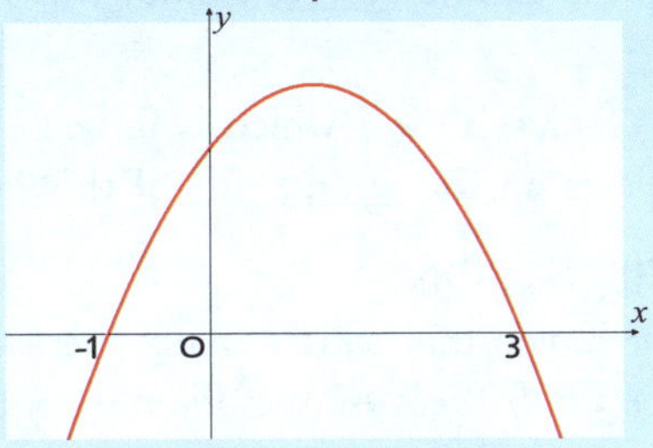

This is all the information you need to make a quick sketch to help you answer the question.

4) Now solve the inequality — you want the bit where the graph is above the x-axis (as it's a >). Reading off the graph, you can see that the solution is $-1 < x < 3$.

You can sketch graphs to help you solve quadratic inequalities

When you're sketching a quadratic graph, remember that if it has a positive x^2 term it will be a u-shaped graph, and if it has a negative x^2 term it will be an n-shaped graph. Now have a go at these Practice Questions.

Q1 Solve these inequalities: a) $p^2 < 49$ [2 marks] b) $-\frac{1}{2}p^2 \leq -32$ [3 marks] GRADE 8

Q2 Write down all the integer values that satisfy the inequality $x^2 - 4x \leq 0$. [3 marks] GRADE 9

Graphical Inequalities

*These questions always involve **shading a region on a graph**. The method sounds very complicated, but once you've seen it in action with an example, you'll see that it's OK...*

Showing **Inequalities** on a **Graph**

GRADE 6

Here's the method to follow:

1) CONVERT each INEQUALITY to an EQUATION
by simply putting an '=' in place of the inequality sign.
2) DRAW THE GRAPH FOR EACH EQUATION — if the inequality sign is < or > draw a dotted line, but if it's ≥ or ≤ draw a solid line.
3) Work out WHICH SIDE of each line you want — put a point (usually the origin) into the inequality to see if it's on the correct side of the line.
4) SHADE THE REGION this gives you.

If using the origin doesn't work (e.g. if the origin lies on the line), just pick another point with easy coordinates and use that instead.

EXAMPLE **Shade the region that satisfies all three of the following inequalities:**
$x + y < 5$ $y \leq x + 2$ $y > 1$.

1) CONVERT EACH INEQUALITY TO AN EQUATION:
$x + y = 5$, $y = x + 2$ and $y = 1$

2) DRAW THE GRAPH FOR EACH EQUATION (see p.66-67)
You'll need a dotted line for $x + y = 5$ and $y = 1$ and a solid line for $y = x + 2$.

3) WORK OUT WHICH SIDE OF EACH LINE YOU WANT
This is the fiddly bit. Put $x = 0$ and $y = 0$ (the origin) into each inequality and see if this makes the inequality true or false.

$x + y < 5$:
$x = 0$, $y = 0$ gives $0 < 5$ which is true.
This means the origin is on the correct side of the line.

$y \leq x + 2$:
$x = 0$, $y = 0$ gives $0 \leq 2$ which is true.
So the origin is on the correct side of this line.

$y > 1$:
$x = 0$, $y = 0$ gives $0 > 1$ which is false.
So the origin is on the wrong side of this line.

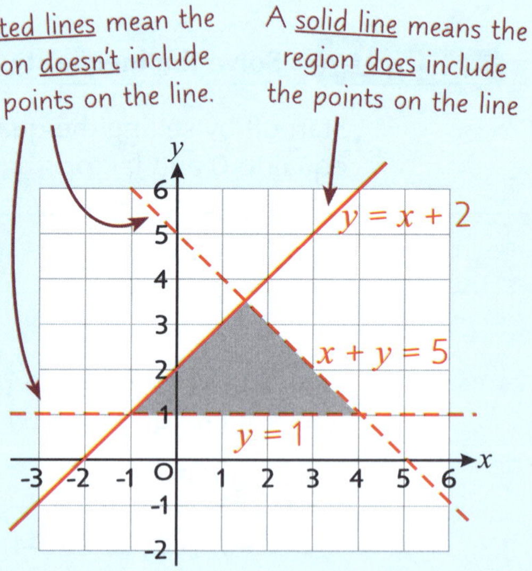

4) SHADE THE REGION
You want the region that satisfies all of these:
— below $x + y = 5$ (because the origin is on this side)
— right of $y = x + 2$ (because the origin is on this side)
— above $y = 1$ (because the origin isn't on this side).

Make sure you read the question carefully — you might be asked to label the region instead of shade it, or just mark on points that satisfy all three inequalities. No point throwing away marks because you didn't read the question properly.

Just draw the graphs and shade the region

Don't panic if you're not quite sure how to sketch the graphs — you'll find out how on p.66-67. Then you just have to make sure you shade the right region — pick a point (usually (0, 0)) and try it.

Q1 On a grid, shade the region that satisfies $x \leq 5$, $y > -1$ and $y < x + 1$. [3 marks] GRADE 6

Q1 Video Solution

Iterative Methods

Iterative methods *are techniques where you keep* ***repeating*** *a calculation in order to get closer and closer to the solution you want. You need to know all about two methods that find a solution using a* ***sign change****...*

Where There's a **Sign Change**, There's a **Solution**

GRADE 7

If you're trying to solve an equation that equals 0, there's one very important thing to remember:

> If there's a sign change (i.e. from positive to negative or vice versa) when you put two numbers into the equation, there's a solution between those numbers.

Think about the equation $x^3 - 3x - 1 = 0$. When $x = -1$, the expression gives $(-1)^3 - 3(-1) - 1 = 1$, which is positive, and when $x = -2$ the expression gives $(-2)^3 - 3(-2) - 1 = -3$, which is negative. This means that the expression will be 0 for some value between $x = -1$ and $x = -2$ (the solution).

Use **Iteration** When an Equation is **Too Hard** to Solve

GRADE 7

Not all equations can be solved using the methods you've seen so far in this section (e.g. factorising, the quadratic formula etc.). But if you know an interval that contains a solution to an equation, you can use an iterative method to find the approximate value of the solution.

This is known as the decimal search method.

EXAMPLE

A solution to the equation $x^3 - 3x - 1 = 0$ lies between –1 and –2. By considering values in this interval, find a solution to this equation to 1 d.p.

1) Try (in order) the values of x with 1 d.p. that lie between –1 and –2. There's a sign change between –1.5 and –1.6, so the solution lies in this interval.
2) Now try values of x with 2 d.p. between –1.5 and –1.6. There's a sign change between –1.53 and –1.54, so the solution lies in this interval.
3) Both –1.53 and –1.54 round to –1.5 to 1 d.p. so a solution to $x^3 - 3x - 1 = 0$ is **$x = -1.5$ to 1 d.p.**

x	$x^3 - 3x - 1$	
–1.0	1	Positive
–1.1	0.969	Positive
–1.2	0.872	Positive
–1.3	0.703	Positive
–1.4	0.456	Positive
–1.5	0.125	Positive
–1.6	–0.296	Negative
–1.51	0.087049	Positive
–1.52	0.048192	Positive
–1.53	0.008423	Positive
–1.54	–0.032264	Negative

To save time in the decimal search method, work out each end of the interval first, and start from the end that gives an answer closer to 0. If it's the top end, count down from there.

This is called the interval bisection method.

EXAMPLE

The equation $2x^3 + x^2 - 2 = 0$ has a solution between $x = 0.6$ and $x = 1.0$. Find this solution to 1 d.p.

1) Find the midpoint of the interval where the solution lies — between $x = 0.6$ and $x = 1.0$, that's 0.8.
2) There's a sign change between $x = 0.8$ and $x = 1.0$, so the solution must lie in this interval.
3) Now try the midpoint of 0.8 and 1.0, $x = 0.9$. There's a sign change between 0.8 and 0.9, so the solution is in this interval.
4) Finally, try 0.85 — there's a sign change between 0.85 and 0.9. Both of these round to 0.9 to 1 d.p. so a solution to $2x^3 + x^2 - 2 = 0$ is **$x = 0.9$ to 1 d.p.**

x	$2x^3 + x^2 - 2$	
0.6	–1.208	Negative
1.0	1.0	Positive
0.8	–0.336	Negative
0.9	0.268	Positive
0.85	–0.04925	Negative

With each sign change, you narrow the interval containing the solution.

Keep looking for the sign change until your answer is accurate enough

You can use either method in the exam, so just use the one you're more comfortable with.

Q1 a) Show that $x^3 - 3x - 1 = 0$ has another solution in the interval $1 < x < 2$. [1 mark]
b) By considering values in this interval, find this solution to 1 d.p. [3 marks]

GRADE 7

Warm-Up and Worked Exam Questions

OK, the topics in this section look a bit nasty — but for all of them, it's just a case of learning the symbols and methods and practising lots of questions...

Warm-Up Questions

1) List all integer values for x where $12 < x < 17$.
2) Find all integer values of n if $-3 \leq n \leq 3$.
3) Find all integer values of x such that $8 < 4x < 20$.
4) Solve the inequality $2q + 2 \leq 12$.
5) Solve the inequality $4p + 12 > 30$.
6) a) On the same axis, draw the graphs of $y = 0$, $y = 2x$, $y = 6 - x$.
   b) R is the region defined by the inequalities $y \leq 2x$, $y \leq 6 - x$, $y \geq 0$. Shade this region and label it R.
7) Show that $x^3 + 6x = 69$ has a solution between 3 and 4.
8) $x^3 - 11x = 100$ has a solution between 5 and 6. Use the decimal search method to find this solution to 1 d.p.

Worked Exam Questions

I'll show you how to do two exam questions, then you're on your own for the questions on the next page. Enjoy.

1 Circle the inequality that is shown on the number line below.

You want x on its own, so divide. Then think about what each circle means.

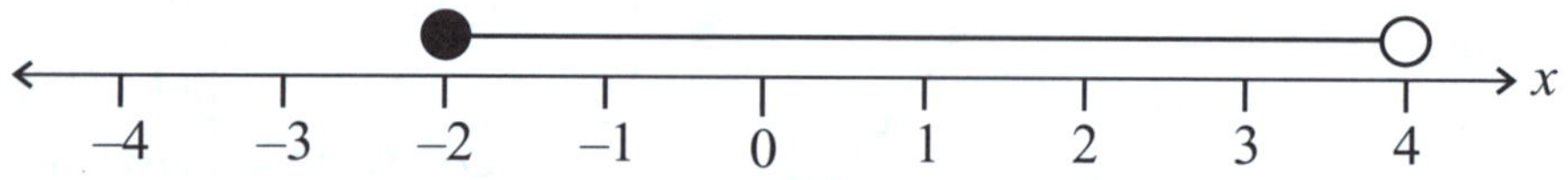

$6 < 3x < 9$ → $2 < x < 3$

(circled) $-4 \leq 2x < 8$ → $-2 \leq x < 4$

$-8 \leq 4x \leq 16$ → $-2 \leq x \leq 4$

$-10 \leq 5x < 15$ → $-2 \leq x < 3$

[1 mark]

2 Look at the grid below.

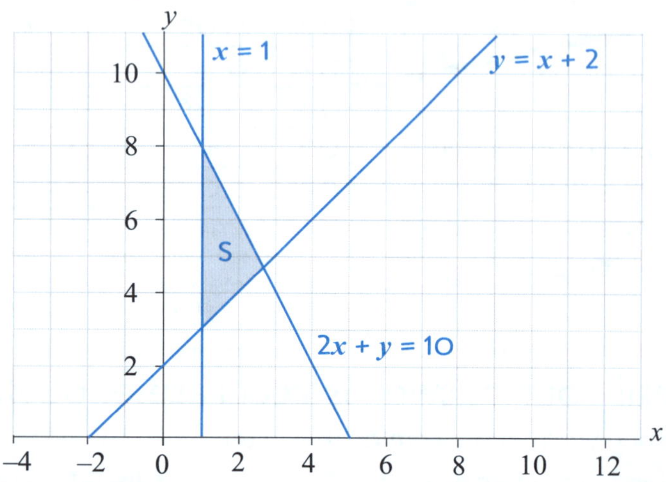

a) Use the grid to draw the graphs of $2x + y = 10$ and $y = x + 2$.

$y = 10 - 2x$

It can help to rearrange the equations into the form $y = mx + c$ first.

[2 marks]

b) Shade and label, using the letter S, the area represented by the inequalities $x \geq 1$, $2x + y \leq 10$, $y \geq x + 2$.

[2 marks]

Exam Questions

3 Find the largest three consecutive even numbers that sum to less than 1000.

....

[3 marks]

4 Find the integer values that satisfy both of the following inequalities:

$5n - 3 \leq 17$ and $2n + 6 > 8$

....

[3 marks]

5 The equation $3x - 2x^3 + 5 = 0$ has one solution.

a) Show that this solution lies in the interval $1.6 < x < 2$.

[2 marks]

b) By filling in the table below, find an approximation to the solution of $3x - 2x^3 + 5 = 0$ to 1 d.p. You might not need to use all the rows.

x	$3x - 2x^3 + 5$	
1.6		
2		
1.8		

Use the interval bisection method here.

$x =$ ....

[4 marks]

6 Solve the following inequalities.

a) $5x^2 < 80$

....

[3 marks]

b) $x^2 + 1 < x + 7$

....

[3 marks]

Simultaneous Equations 1

There are two types of simultaneous equations you could get — EASY ONES (where both equations are linear) and TRICKY ONES (where one's quadratic).

1. $2x = 6 - 4y$ and $-3 - 3y = 4x$
2. $7x + y = 1$ and $2x^2 - y = 3$

1 Six Steps for **Easy Simultaneous Equations**

GRADE 5

EXAMPLE **Solve the simultaneous equations $2x = 6 - 4y$ and $-3 - 3y = 4x$**

1) Rearrange both equations into the form $ax + by = c$, and label the two equations ① and ②.

a, b and c are numbers (which can be negative)

$2x + 4y = 6$ — ①
$4x + 3y = -3$ — ②

2) Match up the numbers in front (the 'coefficients') of either the x's or y's in both equations. You may need to multiply one or both equations by a suitable number. Relabel them ③ and ④.

① × 2: $4x + 8y = 12$ — ③
$4x + 3y = -3$ — ④

3) Add or subtract the two equations to eliminate the terms with the same coefficient.

③ – ④ $0x + 5y = 15$

If the coefficients have the same sign (both +ve or both –ve) then subtract. If the coefficients have opposite signs (one +ve and one –ve) then add.

4) Solve the resulting equation.

$5y = 15 \Rightarrow y = 3$

5) Substitute the value you've found back into equation ① and solve it.

Sub $y = 3$ into ①: $2x + (4 \times 3) = 6 \Rightarrow 2x + 12 = 6 \Rightarrow 2x = -6 \Rightarrow x = -3$

6) Substitute both these values into equation ② to make sure it works. If it doesn't then you've done something wrong and you'll have to do it all again.

Sub x and y into ②: $(4 \times -3) + (3 \times 3) = -12 + 9 = -3$, which is right, so it's worked.

So the solutions are: $x = -3$, $y = 3$

And these are the easy simultaneous equations

It might just be me, but I think simultaneous equations are quite fun... well, maybe not fun... but quite satisfying. Anyway, it doesn't matter whether you like them or not — you have to learn how to do them.

Q1 Issy buys two cups of tea and three slices of cake for £9.
Rudy buys four cups of tea and one slice of cake from the same cafe for £8.
Find the cost of one cup of tea and the cost of one slice of cake. [3 marks] GRADE 5

Q2 Find x and y given that $2x - 10 = 4y$ and $3y = 5x - 18$. [3 marks] GRADE 5

Q1 Video Solution

Simultaneous Equations 2

2 Seven Steps for TRICKY Simultaneous Equations

GRADE 8

EXAMPLE **Solve these two equations simultaneously:** $7x + y = 1$ **and** $2x^2 - y = 3$

1) Rearrange the quadratic equation so that you have the non-quadratic unknown on its own. Label the two equations ① and ②.

$7x + y = 1$ — ① $y = 2x^2 - 3$ — ②

You could also rearrange the linear equation and substitute it into the quadratic.

2) Substitute the quadratic expression into the other equation. You'll get another equation — label it ③.

$7x + y = 1$ — ①
$y = 2x^2 - 3$ — ②
$\Rightarrow 7x + (2x^2 - 3) = 1$ — ③

Put the expression for y into equation ① in place of y.

3) Rearrange to get a quadratic equation. And guess what... You've got to solve it.

$2x^2 + 7x - 4 = 0$
$(2x - 1)(x + 4) = 0$
So $2x - 1 = 0$ OR $x + 4 = 0$
$x = 0.5$ OR $x = -4$

Remember — if it won't factorise, you can either use the formula or complete the square. Have a look at p.40-42 for more details.

4) Stick the first value back in one of the original equations (pick the easy one).

① $7x + y = 1$
Substitute in $x = 0.5$: $3.5 + y = 1$, so $y = 1 - 3.5 = -2.5$

5) Stick the second value back in the same original equation (the easy one again).

① $7x + y = 1$
Substitute in $x = -4$: $-28 + y = 1$, so $y = 1 + 28 = 29$

6) Substitute both pairs of answers back into the other original equation to check they work.

② $y = 2x^2 - 3$
Substitute in $x = 0.5$: $y = (2 \times 0.25) - 3 = -2.5$
Substitute in $x = -4$: $y = (2 \times 16) - 3 = 29$

7) Write the pairs of answers out again, clearly, at the bottom of your working.

The two pairs of solutions are: $x = 0.5, y = -2.5$ and $x = -4, y = 29$

The solutions to simultaneous equations are actually the coordinates of the points where the graphs of the equations cross — so in this example, the graphs of $7x + y = 1$ and $2x^2 - y = 3$ will cross at (0.5, –2.5) and (–4, 29). There's more on this on p.77.

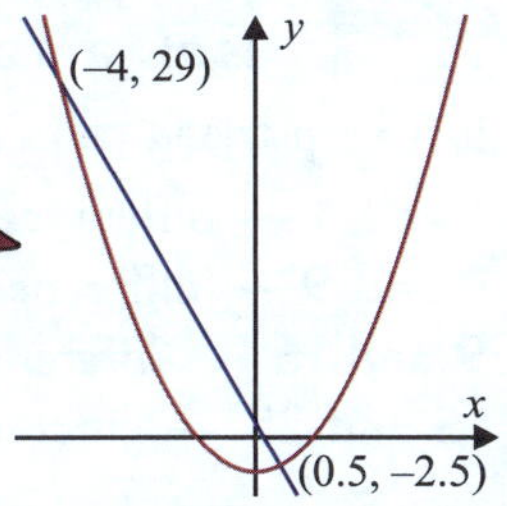

Remember to write the two pairs out clearly

Don't get confused — you end up with 2 pairs of solutions, not 4 separate solutions.

Q1 Solve the simultaneous equations $y = 2 - 3x$ and $y + 2 = x^2$. [4 marks] GRADE 8

Q2 Find the coordinates of A and B, the points where the graphs of $y = x^2 + 4$ and $y - 6x - 4 = 0$ intersect, and use your answer to find the exact length of AB. [5 marks] GRADE 8

Q1 Video Solution

Proof

*I'm not going to lie — **proof questions** can look a bit terrifying. There are **loads** of things you could be asked to prove — I'll start with some **algebraic** proofs on this page, then move on to **wild and wonderful** topics.*

Show Things Are **Odd, Even** or **Multiples** by **Rearranging**

Before you get started, there are a few things you need to know — they'll come in very handy when you're trying to prove things.

- Any even number can be written as $2n$ — i.e. 2 × something.
- Any odd number can be written as $2n + 1$ — i.e. 2 × something + 1.
- Consecutive numbers can be written as n, $n + 1$, $n + 2$ etc. — you can apply this to e.g. consecutive even numbers too (they'd be written as $2n$, $2n + 2$, $2n + 4$). (In all of these statements, n is just any integer.)
- The sum, difference and product of integers is always an integer.

This can be extended to multiples of other numbers too — e.g. to prove that something is a multiple of 5, show that it can be written as 5 × something.

EXAMPLE **Prove that the sum of any three odd numbers is odd.** (GRADE 6)

So what you're trying to do here is show that the sum of three odd numbers can be written as (2 × integer) + 1.

Take three odd numbers: $2a + 1$, $2b + 1$ and $2c + 1$ (they don't have to be consecutive)

Add them together:

$2a + 1 + 2b + 1 + 2c + 1 = 2a + 2b + 2c + 2 + 1$
$= 2(a + b + c + 1) + 1$
$= 2n + 1$ where n is an integer $(a + b + c + 1)$

You'll see why I've written 3 as 2 + 1 in a second.

So the sum of any three odd numbers is odd.

EXAMPLE **Prove that $(n + 3)^2 - (n - 2)^2 \equiv 5(2n + 1)$.** (GRADE 5)

Take one side of the equation and play about with it until you get the other side:

LHS: $(n + 3)^2 - (n - 2)^2 \equiv n^2 + 6n + 9 - (n^2 - 4n + 4)$
$\equiv n^2 + 6n + 9 - n^2 + 4n - 4$
$\equiv 10n + 5$
$\equiv 5(2n + 1) = \text{RHS}$ ✓

$\equiv$ is the identity symbol, and means that two things are identically equal to each other. So $a + b \equiv b + a$ is true for all values of a and b (unlike an equation, which is only true for certain values).

Disprove Things by Finding a **Counter Example**

If you're asked to prove a statement isn't true, all you have to do is find one example that the statement doesn't work for — this is known as disproof by counter example.

EXAMPLE **Amy says "the difference between any two consecutive square numbers is always a prime number". Prove that Amy is wrong.** (GRADE 4)

Just keep trying pairs of consecutive square numbers (e.g. 1^2 and 2^2) until you find one that doesn't work:

1 and 4 — difference = 3 (a prime number)
4 and 9 — difference = 5 (a prime number)
9 and 16 — difference = 7 (a prime number)
16 and 25 — difference = 9 (NOT a prime number) so Amy is wrong.

You don't have to go through loads of examples if you can spot one that's wrong straight away — you could go straight to 16 and 25.

Proof questions aren't as bad as they look

The only way to get on top of proof questions is practice — so start with these:

Q1 Prove that the sum of two consecutive even numbers is even. [3 marks] (GRADE 6)

Q2 $4x + 2 = 3(3a + x)$. For odd integer values of a, prove that x is never a multiple of 8. [3 marks] (GRADE 7)

Proof

*There's **no set method** for proof questions — you have to think about all the things you're **told** in the question (or that you **know** from other areas of maths) and **jiggle them around** until you've come up with a proof.*

Proofs Will Test You on **Other Areas of Maths**

You could get asked just about anything in a proof question, from power laws...

EXAMPLE **Show that the difference between 10^{18} and 6^{21} is a multiple of 2.** GRADE 8

$10^{18} - 6^{21} = (10 \times 10^{17}) - (6 \times 6^{20})$

$= (2 \times 5 \times 10^{17}) - (2 \times 3 \times 6^{20}) = 2[(5 \times 10^{17}) - (3 \times 6^{20})]$

which can be written as $2x$ where $x = [(5 \times 10^{17}) - (3 \times 6^{20})]$ so is a multiple of 2.

... to questions on mean, median, mode or range (see p.174)...

EXAMPLE **The range of a set of positive numbers is 5. Each number in the set is doubled. Show that the range of the new set of numbers also doubles.** GRADE 6

Let the smallest value in the first set of numbers be n.

Then the largest value in this set is n + 5 (as the range for this set is 5).

When the numbers are doubled, the smallest value in the new set is 2n and the largest value is 2(n + 5) = 2n + 10.

To find the new range, subtract the smallest value from the largest:

2n + 10 – 2n = 10 = 2 × 5, which is double the original range.

... or ones where you have to use inequalities (see p.50-51)...

EXAMPLE **Ellie says, "If $x > y$, then $x^2 > y^2$". Is she correct? Explain your answer.** GRADE 5

Try some different values for x and y:

$x = 2, y = 1$: $x > y$ and $x^2 = 4 > 1 = y^2$

$x = 5, y = 2$: $x > y$ and $x^2 = 25 > 4 = y^2$

At first glance, Ellie seems to be correct. BUT... $x = -1, y = -2$: $x > y$ but $x^2 = 1 < 4 = y^2$,

so Ellie is wrong as the statement does not hold for all values of x and y.

This is an example of finding a counter example — see previous page.

... or even geometric proofs (see section 5 for more on geometry).

EXAMPLE **Prove that the sum of the exterior angles of a triangle is 360°.** GRADE 5

First sketch a triangle with angles a, b and c:

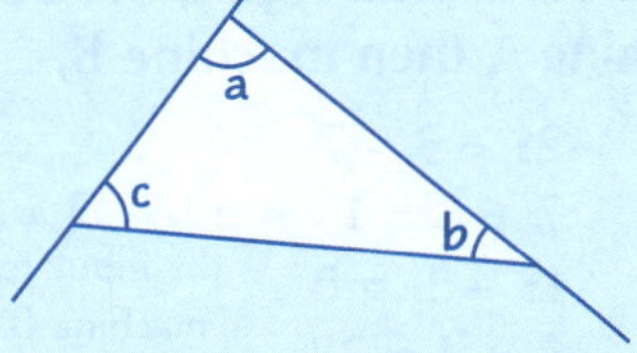

Then the exterior angles are:

180° – a, 180° – b and 180° – c

So their sum is:

(180° – a) + (180° – b) + (180° – c)

= 540° – (a + b + c) = 540° – 180° (as the angles in a triangle add up to 180°)

= 360°

It's hard to revise for proof questions, but practice definitely helps

Proof questions can cover any topic — you could even get a proof hidden inside another question.

Q1 Prove that the difference between two consecutive square numbers is always odd. [3 marks] GRADE 6

Q2 Triangle numbers are formed from the expression $\frac{1}{2}n(n + 1)$. Prove that the ratio between the nth and $(n + 1)$th triangle numbers is always $n : n + 2$. [2 marks] GRADE 7

Q2 Video Solution

Functions

A ***function*** *takes an* ***input*** *(x),* ***processes*** *it and* ***outputs*** *a value (y). They're often written as* ***function machines****.*
You need to know how to use them ***forwards*** *and* ***backwards****.*

Evaluating Functions — Just Put the Numbers In

This is easy — just shove the numbers into the function machine and you're away.

EXAMPLE **Find the value of y when $x = 36$ is put into the function machine below.**

x → $\div 3$ → $+ 6$ → y

Machine A

$36 \div 3 = 12$
$12 + 6 = 18$

The function represented by the machine is $y = (x \div 3) + 6$.

Inverse Functions — Work **Backwards**

The inverse of a function is another function which reverses the original function — it takes a y-value and processes it to produce an x-value. To find it, just work backwards through the machine — make sure you reverse the operations themselves and the order in which you do them.
So for machine A, above, the inverse function is $x = (y - 6) \times 3$.

EXAMPLE **When an x-value is put into the function machine below, the output is $y = 56$. What x-value was used as the input?**

x → $- 5$ → $\times 4$ → y

Machine B

The inverse function is $x = (y \div 4) + 5$.
$56 \div 4 = 14$
$14 + 5 = 19$

You can check your answer by seeing if the input you've found produces the same output: $19 - 5 = 14$, $14 \times 4 = 56$, as expected.

Combining Functions — Do One Function Then Another

1) You might get a question with two functions — you have to do one function then the other to combine them into a single function (called a composite function).
2) All you have to do is put your input into the first function, find the output, then use the output as the input of the second function.
3) The order of the functions matters — you will usually get a different result if you apply the functions in different orders, so be careful.

The first function takes an x-value and turns it into a y-value. The second function takes the y-value and produces a z-value. So the composite function takes the x-value and produces the z-value.

EXAMPLE **The function machines above are labelled A and B.**
For an x-value of 21, find the output when it is put into:
a) machine A then machine B, **b) machine B then machine A.**

a) Machine A: $21 \div 3 = 7$
$7 + 6 = 13$ ← Use 13 as the input for machine B.
Machine B: $13 - 5 = 8$
$8 \times 4 = 32$

b) Machine B: $21 - 5 = 16$
$16 \times 4 = 64$
Machine A: $64 \div 3 = 21\frac{1}{3}$
$21\frac{1}{3} + 6 = 27\frac{1}{3}$

You can see that the results are different when you do the machines in a different order.

Keep track of inputs and outputs

When working with composite functions, the number that is output from the first function becomes the input for the second function. So jotting down what comes out of each function will make your life much easier.

Q1 A number is input into function machine B above. The output is then put back into the same machine. The final output is 220. What was the original number? [2 marks]

Warm-Up and Worked Exam Questions

There was a lot of maths to take in over the last few pages — but you can breathe a sigh of relief as you've made it to the end of the algebra section. Just a few pages of questions to go.

Warm-Up Questions

1) Find x and y given that $5 - 2y = 3x$ and $2x = 5y + 16$.
2) Solve the simultaneous equations $2x + 3y = 19$ and $2x + y = 9$.
3) Solve the simultaneous equations $2x + y = 1$ and $y = 4 - x^2$.
4) Prove that the product of two odd numbers is odd.
5) Prove that $(2x - 5)^2 - (x - 1)^2$ is a multiple of 3 for all integer values of x.
6) Calculate the x-value that was input into the function machine below.

x → [÷ 4] → [+ 12] → $y = 21$

Worked Exam Questions

Here are a couple of worked examples for you. Make the most of them, because the next page is full of exam-type questions for you to have a go at on your own.

1 Solve the following pair of simultaneous equations. (Grade 8)

$$x^2 + y = 4$$
$$y = 4x - 1$$

$y = 4x - 1$ ① $y = 4 - x^2$ ②

Rearrange the quadratic equation to get y on its own, then sub the expression for y into the other equation. Then you just need to solve it.

Substitute ② into ①:

$(4 - x^2) = 4x - 1$

$x^2 + 4x - 5 = 0$

$(x + 5)(x - 1) = 0$

$x = -5$ or $x = 1$

When $x = 1$, $y = (4 \times 1) - 1 = 3$

When $x = -5$, $y = (4 \times -5) - 1 = -21$

So the solutions are $x = 1$, $y = 3$ and $x = -5$, $y = -21$

$x =$ ....1...., $y =$ ....3....

and $x =$ ....−5...., $y =$ ....−21....

[5 marks]

2 Prove that the difference between the squares of two consecutive even numbers is always a multiple of 4. (Grade 7)

n is an integer. 2n represents any even number, so the difference between the squares of two consecutive even numbers will be given by:

$$(2n + 2)^2 - (2n)^2 = (4n^2 + 8n + 4) - 4n^2$$
$$= 8n + 4 = 4(2n + 1)$$
$$= 4x \text{ (where } x \text{ is an integer given by } x = 2n + 1)$$

Any integer multiplied by 4 is a multiple of 4, so $4x$ must be a multiple of 4 and therefore the difference between the squares of two consecutive even numbers will always be a multiple of 4.

[3 marks]

Exam Questions

3 Solve this pair of simultaneous equations. (Grade 5)

$$2x + 3y = 12$$
$$5x + 4y = 9$$

$x =$ .... $y =$ ....

[4 marks]

4 Meena says "If $a < b < c < d$ (where b and d are not zero), then $\frac{a}{b} < \frac{c}{d}$". (Grade 6)
Is she correct? Explain your answer.

[3 marks]

5 Two function machines, A and B, are shown below. (Grade 6)

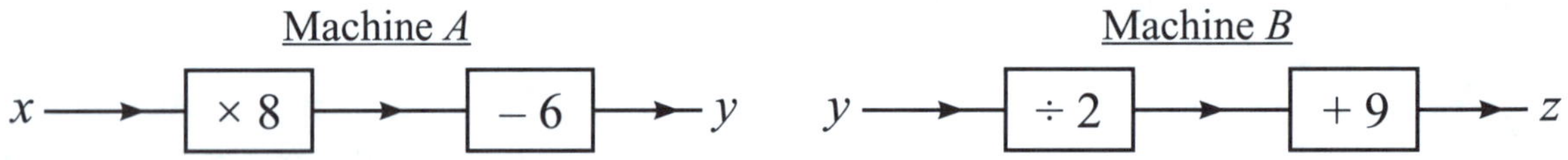

a) Sarah wants to combine machine A and machine B into one machine, C, which gives an output of z when you input x.

Fill in the empty boxes below to show a possible simplified function machine C.

[3 marks]

b) Amy wants to reverse Sarah's machine.
Write the equation that will allow Amy to find the value of x, given a value of z.

....

[2 marks]

6 Show that the number $2^{64} - 1$ is not prime. (Grade 9)

[3 marks]

Revision Questions for Section Two

It's time to check how much of Section Two you've learned...

- Try these questions and tick off each one when you get it right.
- When you're completely happy with a topic, tick that off too.

For even more practice, try the quiz for Section Two — just scan this QR code!

Section Two Quiz

Algebra (p24-29)

1) Simplify by collecting like terms: $3x + 2y - 5 - 6y + 2x$
2) Simplify the following: a) $x^3 \times x^6$ b) $y^7 \div y^5$ c) $(z^3)^4$
3) Multiply out these brackets: a) $3(2x + 1)$ b) $(x + 2)(x - 3)$ c) $(x - 1)(x + 3)(x + 5)$
4) Factorise: a) $8x^2 - 2y^2$ b) $49 - 81p^2q^2$ c) $12x^2 - 48y^2$
5) Simplify the following: a) $\sqrt{27}$ b) $\sqrt{125} \div \sqrt{5}$
6) Write $\sqrt{98} + 3\sqrt{8} - \sqrt{200}$ in the form $a\sqrt{2}$, where a is an integer.

Solving Equations and Rearranging Formulas (p32-35)

7) Solve these equations: a) $5(x + 2) = 8 + 4(5 - x)$ b) $x^2 - 21 = 3(5 - x^2)$
8) Make p the subject of these: a) $\frac{p}{p+y} = 4$ b) $\frac{1}{p} = \frac{1}{q} + \frac{1}{r}$

Quadratics (p38-42)

9) Solve the following by factorising them first: a) $x^2 + 9x + 18 = 0$ b) $5x^2 - 17x - 12 = 0$
10) Write down the quadratic formula.
11) Find the solutions of these equations (to 2 d.p.) using the quadratic formula:
a) $x^2 + x - 4 = 0$ b) $5x^2 + 6x = 2$ c) $(2x + 3)^2 = 15$
12) Find the exact solutions of these equations by completing the square:
a) $x^2 + 12x + 15 = 0$ b) $x^2 - 6x = 2$
13) The graph of $y = x^2 + px + q$ has a turning point at (2, 5). Find the values of p and q.

Algebraic Fractions (p45)

14) Write $\frac{2}{x+3} + \frac{1}{x-1}$ as a single fraction.

Sequences (p46-47)

15) Find the expression for the nth term in the following sequences:
a) 7, 9, 11, 13 b) 11, 8, 5, 2 c) 5, 9, 15, 23.
16) The nth term of a sequence is given by $n^2 + 7$. Is 32 a term in this sequence?

Inequalities (p50-52)

17) Solve the following inequalities: a) $4x + 3 \le 6x + 7$ b) $5x^2 > 180$
18) Show on a graph the region described by these conditions: $x + y \le 6$, $y > 0.5$, $y \le 2x - 2$

Iterative Methods (p53)

19) Show that the equation $x^3 - 4x^2 + 2x - 3 = 0$ has a solution between $x = 3$ and $x = 4$.

Simultaneous Equations (p56-57)

20) Solve the following pair of simultaneous equations: $4x + 5y = 23$ and $3y - x = 7$
21) Solve these simultaneous equations: $y = 3x + 4$ and $x^2 + 2y = 0$

Proof and Functions (p58-60)

22) Prove that the product of an odd number and an even number is even.
23) A function machine takes an x-value, squares it then subtracts 8. What is the output when $x = 7$?

Straight Lines and Gradients

To start off with, here's some basic stuff about straight-line graphs. Get stuck into it — it'll get you off to a running start.

Learn to Spot These Straight-Line Equations

GRADE 3

If an equation has a y and/or x but no higher powers (like x^2 or x^3), then it's a straight-line equation.

Vertical and horizontal lines: 'x = a' and 'y = a'

'x = a' is a vertical line through 'a' on the x-axis.

'y = a' is a horizontal line through 'a' on the y-axis.

The equation of the x-axis is $y = 0$. The equation of the y-axis is $x = 0$.

The main diagonal through the origin: '$y = x$'

'$y = x$' is the main diagonal that goes UPHILL from left to right.

$y = -x$ is the main diagonal sloping downhill.

Other sloping lines through the origin: '$y = ax$'

The value of 'a' is the gradient — see below.

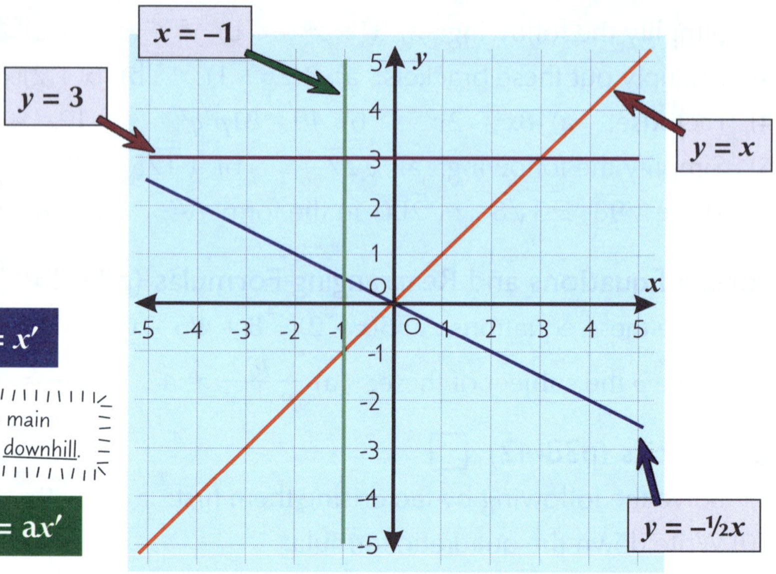

The Gradient is the Steepness of the Line

GRADE 3

The gradient of the line is how steep it is — the larger the gradient, the steeper the slope.
A negative gradient tells you it slopes downhill. You find it by dividing the change in y by the change in x.

EXAMPLE Find the gradient of the straight line to the right.

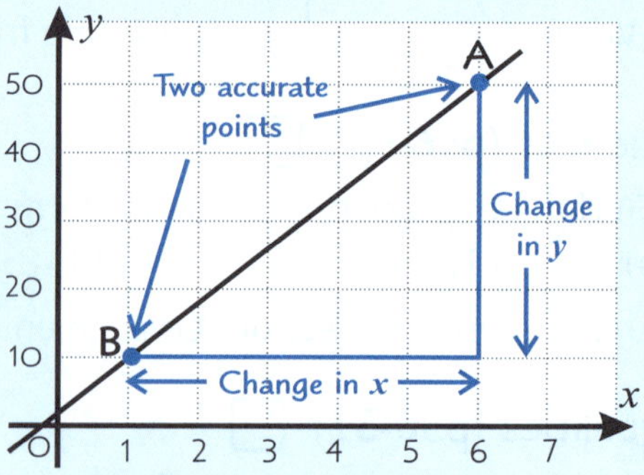

1. Choose two accurate points on the line. A: (6, 50) B: (1, 10)
2. Find the change in y and change in x.

   Change in y = 50 – 10 = 40
   Change in x = 6 – 1 = 5

   Make sure you subtract the y and x-coordinates in the same order. E.g. $y_A - y_B$ and $x_A - x_B$
3. Use this formula: $\text{GRADIENT} = \dfrac{\text{CHANGE IN Y}}{\text{CHANGE IN X}}$ Gradient = $\frac{40}{5}$ = 8

Always check the sign of your gradient. Remember, uphill = positive and downhill = negative.

Learn these simple straight lines

Vertical line: $x = a$, horizontal line: $y = a$, main diagonals: $y = x$ and $y = -x$.
Remember — to find the gradient, mark two points on the line then divide the change in y by the change in x. Try this Practice Question.

Q1 Find the gradient of the line shown on the right. [2 marks] GRADE 3

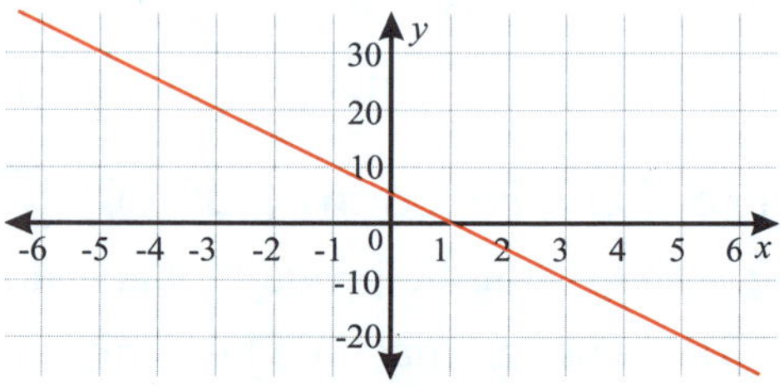

y = mx + c

*Using **'y = mx + c'** is the most straightforward way of dealing with straight-line equations, and it's very useful in exams. The first thing you have to do though is **rearrange** the equation into the standard format like this:*

Straight line:		Rearranged into 'y = mx +c'	
$y = 2 + 3x$	→	$y = 3x + 2$	(m = 3, c = 2)
$x - y = 0$	→	$y = x + 0$	(m = 1, c = 0)
$4x - 3 = 5y$	→	$y = \frac{4}{5}x - \frac{3}{5}$	(m = $\frac{4}{5}$, c = $-\frac{3}{5}$)

where:
'm' = gradient of the line.
'c' = 'y-intercept' (where it hits the y-axis)

WATCH OUT: people mix up 'm' and 'c' when they get something like $y = 5 + 2x$.
Remember, 'm' is the number in front of the 'x' and 'c' is the number on its own.

Finding the **Equation** of a Straight-Line **Graph**

GRADE 4

When you're given the graph itself, it's quick and easy to find the equation of the straight line.

EXAMPLE

Find the equation of the line on the graph in the form $y = mx + c$.

1) Find 'm' (gradient) and 'c' (y-intercept).

$$\text{'m'} = \frac{\text{change in } y}{\text{change in } x} = \frac{15}{30} = \frac{1}{2}$$

'c' = 15

2) Use these to write the equation in the form $y = mx + c$.

$$y = \frac{1}{2}x + 15$$

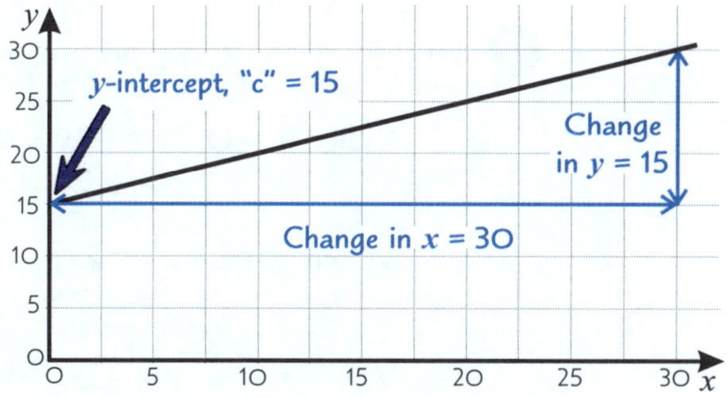

Finding the **Equation** of a Line **Through Two Points**

GRADE 5

If you're given two points on a line you can find the gradient, then you can use the gradient and one of the points to find the equation of the line. This is super handy, so practise it until you can do it in your sleep.

EXAMPLE

Find the equation of the straight line that passes through (–2, 9) and (3, –1). Give your answer in the form $y = mx + c$.

1) Use the two points to find 'm' (gradient).

$$m = \frac{\text{change in } y}{\text{change in } x} = \frac{-1 - 9}{3 - (-2)} = \frac{-10}{5} = -2$$

So $y = -2x + c$

2) Substitute one of the points into the equation you've just found.

Substitute (–2, 9) into eqn: $9 = -2(-2) + c$
$9 = 4 + c$

3) Rearrange the equation to find 'c'.

$c = 9 - 4 = 5$

4) Substitute back into $y = mx + c$:

$y = -2x + 5$

Sometimes you'll be asked to give your equation in other forms such as $ax + by + c = 0$.
Just rearrange your $y = mx + c$ equation to get it in this form. It's no biggie.

m is the gradient and c is the y-intercept

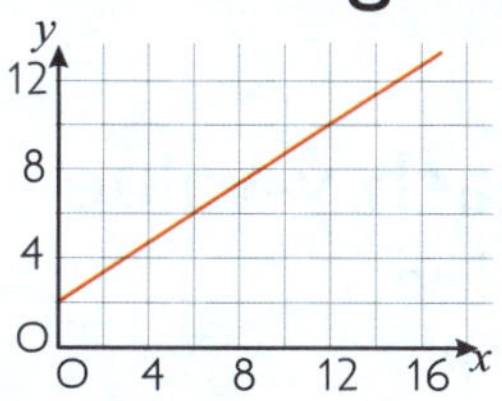

Remember the steps for finding equations and try out your new-found graph skills.

Q1 Find the equation of the line on the graph to the left. [2 marks] (GRADE 4)

Q2 Line Q goes through (0, 5) and (4, 7).
Find the equation of Line Q in the form $y = mx + c$. [3 marks] (GRADE 5)

Q2 Video Solution

Drawing Straight-Line Graphs

You've got three methods for ***drawing straight-line graphs*** *on these two pages.*
Make sure you're happy with ***all three****.*

The '**Table of 3 Values**' Method

GRADE 3

EXAMPLE **Draw the graph of $y = -2x + 4$ for values of x from –1 to 4.**

1. Draw up a table with three suitable values of x.

x	0	2	4
y			

2. Find the y-values by putting each x-value into the equation:

When $x = 0$, $y = -2x + 4$
$= (-2 \times 0) + 4 = 4$

When $x = 2$, $y = -2x + 4$
$= (-2 \times 2) + 4 = 0$

When $x = 4$, $y = -2x + 4$
$= (-2 \times 4) + 4 = -4$

x	0	2	4
y	4	0	–4

3. Plot the points and draw the line.

The table gives the points (0, 4), (2, 0) and (4, –4)

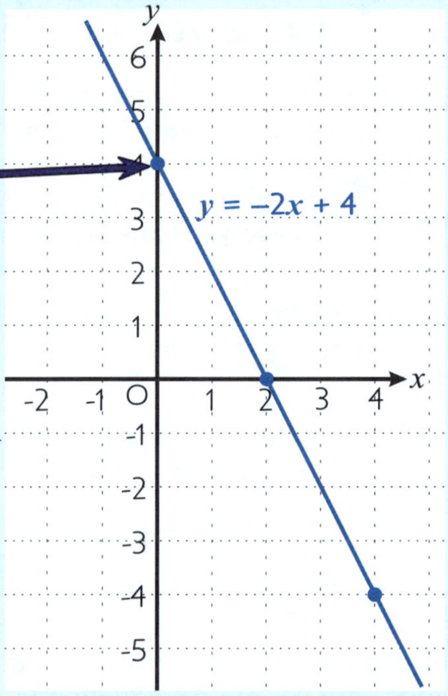

If it's a straight-line equation, the 3 points will be in a dead straight line with each other. If they aren't, you need to go back and CHECK YOUR WORKING.

Put the x-values from your table into the equation to get each y-value

Make sure you use a ruler when drawing straight-line graphs. Also, it's good to check your line afterwards by looking at the equation. Remember, 'm' is the gradient and 'c' is the y-intercept.

Drawing Straight-Line Graphs

Using y = mx + c

EXAMPLE **Draw the graph of $4y - 2x = -4$.**

1. Get the equation into the form $y = mx + c$. $4y - 2x = -4 \rightarrow y = \frac{1}{2}x - 1$

2. Put a dot on the <u>y-axis</u> at the <u>value of c</u>. 'c' = –1

3. Using <u>m</u>, go across and up or down a certain number of units. Make another dot, then repeat this step a few times in both directions. Go <u>2 along</u> and <u>1 up</u> because 'm' = $+\frac{1}{2}$.

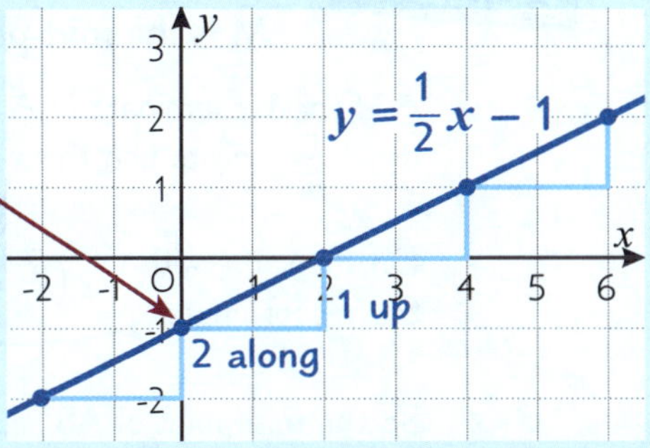

4. When you have 4 or 5 dots, draw a <u>straight line</u> through them.

5. Finally check that the <u>gradient</u> looks right. A gradient of $+\frac{1}{2}$ should be <u>quite gentle</u> and <u>uphill</u> left to right — which it is, so it looks OK.

The 'x = 0, y = 0' Method

Here's a third method for drawing <u>straight lines</u>. This one's really handy if you just want to do a <u>sketch</u>.

EXAMPLE **Sketch the straight line $y = 3x - 5$ on the diagram.**

Don't forget to label your line.

1) <u>Set $x = 0$</u> in the equation, and <u>find y</u> — this is where it <u>crosses the y-axis</u>.
$y = 3x - 5$. When $x = 0$, $y = -5$.

2) <u>Set $y = 0$</u> in the equation and <u>find x</u> — this is where it <u>crosses the x-axis</u>.
When $y = 0$, $0 = 3x - 5$. So $x = \frac{5}{3}$.

3) Mark on the two <u>points</u> and <u>draw a line</u> passing <u>through</u> them.

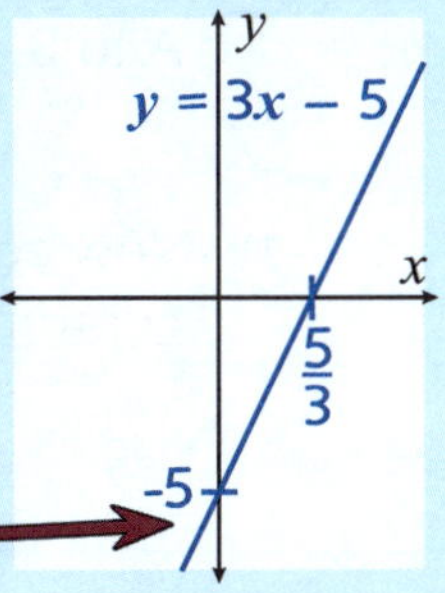

Drawing straight-line graphs isn't as scary with these simple methods

These pages give you three simple methods for drawing straight-line graphs.
Learn them all, just in case you have to use a specific one in the exam. Then try this Practice Question.

Q1 Sketch the graph of $5y + 2x = 20$. [3 marks]

Coordinates and Ratio

Now you're all clued up on the equations of straight lines, it's time to move on to ***line segments****.*
Instead of going on forever, a line segment is the ***part of a line*** *between two* ***end points****.*

Find the **Mid-Point** Using the **Average** of the **End Points**

GRADE 5

To find the mid-point of a line segment, just add the x-coordinates and divide by two, then do the same for the y-coordinates.

EXAMPLE **Points A and B are given by the coordinates (7, 4) and (–1, –2) respectively. M is the mid-point of the line segment AB. Find the coordinates of M.**

Add the x-coordinate of A to the x-coordinate of B and divide by two to find the x-coordinate of the midpoint.

Do the same with the y-coordinates.

$$\left(\frac{7+-1}{2}, \frac{4+-2}{2}\right) = \left(\frac{6}{2}, \frac{2}{2}\right) = (3, 1)$$

So the mid-point of AB has coordinates (3, 1)

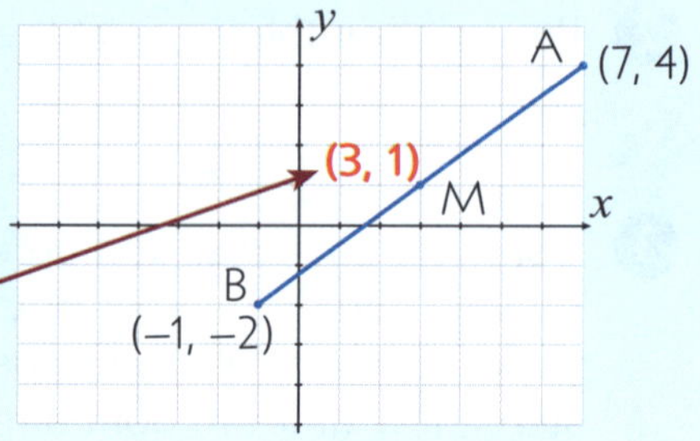

Use **Ratios** to Find **Coordinates**

GRADE 7

Ratios can be used to express where a point is on a line.
You can use a ratio to find the coordinates of a point.

EXAMPLE **Point A has coordinates (–3, 5) and point B has coordinates (18, 33). Point C lies on the line segment AB, so that AC : CB = 4 : 3**
Find the coordinates of C.

1) First find the difference between the coordinates of A and B:
   Difference in x-coordinates: 18 – –3 = 21
   Difference in y-coordinates: 33 – 5 = 28

2) Now look at the ratio you've been given: AC : CB = 4 : 3

3) The ratio tells you C is $\frac{4}{7}$ of the way from A to B — so find $\frac{4}{7}$ of each difference.
   x: $\frac{4}{7} \times 21 = 12$
   y: $\frac{4}{7} \times 28 = 16$

4) Now add these to the coordinates of A to find C.
   x-coordinate: –3 + 12 = 9
   y-coordinate: 5 + 16 = 21
   Coordinates of C are (9, 21)

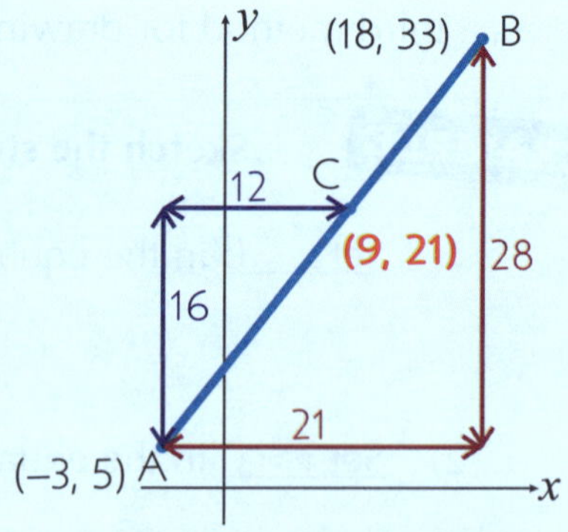

Midpoints — add the x's together and halve, then do the same for the y's

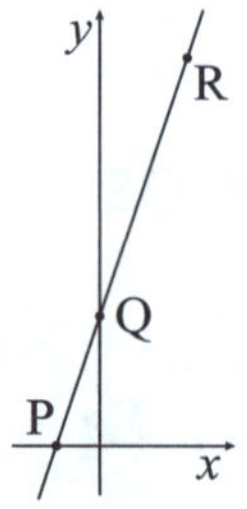

If you get a wordy line segments question, sketch a quick diagram to help you get your head around it. Have a go at these questions to see if this stuff has sunk in yet:

Q1 A (–4, –1) and B (8, –3) are points on the circumference of a circle. AB is a diameter. Find the coordinates of the centre of the circle. [2 marks] GRADE 6

Q2 P, Q and R lie on the straight line with equation $y - 3x = 6$, as shown on the left. PQ : QR = 1 : 2. Find the coordinates of R. [4 marks] GRADE 7

Q2 Video Solution

Parallel and Perpendicular Lines

*On p.65 you saw how to write the **equation of a straight line**. Well, you also have to be able to write the equation of a line that's **parallel** or **perpendicular** to the straight line you're given.*

Parallel Lines Have the **Same Gradient**

Parallel lines all have the same gradient, which means their $y = mx + c$ equations all have the same value of m.
So the lines: $y = 2x + 3$, $y = 2x$ and $y = 2x - 4$ are all parallel.

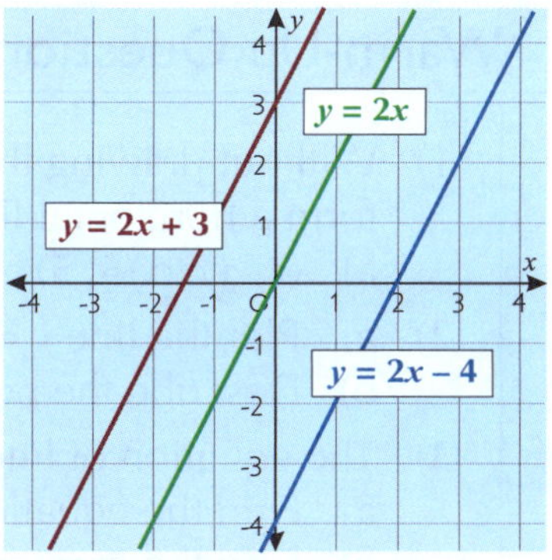

EXAMPLE

Line J has a gradient of –0.25. Find the equation of Line K, which is parallel to Line J and passes through the point (2, 3).

Lines J and K are parallel so their gradients are the same $\Rightarrow$ $m = -0.25$

$y = -0.25x + c$

when $x = 2$, $y = 3$:
$3 = (-0.25 \times 2) + c \Rightarrow 3 = -0.5 + c$
$c = 3.5$

$y = -0.25x + 3.5$

1) First find the 'm' value for Line K.
2) Substitute the value for 'm' into $y = mx + c$ to give you the 'equation so far'.
3) Substitute the x and y values for the given point on Line K and solve for 'c'.
4) Write out the full equation.

Perpendicular Line Gradients

Perpendicular lines cross at a right angle, and if you multiply their gradients together you'll get –1. Pretty nifty that.

If the gradient of the first line is m, the gradient of the other line will be $-\frac{1}{m}$, because $m \times -\frac{1}{m} = -1$.

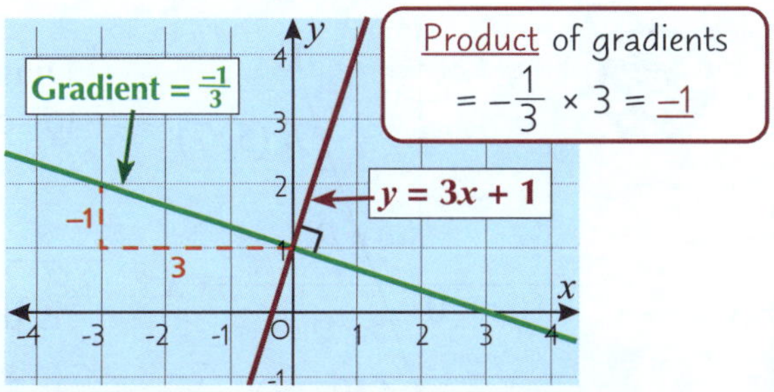

EXAMPLE

Lines A and B are perpendicular and intersect at (3, 3).
If Line A has the equation $3y - x = 6$, what is the equation of Line B?

1) Find 'm' (the gradient) for Line A. $3y - x = 6 \Rightarrow 3y = x + 6$
$\Rightarrow y = \frac{1}{3}x + 2$, so $m_A = \frac{1}{3}$

2) Find the 'm' value for the perpendicular line (Line B). $m_B = -\frac{1}{m_A} = -1 \div \frac{1}{3} = -3$

3) Put this into $y = mx + c$ to give the 'equation so far'. $y = -3x + c$

4) Put in the x and y values of the point and solve for 'c'. $x = 3$, $y = 3$ gives:
$3 = (-3 \times 3) + c$
$\Rightarrow 3 = -9 + c \Rightarrow c = 12$

5) Write out the full equation. $y = -3x + 12$

Parallel lines have the same gradient

Perpendicular lines can be a little more tricky, so remember that their gradients multiply together to give –1.

Q1 Find the equation of the line parallel to $2x + 2y = 3$ which passes through the point (1, 4). Give your answer in the form $y = mx + c$. [3 marks]

Q2 Show that the lines $y + 5x = 2$ and $5y = x + 3$ are perpendicular. [3 marks]

Q1 Video Solution

Warm-Up and Worked Exam Questions

On the day of the exam you'll have to know straight-line graphs like the back of your hand. If you struggle with any of the warm-up questions, go back over the section again before you go any further.

Warm-Up Questions

1) Without drawing them, state whether the lines passing through the following points form a horizontal line, a vertical line, the line $y = x$ or the line $y = -x$.
   a) (1, 1) to (5, 5) b) (0, 4) to (–3, 4) c) (–1, 3) to (–1, 7) d) (4, –4) to (–3, 3).
2) a) Plot the line $y = 3x - 4$.
   b) Describe the position of the line with equation $y = 3x + 2$ in relation to $y = 3x - 4$.
3) The equation of line S is $y = 2x - 3$.
   a) Find the equation of the line which is parallel to line S and passes through the point (0, 4).
   b) Find the gradient of a line which is perpendicular to S.

Worked Exam Question

You know the routine by now — work carefully through this example and make sure you understand it. Then it's on to the real test of doing some exam questions for yourself.

1 Line **L** passes through the points *A* (0, –3) and *B* (5, 7), as shown below.

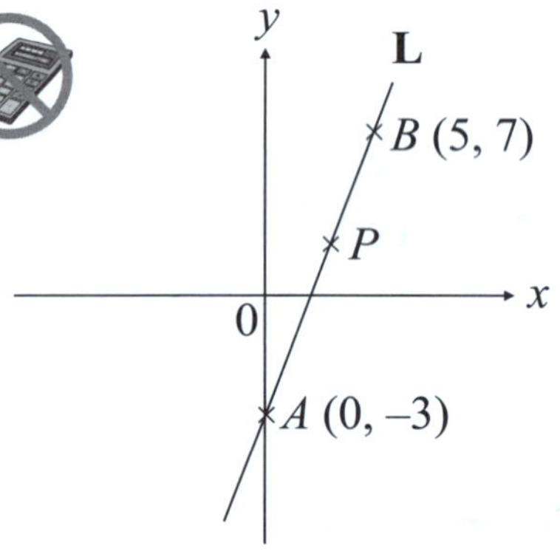

Diagram **NOT** to scale

a) Find the equation of line **L**. (Grade 5)

The equation is $y = mx + c$, where m is the gradient, and c is the y-intercept.

$m = \dfrac{(7 - (-3))}{(5 - 0)}$

m = 2

The line passes through (0, –3), so c = –3.

$y = 2x - 3$

[3 marks]

b) Write down the equation of the line which is parallel to line **L** and passes through the point (2, 10). (Grade 5)

A line which is parallel to line L will have the same gradient.

m = 2, so $y = 2x + c$

$10 = (2 \times 2) + c$ ← Substitute in the x and y values of the point to find c.

$c = 10 - 4 = 6$

So $y = 2x + 6$

$y = 2x + 6$

[2 marks]

c) Point *P* lies on the line segment *AB*, such that *AP* : *PB* = 2 : 3 (Grade 7)
What are the coordinates of *P*?

Difference in x-coordinate from A to B: $5 - 0 = 5$

Difference in y-coordinate from A to B: $7 - (-3) = 10$

So the x-coordinate of P = $0 + \dfrac{2}{5} \times 5 = 2$ ← Convert the ratio into a fraction, then multiply the x and y differences by it.

and the y-coordinate of P = $-3 + \dfrac{2}{5} \times 10 = 1$

(2, 1)

[3 marks]

Exam Questions

2 The lines $y = 3x + 4$ and $y = 2x + 6$ intersect at the point M. (GRADE 6)

Line **N** goes through point M and is parallel to the line $y = \frac{1}{2}x + 6$.
Find the equation of line **N**.

....
[5 marks]

3 A straight line, **S**, passes through the points (a, b) and (c, d). (GRADE 7)
It is given that: $2a + 4 = 2c$
$b - 6 = d$

a) What is the gradient of **S**?

Gradient = ...
[3 marks]

b) Line **R** is perpendicular to Line **S** and passes through (6, 3). Find the equation of the line.

....
[2 marks]

4 Eshe plots the points A (5, 7), B (1, –1), C (13, 4) and D (3, –2). She claims she can draw a line perpendicular to AB that passes through the midpoint of both AB and CD. (GRADE 8)
Is she correct? Explain your answer.

[4 marks]

Quadratic Graphs

Quadratic functions take the form ***y = anything with x^2*** *(but no higher powers of x).*
x^2 graphs all have the same ***symmetrical*** *bucket shape.*

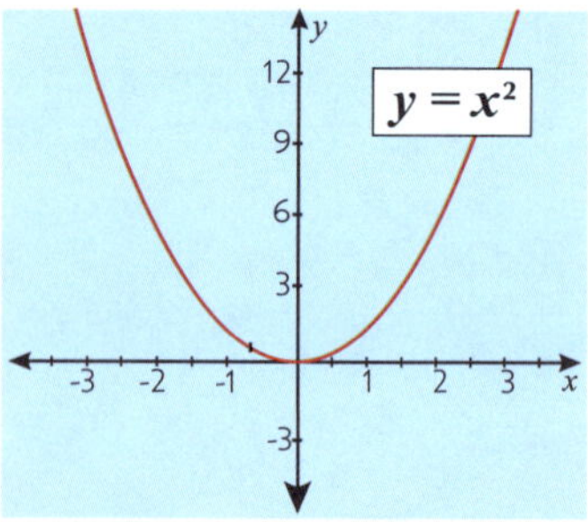

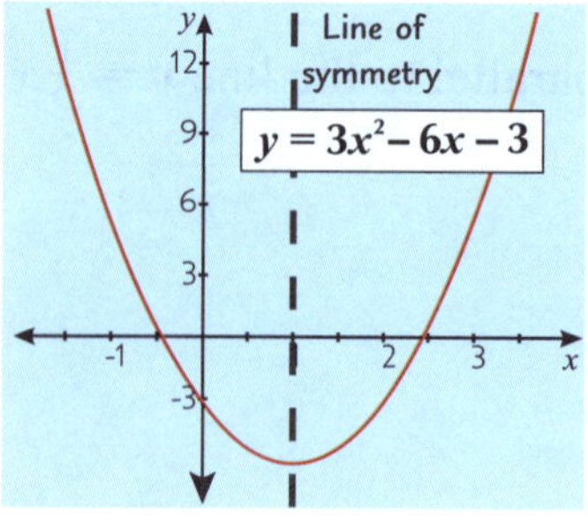

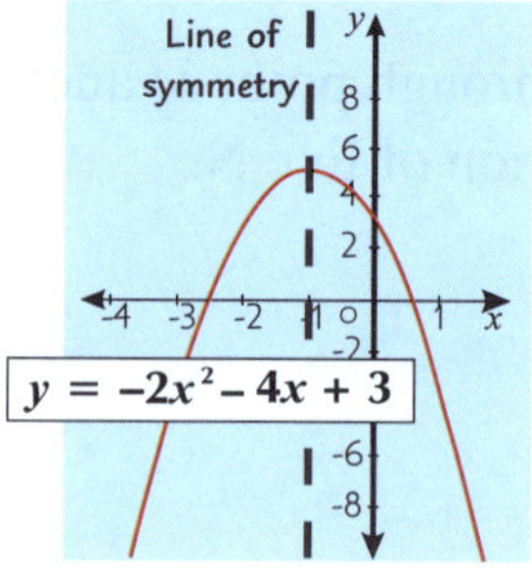

If the x^2 bit has a '–' in front of it then the bucket is upside down.

Plotting **Quadratics** (Grade 5)

EXAMPLE

Complete the table of values for the equation $y = x^2 + 2x - 3$ and then plot the graph.

x	–4	–3	–2	–1	0	1	2
y	5	0	–3	–4	–3	0	5

1) Substitute each x-value into the equation to get each y-value.
   E.g. $y = (-4)^2 + (2 \times -4) - 3 = 5$
2) Plot the points and join them with a completely smooth curve.

NEVER EVER let one point drag your graph off in some ridiculous direction. When a graph is generated from an equation, you never get spikes or lumps.

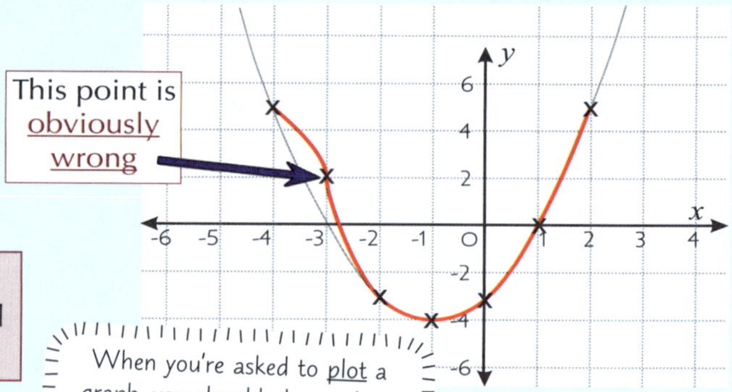

When you're asked to plot a graph, you should always draw it accurately using this method.

Sketching Quadratics (Grade 7)

If you're asked to sketch a graph, you won't have to use graph paper or be dead accurate — just find and label the important points and make sure the graph is roughly in the correct position on the axes.

EXAMPLE

Sketch the graph of $y = -x^2 - 2x + 8$, labelling the turning point and x-intercepts with their coordinates.

1. Find all the information you're asked for.
   Solve $-x^2 - 2x + 8 = 0$ to find the x-intercepts (see p.51).
   $-x^2 - 2x + 8 = -(x + 4)(x - 2) = 0$ so $x = -4$, $x = 2$
   Use symmetry to find the turning point of the curve:
   The x-coordinate of the turning point is halfway between -4 and 2.
   $x = \frac{-4+2}{2} = -1$
   $y = -(-1)^2 - 2(-1) + 8 = 9$
   So the turning point is (–1, 9).
2. Use the information you know to sketch the curve and label the important points.

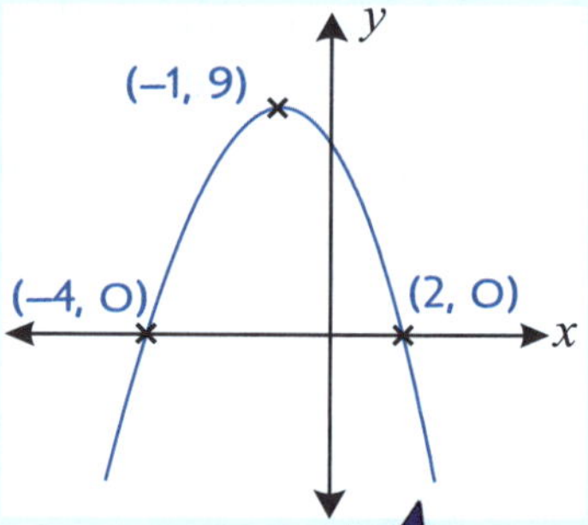

The x^2 is negative, so the curve is n-shaped.

Tables of values, plotting — easy marks, as long as you're accurate

Filling in tables of values and plotting graphs are easy questions, but too many people rush them and make silly errors. Take your time and get them right — if your curve isn't smooth, check the points in your table.

Q1 Plot the graph of $y = x^2 - 4x - 1$ for values of x between –2 and 6. [4 marks] (Grade 5)

Harder Graphs 1

Graphs come in all sorts of shapes, sizes and wiggles — there are 7 more types you need to know. Here are the first two:

Circles: $x^2 + y^2 = r^2$

Grade 7

The equation for a circle with centre (0, 0) and radius r is:

$$x^2 + y^2 = r^2$$

$x^2 + y^2 = 25$ is a circle with centre (0, 0).
$r^2 = 25$, so the radius, r, is 5.

$x^2 + y^2 = 100$ is a circle with centre (0, 0).
$r^2 = 100$, so the radius, r, is 10.

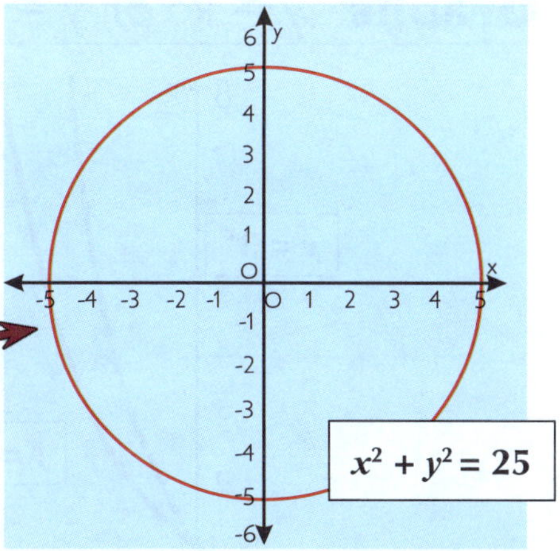

EXAMPLE

Find the equation of the tangent to $x^2 + y^2 = 100$ at the point (8, –6).

Grade 9

1) Find the gradient of the line from the origin to (8, –6). This is a radius of the circle.

$$\text{Gradient} = \frac{\text{Change in } y}{\text{Change in } x} = \frac{-6 - 0}{8 - 0} = \frac{-3}{4}$$

2) A tangent meets a radius at 90°, (see p.116) so they are perpendicular — so the gradient of the tangent is $-\frac{1}{m}$.

$$\text{Gradient of tangent} = -\frac{1}{m} = -\frac{1}{\frac{-3}{4}} = \frac{4}{3}$$

3) Find the equation of the tangent by substituting (8, –6) into $y = mx + c$.

$$y = mx + c \Rightarrow (-6) = \frac{4}{3}(8) + c$$

$$-6 = \frac{32}{3} + c$$

$$c = -\frac{50}{3}$$

$$y = \frac{4}{3}x - \frac{50}{3}$$

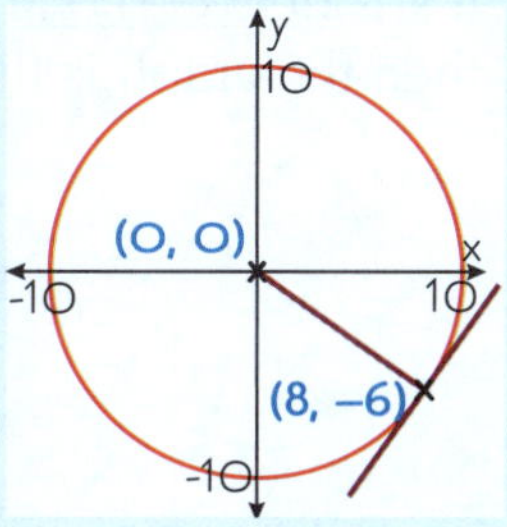

1/x (Reciprocal) Graphs: $y = A/x$ or $xy = A$

Grade 6

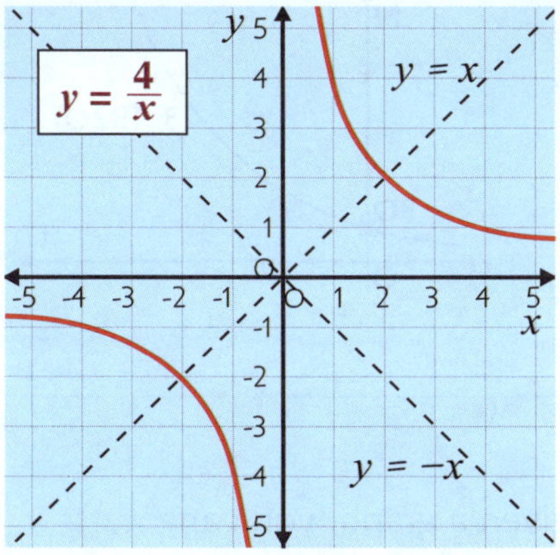

These are all the same basic shape, except the negative ones are in opposite quadrants to the positive ones (as shown). The two halves of the graph don't touch.
The graphs don't exist for $x = 0$.
They're all symmetrical about the lines $y = x$ and $y = -x$.

(You get this type of graph with inverse proportion — see p.95.)

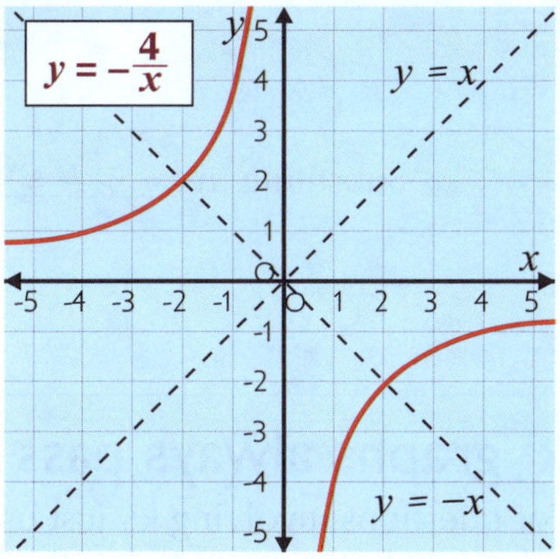

Learn all the different types of graph

Circle graphs and reciprocals may come up in your exam, so make sure you can recognise them and are able to sketch them. For circle graphs, remember to square-root r^2 to find the radius.

Q1 The point (5, 12) lies on a circle with centre (0, 0).
Find the radius and equation of the circle. [3 marks] Grade 7

Q1 Video Solution

Harder Graphs 2

The stars of this page are the k^x graphs. k^x graphs are sometimes called 'exponential graphs', and never touch the x-axis.

k^x Graphs: $y = k^x$ or $y = k^{-x}$ (k is some positive number)

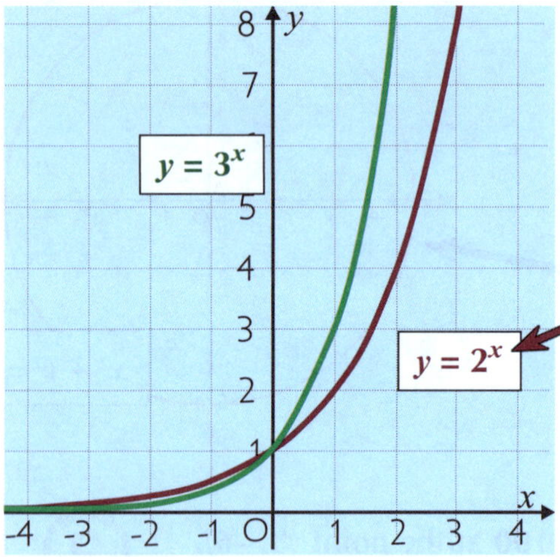

1) These 'exponential' graphs are always above the x-axis, and always go through the point (0, 1).
2) If $k > 1$ and the power is +ve, the graph curves upwards.
3) If k is between 0 and 1 OR the power is negative, then the graph is flipped horizontally.

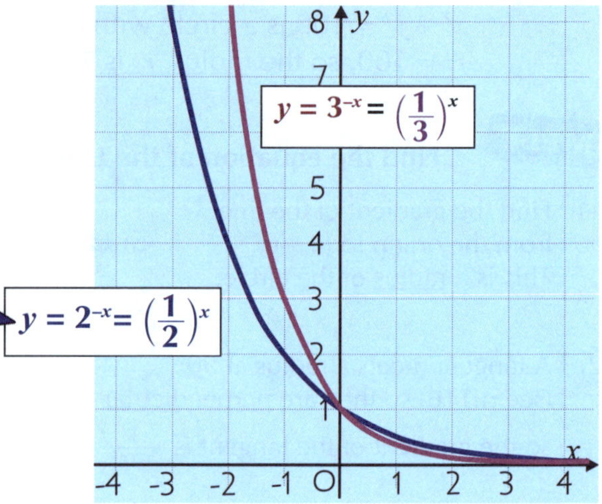

EXAMPLE **This graph shows how the number of victims of an alien virus (N) increases in a science fiction film. The equation of the graph is $N = fg^t$, where t is the number of days into the film. f and g are positive constants. Find the values of f and g.**

When $t = 0$, $N = 30$ so substitute these values into the equation:

$g^0 = 1$, so you can find f.

$$30 = fg^0 \Rightarrow 30 = f \times 1 \Rightarrow f = 30$$

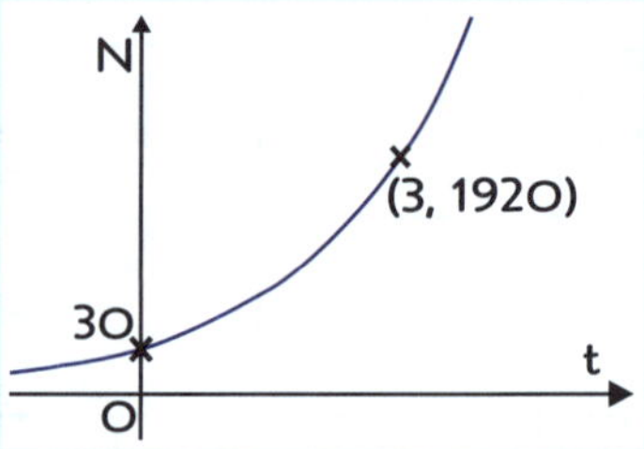

Substitute in $t = 3$, $N = 1920$: $N = 30g^t \Rightarrow 1920 = 30g^3$

$$g = \sqrt[3]{64} \Rightarrow g = 4$$

k^x graphs always pass through (0, 1)

For questions involving k^x, just use the information you're given. If you're given a graph, you can substitute known values into the equation to find the missing constants, like in the example above.

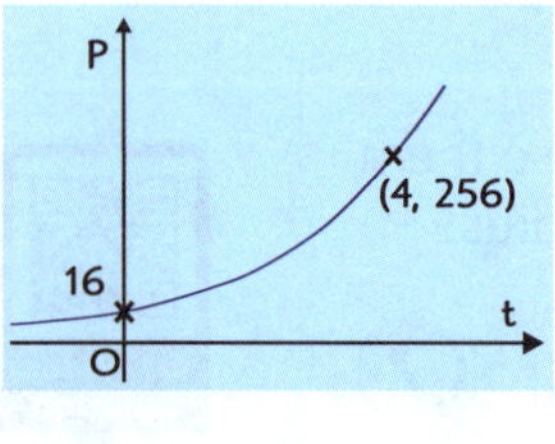

Q1 The increasing population of rats over time (shown on the graph) is modelled by the equation $P = ab^t$, where P = population, t = number of months and a and b are positive constants.

a) Find a and b. [4 marks]

b) Estimate the population after 7 months. [2 marks]

Q1 Video Solution

Harder Graphs 3

Nearly there — just a couple more pages of tough graphs. You need to be able to plot or sketch these graphs. Knowing what you're aiming for really helps.

x^3 Graphs: $y = ax^3 + bx^2 + cx + d$ (b, c and d can be zero)

GRADE 5

All x^3 graphs (also known as cubic graphs) have a wiggle in the middle — sometimes it's a flat wiggle, sometimes it's more pronounced. $-x^3$ graphs always go down from top left, $+x^3$ ones go up from bottom left.

Note that x^3 must be the highest power and there must be no other bits like $1/x$ etc.

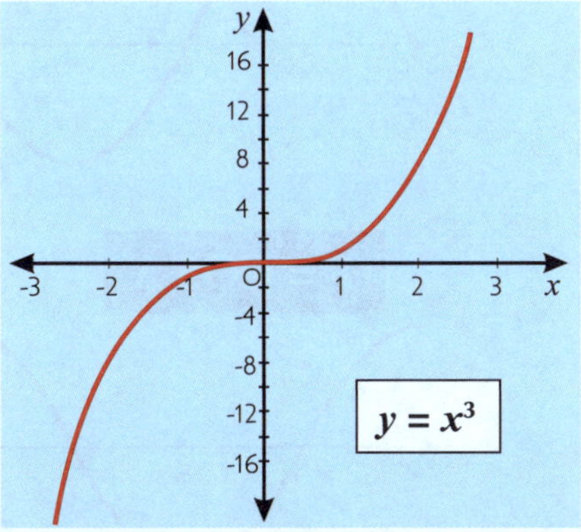

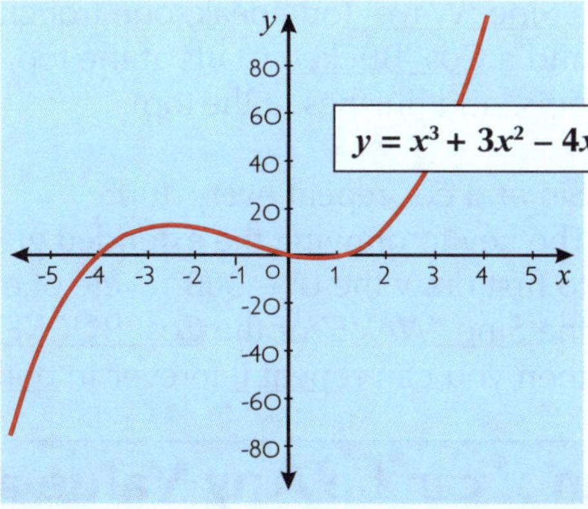

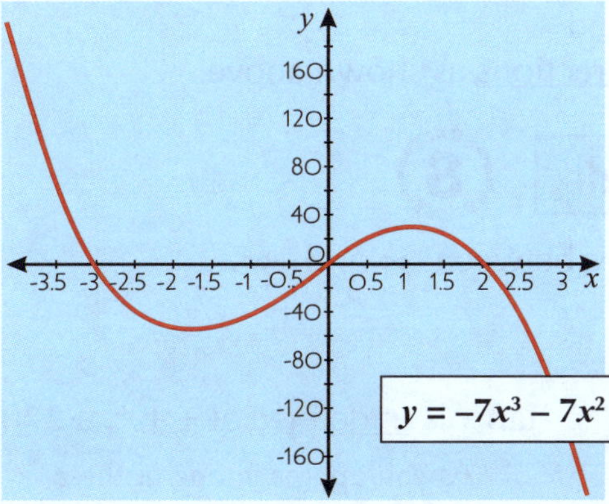

EXAMPLE

Draw the graph of $y = x^3 + 4x^2$ for values of x between –4 and +2.

Start by making a table of values.

x	–4	–3	–2	–1	0	1	2
$y = x^3 + 4x^2$	0	9	8	3	0	5	24

Plot the points and join them with a lovely smooth curve. DON'T use your ruler — that would be a trifle daft.

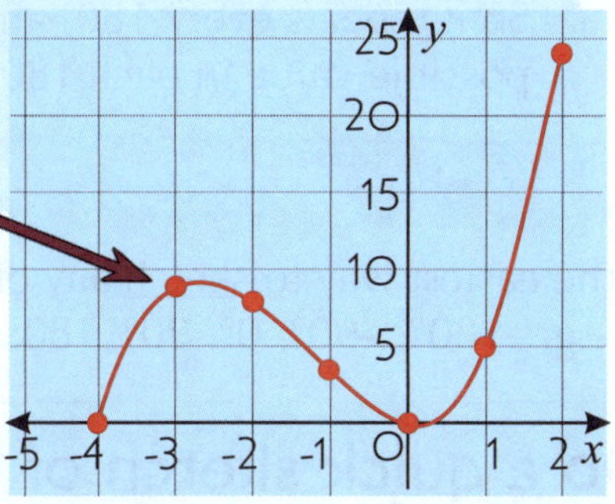

x^3 graphs always have a wiggle in the middle

Cubic graphs look nasty, but as long as you have the equation you can make a table of values. It's a dead easy method, but make sure the curve looks right when you plot it or it's likely you've made a mistake.

Harder Graphs 4

Before you leave this page, you should be able to close your eyes and picture these three graphs in your head, ***properly labelled*** *and everything. If you can't, you need to learn them more. I'm not kidding.*

Sine 'Waves' and Cos 'Buckets'

1) The underlying shape of the sin and cos graphs is identical — they both bounce between y-limits of exactly +1 and –1.
2) The only difference is that the sin graph is shifted right by 90° compared to the cos graph.
3) For 0° – 360°, the shapes you get are a Sine 'Wave' (one peak, one trough) and a Cos 'Bucket' (starts at the top, dips, and finishes at the top).
4) Sin and cos repeat every 360°. The key to drawing the extended graphs is to first draw the 0° – 360° cycle of either the Sine 'WAVE' or the Cos 'BUCKET' and then you can repeat it forever in both directions as shown above.

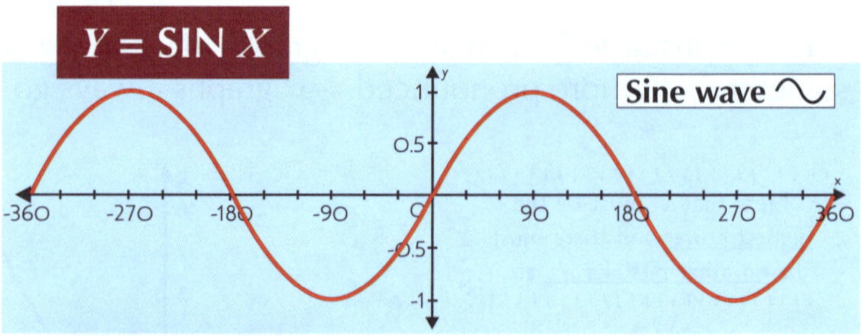

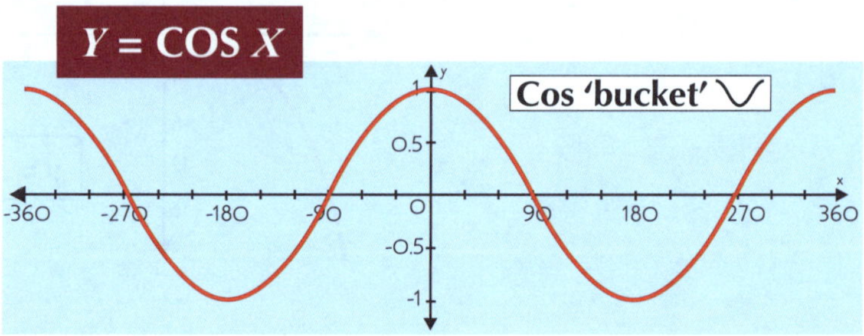

tan x can be Any Value at all

tan x is different from sin x or cos x — it goes between -∞ and +∞.

Tan x repeats every 180°

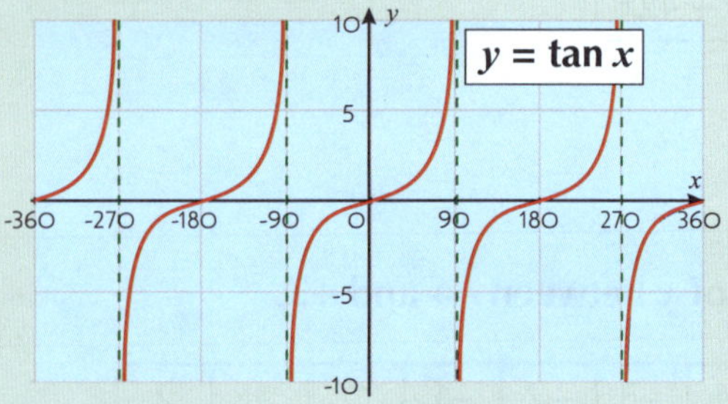

tan x goes from –∞ to +∞ every 180°.

So it repeats every 180° and takes every possible value in each 180° interval.

tan x is undefined at ±90°, ±270°,...

As you approach one of these undefined points from the left, tan x just shoots up to infinity.

As you approach from the right, it drops to minus infinity.

The graph never ever touches these lines. But it does get infinitely close, if you see what I mean...

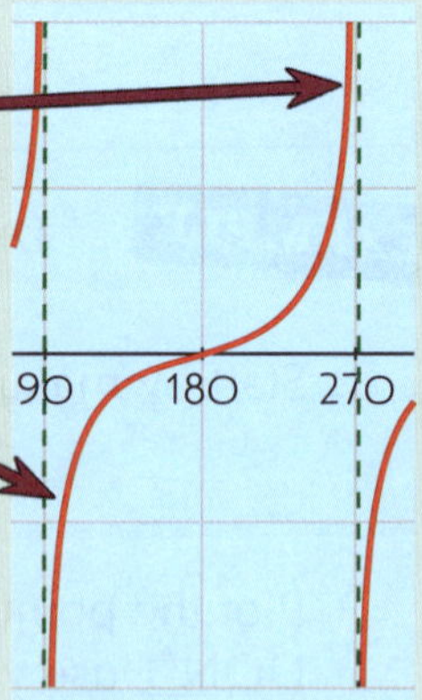

The easiest way to sketch any of these graphs is to plot the important points which happen every 90° (e.g. –180°, –90°, 0°, 90°, 180°, 270°, 360°, 450°, 540°...) and then just join the dots up.

Do a quick sketch of each graph

Sin$x$ and cos$x$ look pretty similar so they're easy to mix up. The trick is to learn the key points and shape of each graph. When you're ready, cover up the page and try out these practice questions.

Q1 a) Sketch the graph of $y = \cos x$ for values of x between –360° and 360°. [4 marks]

b) Sketch the graph of $y = \sin x$ for values of x between 0 and 720°. [4 marks]

Solving Equations Using Graphs

*You can plot graphs to find **approximate solutions** to simultaneous equations or other awkward equations. Plot the equations you want to solve and the solution lies where the lines **intersect**.*

Plot **Both Graphs** and See Where They **Cross**

EXAMPLE **By plotting the graphs, solve the simultaneous equations $x^2 + y^2 = 16$ and $y = 2x + 1$.**

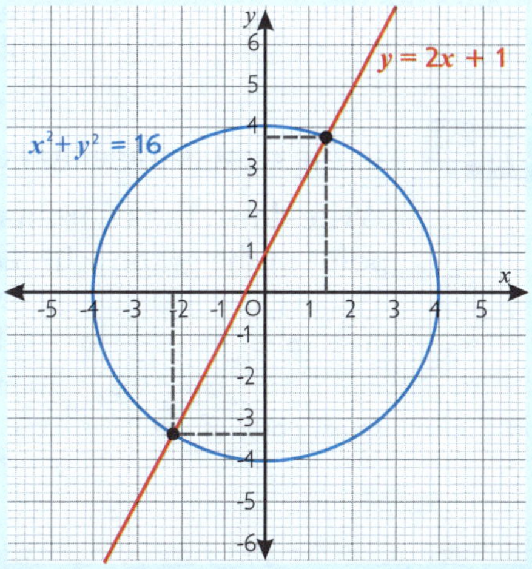

1) DRAW BOTH GRAPHS.
$x^2 + y^2 = 16$ is the equation of a circle with centre (0, 0) and radius 4 (see p.73). Use a pair of compasses to draw it accurately.

2) LOOK FOR WHERE THE GRAPHS CROSS.
The straight line crosses the circle at two points. Reading the x and y values of these points gives the solutions $x = 1.4$, $y = 3.8$ and $x = -2.2$, $y = -3.4$ (all to 1 decimal place).

Using Graphs to Solve **Harder Equations**

EXAMPLES

1. The graph of $y = \sin x$ is shown to the right. Use the graph to estimate the solutions to $\sin x = 0.7$ between –180° and 180°

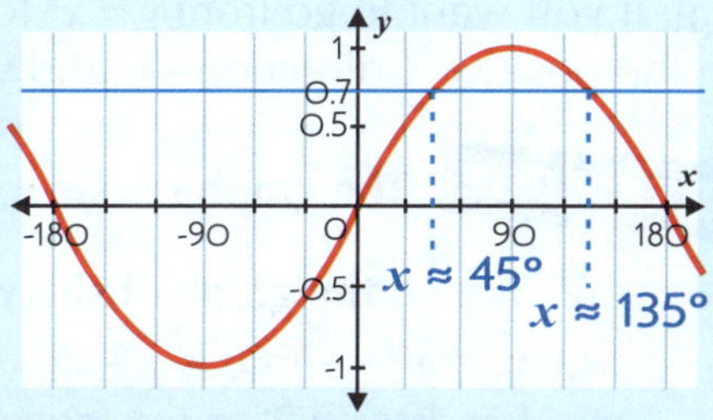

Draw the line $y = 0.7$ on the graph, then read off where it crosses $\sin x$.

The solutions are $x \approx 45°$ and $x \approx 135°$.

2. The graph of $y = 2x^2 - 3x$ is shown on the right.

a) Use the graph to estimate both roots of $2x^2 - 3x = 7$.

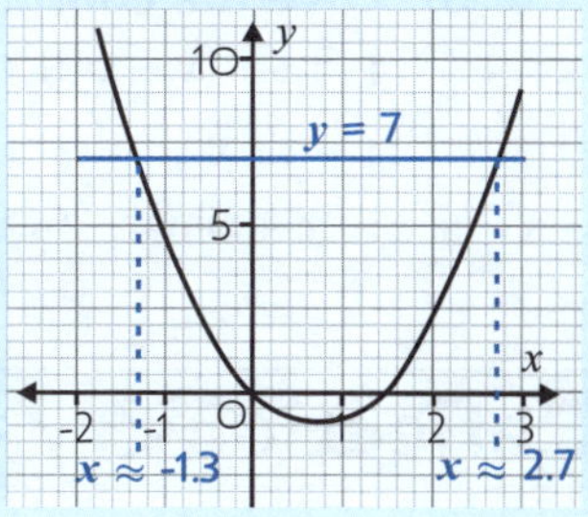

$2x^2 - 3x = 7$ is what you get when you put $y = 7$ into the equation:

1) Draw a line at $y = 7$.
2) Read the x-values where the curve crosses this line — these are the solutions or roots.

Quadratic equations usually have 2 roots.

The roots are around $x = -1.3$ and $x = 2.7$.

b) Find the equation of the line you would need to draw on the graph to solve $2x^2 - 5x + 1 = 0$

This is a bit nasty — the trick is to rearrange the given equation $2x^2 - 5x + 1 = 0$ so that you have $2x^2 - 3x$ (the graph) on one side.

$$2x^2 - 5x + 1 = 0$$

Adding $2x - 1$ to both sides: $2x^2 - 3x = 2x - 1$

So the line needed is $y = 2x - 1$.

The sides of this equation represent the two graphs $y = 2x^2 - 3x$ and $y = 2x - 1$. Finding the points where these graphs cross will give the solutions to $2x^2 - 5x + 1 = 0$.

The solutions lie where the graphs intersect

Get your graph-plotting pencil ready and have a go at these Practice Questions:

Q1 By plotting graphs, find approximate solutions to $y = x^2 + 2x - 4$ and $y = 6 - x$. [4 marks]

Q2 By plotting graphs, find approximate solutions to $x^2 + y^2 = 25$ and $y = x + 1$. [4 marks]

Q2 Video Solution

Graph Transformations

*Remember that a **function** of x is just some equation involving x that takes an input, processes it and produces an output. So for a graph of, say, $y = x^2$, x^2 is the function (this will be handy for **transformations**).*

Translations on the **y-axis**: y = (function of x) ± a

GRADE 7

This is where the whole graph is slid up or down the y-axis, and is achieved by adding a number to or subtracting a number from the original function: e.g. $y = x^2 + a$.

You must describe this as a 'translation' — don't just say 'slide'.

EXAMPLE **To the right is the graph of $y = x^2 - 4x + 6$.**

Write down the coordinates of the minimum point of the graph with equation $y = x^2 - 4x + 11$.

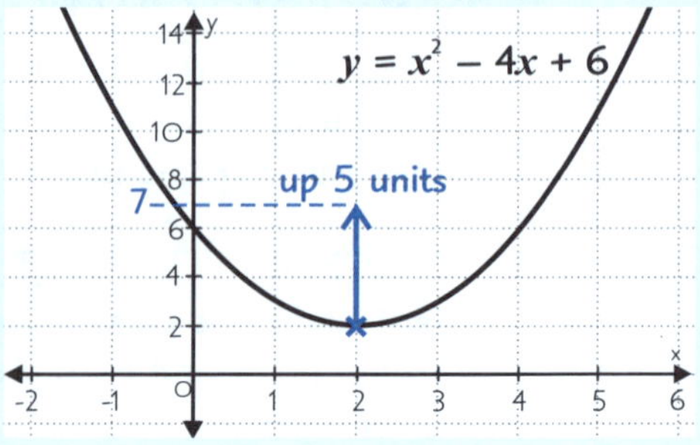

The minimum point of $y = x^2 - 4x + 6$ has coordinates (2, 2).
$y = x^2 - 4x + 11$ is the same shape graph, translated 5 units upwards.
So the minimum point of $y = x^2 - 4x + 11$ is at **(2, 7)**.

Translations on the **x-axis**: y = function of (x ± a)

GRADE 8

This is where the whole graph slides to the left or right and it only happens when you replace 'x' everywhere in the equation with '$x + a$' or '$x - a$'. These are tricky because they go 'the wrong way'. E.g. if you want to go from $y = x^2$ to $y = (x - a)^2$ you must move the whole graph a distance 'a' in the positive x-direction → (and vice versa).

EXAMPLE **The graph $y = \sin x$ is shown below, for $-360° \le x \le 360°$.**

a) **Sketch the graph of $\sin(x - 60)$.**
$y = \sin(x - 60)$ is $y = \sin x$ translated 60° in the positive x-direction.

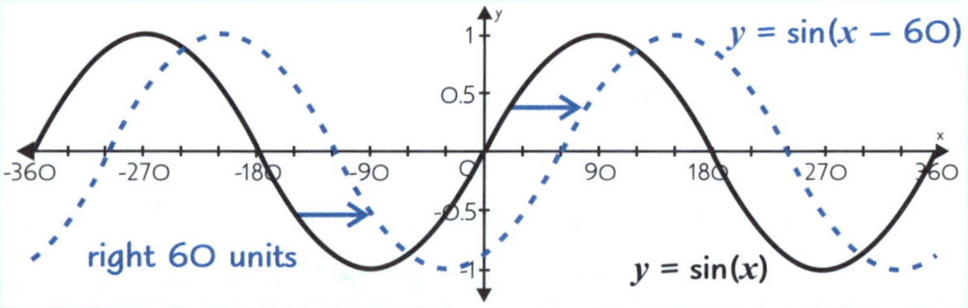

b) **Give the coordinates of a point where $y = \sin(x - 60)$ crosses the x-axis.**
$y = \sin x$ crosses the x-axis at (0, 0),
so $y = \sin(x - 60)$ will cross at **(60, 0)**.

Reflections: y = –(function of x) and y = function of (–x)

GRADE 8

The reflection of $y = x^2 - x - 2$ in the x-axis has the equation $y = -(x^2 - x - 2)$
$= -x^2 + x + 2$.

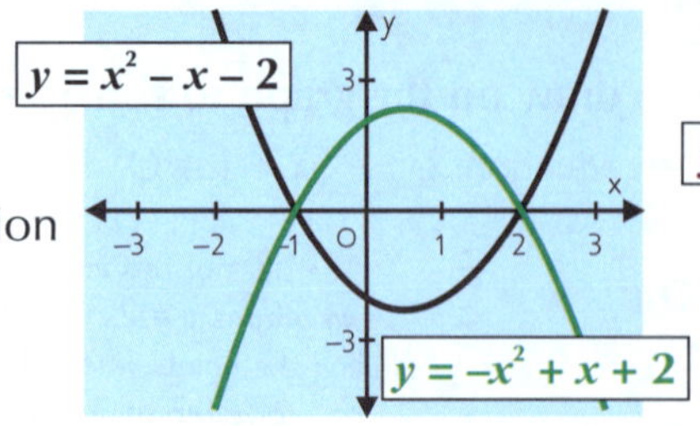

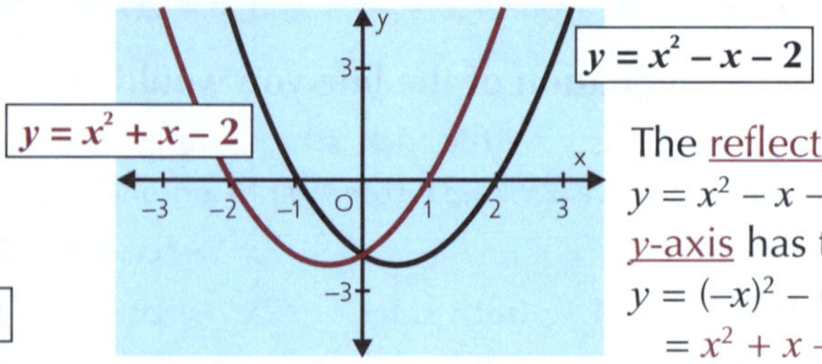

The reflection of $y = x^2 - x - 2$ in the y-axis has the equation $y = (-x)^2 - (-x) - 2$
$= x^2 + x - 2$.

Learn these three types of graph transformations

Graphs can be slid up or down by adding or subtracting a number to/from the whole function, and they can be slid left or right when x is replaced with $x \pm a$. $(x + a)$ slides the graph left, $(x - a)$ slides the graph right.

Q1 The minimum point of the graph of $y = 2x^2 - 4x + 3$ is (1, 1).
Find the coordinates of the minimum point of:
a) $y = -2x^2 + 4x - 3$,
b) $y = 2(x - 3)^2 - 4(x - 3) + 3$,
c) $y = 2(x + 2)^2 - 4(x + 2) + 5$. [3 marks]

GRADE 8

Warm-Up and Worked Exam Questions

The warm-up questions run quickly over the basic facts you'll need in the exam. The exam questions come later — but unless you've learnt the facts first you'll find the exams tougher than stale bread.

Warm-Up Questions

1) a) Complete the table of values for $y = x^2 - 2x - 1$.

x	–2	–1	0	1	2	3	4	5
y								

b) Plot the x and y values from the table and join the points up to form a smooth curve.

c) Use your curve to find the value of y when $x = 3.5$.

d) Find the two values of x when $y = 5$.

2) Sketch the graph of $y = x^2 - 3x$, labelling the turning point and x-intercepts with their coordinates.

3) To the right is the graph of $y = \cos x$ for $0° \leq x \leq 360°$. As shown on the graph, $\cos 50° = 0.643$. Give another value of x, found on this graph, where $\cos x = 0.643$.

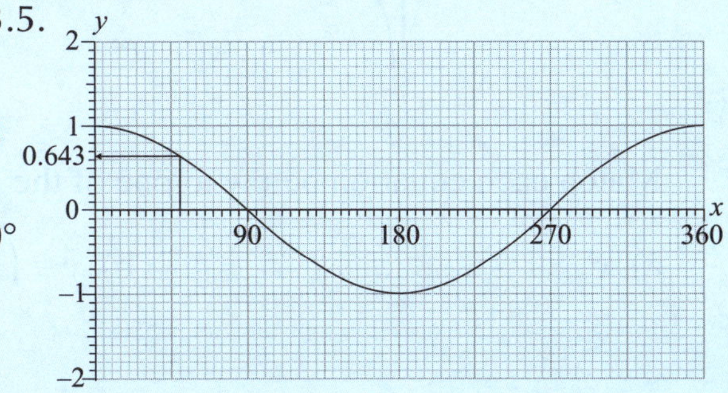

4) Graphs are drawn showing the functions $y = x^2 - 6$ and $y = x^2 - 4$. Describe how the shape and position of the two graphs are related.

Worked Exam Question

Wow, an exam question — with the answers helpfully written in. It must be your birthday.

1 The temperature (T) of a piece of metal changes over time (t) as it is rapidly heated and then cooled again. It is modelled by the equation $T = -5t^2 + 40t - 35$. (Grade 5)

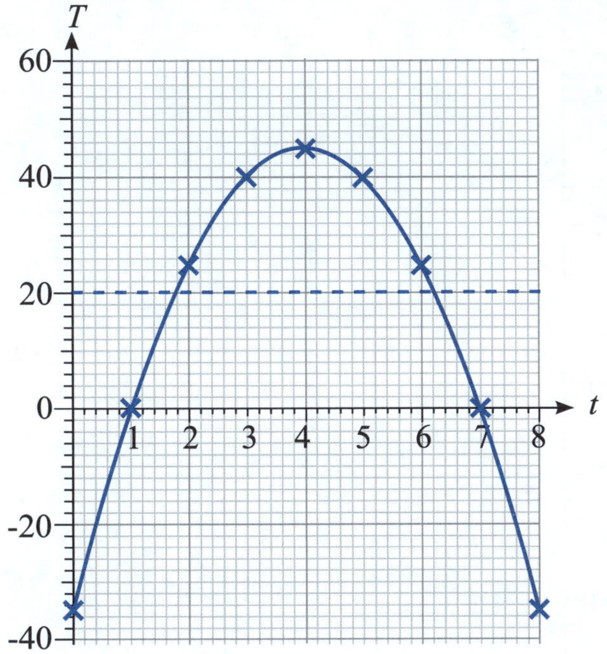

a) Plot the graph of $T = -5t^2 + 40t - 35$ on the grid.

Make a table of values:

t	0	1	2	3	4	5	6	7	8
T	–35	0	25	40	45	40	25	0	–35

Plot these points on the graph and draw a smooth curve through them.

[3 marks]

b) At what time did the metal reach its highest temperature?

$t =$ ......4......

[1 mark]

c) Using your graph, solve the equation $-5t^2 + 40t - 35 = 20$.

Draw the line $y = 20$ and read off where it crosses the curve.

$t =$ ......1.8...... and $t =$ ......6.2......

[2 marks]

Exam Questions

2 Sketches of different graphs are shown below. (Grade 6)

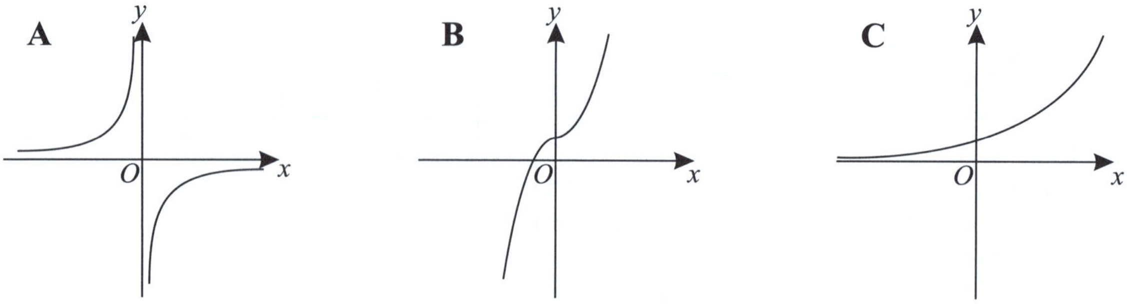

Match each equation below to one of the graphs above.

a) $x^3 + 1$ ....

b) $y = \left(\frac{3}{2}\right)^x$ ....

c) $y = -\frac{1}{x}$ ....

[3 marks]

3 This question is about the function $y = x^3 - 4x^2 + 4$. (Grade 6)

a) Complete the table below.

x	–1	–0.5	0	0.5	1	1.5	2	2.5	3	3.5	4
y	–1	2.875	4	3.125	1	–1.625	–4				

[2 marks]

b) Use your table to draw the graph of $y = x^3 - 4x^2 + 4$ on the grid, for values of x in the range $-1 \leq x \leq 4$.

[2 marks]

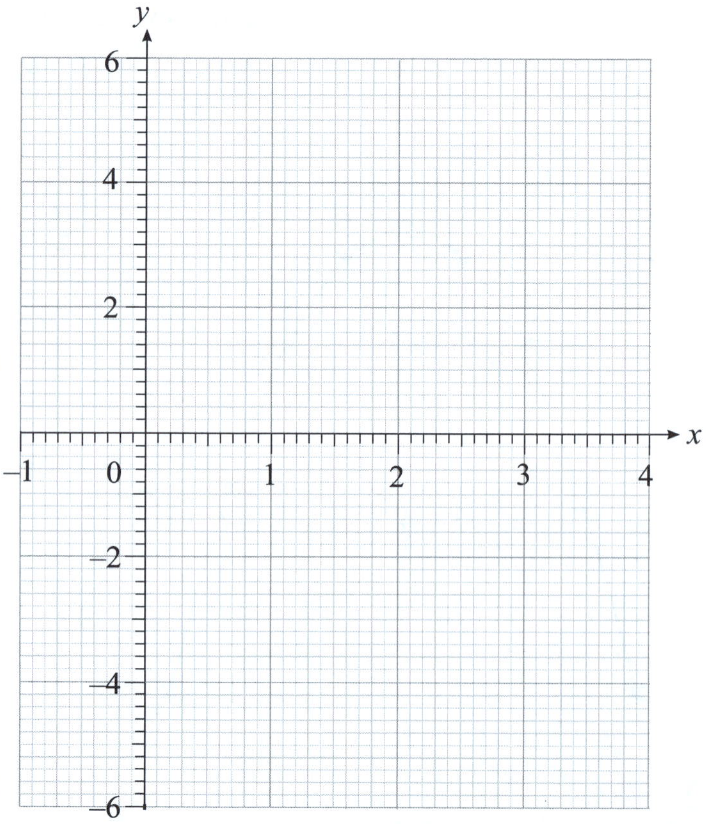

c) Estimate the solutions to the equation $x^3 - 4x^2 + 4 = 0$.

....

[1 mark]

Exam Questions

4 The graph of the curve $y = x^2 + 2x - 5$ is shown below.
By drawing a suitable line on the graph, find the solutions of $x^2 + x = 6$

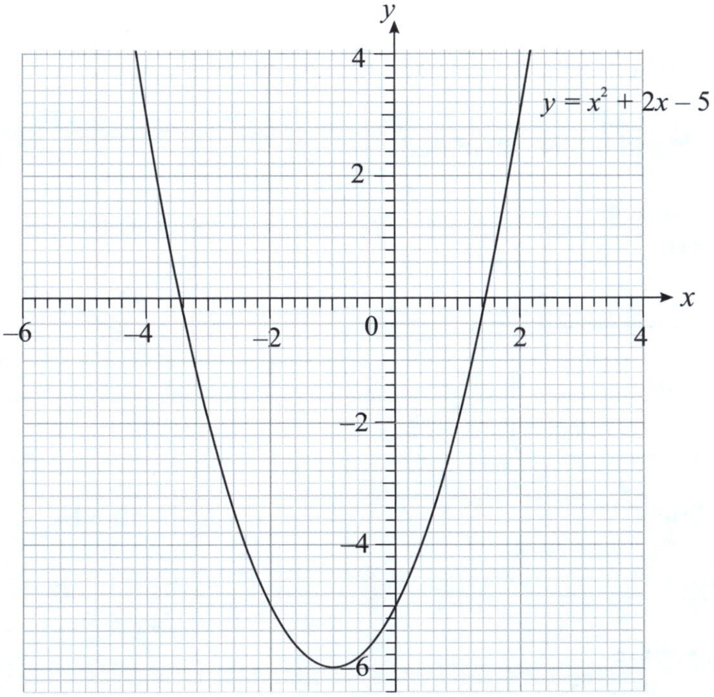

$x =$ ....

$x =$ ....

[4 marks]

5 The graph of $y = \cos x$ for $0° \leq x \leq 360°$ is shown below.

a) Draw the graph of $y = \cos(-x) + 1$ on the grid.

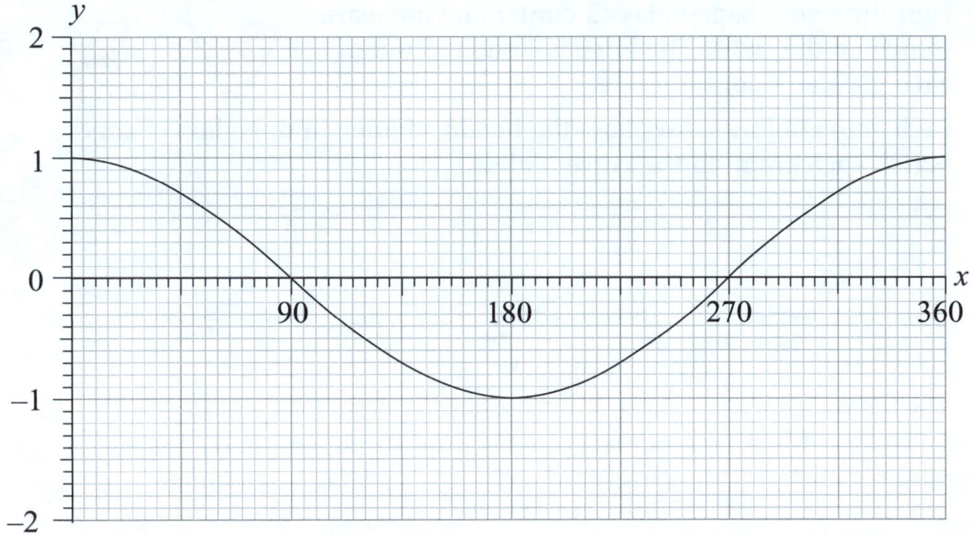

[2 marks]

b) Write down the x-values of the points where $y = -\cos(x + 30)$ crosses the x-axis.

....

[2 marks]

Real-Life Graphs

Now and then, graphs mean something more interesting than just $y = x^3 + 4x^2 - 6x + 4$*...*

Graphs Can Show **Billing Structures**

Grade 3

Many bills are made up of two charges — a fixed charge and a cost per unit. E.g. You might pay £11 each month for your phone line, and then be charged 3p for each minute of calls you make.

EXAMPLE **This graph shows how a broadband bill is calculated.**

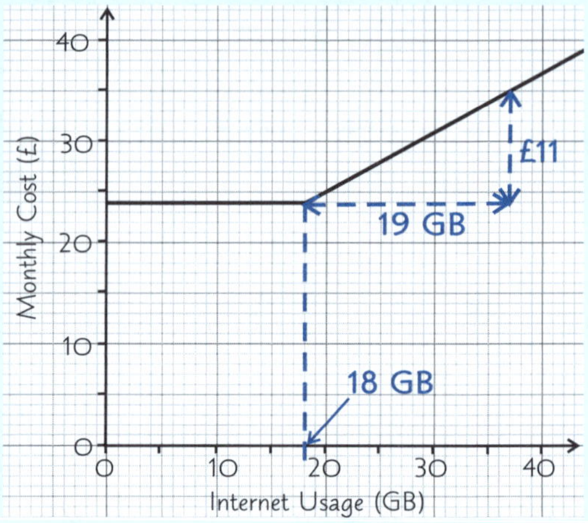

a) **How many gigabytes (GB) of Internet usage are included in the basic monthly cost?**

18 GB

> The first section of the graph is horizontal. You're charged £24 even if you don't use the Internet during the month. It's only after you've used 18 GB that the bill starts rising.

b) **What is the cost for each additional gigabyte (to the nearest 1p)?**

Gradient of sloped section = cost per GB

$\frac{\text{vertical change}}{\text{horizontal change}} = \frac{11}{19} = £0.5789...$ per GB

To the nearest 1p this is £0.58

> No matter what the graph, the gradient is always the y-axis unit PER the x-axis unit (see p.85).

Graphs Can Show **Changes with Time**

Grade 6

EXAMPLE **Four different-shaped glasses containing juice are shown on the right. The juice is siphoned out of each glass at a constant rate.**

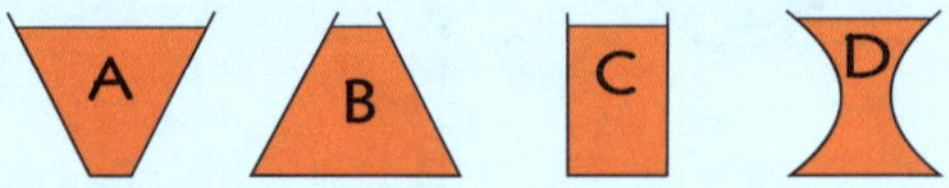

Each graph below shows how the height of juice in one glass changes. Match each graph to the correct glass.

> A steeper slope means that the juice height is changing faster.

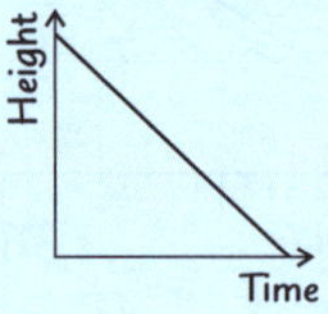

Glass C

Glass C has straight sides, so the juice height falls steadily.

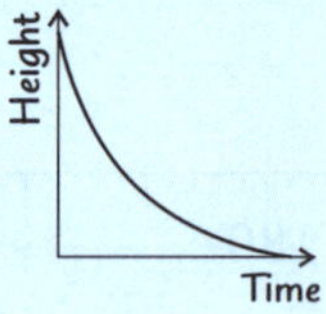

Glass B

Glass B is narrowest at the top, so the juice height falls fastest at first.

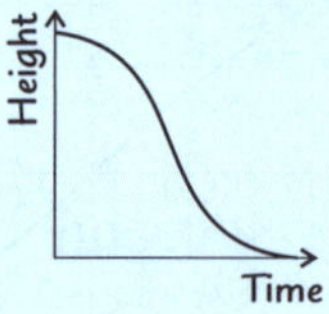

Glass D

Glass D is narrowest in the middle, so the height will fall fastest in the middle part of the graph.

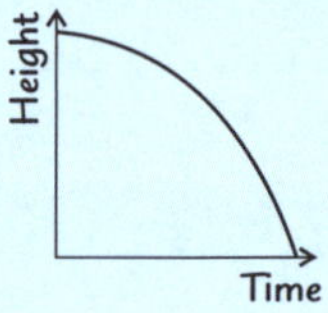

Glass A

Glass A is narrowest at the bottom, so the height will fall fastest at the end of the graph.

Always think carefully about what the gradient means

There are loads of different types of graph you could be given in the exam, so make sure you know exactly what you're dealing with — it's easy to get the axes mixed up. Now have a go at this Practice Question.

Q1 A taxi charges a minimum fare of £4.50, which includes the first three miles.
It then charges 80p for each additional mile.
Draw a graph to show the cost of journeys of up to 10 miles. [4 marks] Grade 4

Distance-Time Graphs

Distance-time graphs just show how distance changes over time.

Distance-Time Graphs

Distance-time graphs can look a bit awkward at first, but they're not too bad once you get your head around them.

Just remember these 4 important points:

1) At any point, GRADIENT = SPEED.
2) The STEEPER the graph, the FASTER it's going.
3) FLAT SECTIONS are where it is STOPPED.
4) If the gradient's negative, it's COMING BACK.

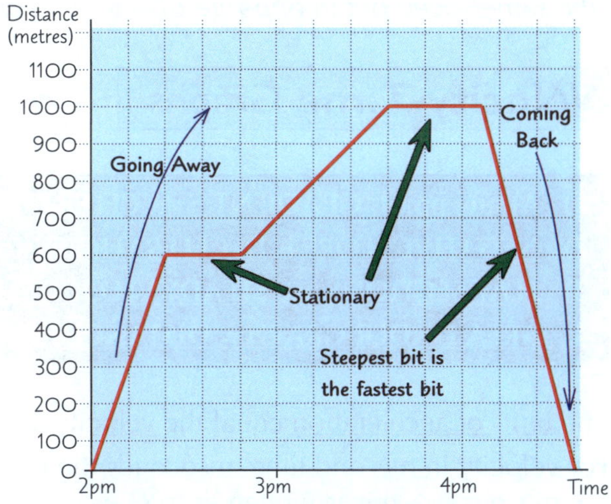

EXAMPLE

Henry went out for a ride on his bike. After a while he got a puncture and stopped to fix it. This graph shows the first part of Henry's journey.

a) **What time did Henry leave home?**

He left home at the point where the line starts. At 8:15

b) **How far did Henry cycle before getting a puncture?**

The horizontal part of the graph is where Henry stopped.

12 km

c) **What was Henry's speed before getting a puncture?**

Using the speed formula is the same as finding the gradient.

$$\text{speed} = \frac{\text{distance}}{\text{time}} = \frac{12 \text{ km}}{0.5 \text{ hours}}$$
$$= 24 \text{ km/h}$$

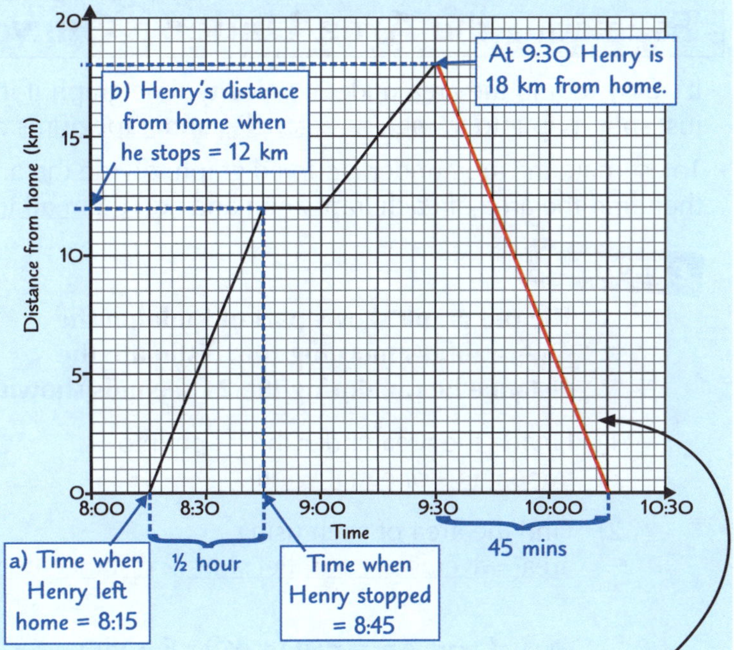

d) **At 9:30 Henry turns round and cycles home at 24 km/h. Complete the graph to show this.**

You have to work out how long it will take Henry to cycle the 18 km home:

$$\text{time} = \frac{\text{distance}}{\text{speed}} = \frac{18 \text{ km}}{24 \text{ km/h}} = 0.75 \text{ hours}$$
$$0.75 \times 60 \text{ mins} = 45 \text{ mins}$$

Decimal times are yuck, so convert it to minutes.

45 minutes after 9:30 is 10:15, so that's the time Henry gets home. Now you can complete the graph.

The gradient of a distance-time graph = speed

Have a quick look at the axes before starting questions like this. Once you notice that it's a plain old distance-time graph, use the 4 key points from the blue box. Now try this Practice Question.

Q1 a) Using the graph above, how long did Henry stop for? [1 mark]

b) What was Henry's speed after he had repaired the puncture, before he turned back home? [2 marks]

Q1 Video Solution

Velocity-Time Graphs

*Velocity is **speed** in a **particular direction**. So two objects with velocities of 20 m/s and –20 m/s are moving at the same speed but in opposite directions. For the purpose of these graphs, velocity is just **speed**.*

Velocity-Time Graphs

GRADE 6

1) At any point, GRADIENT = ACCELERATION.
2) NEGATIVE SLOPE is DECELERATION (slowing down).
3) FLAT SECTIONS are STEADY VELOCITY.
4) AREA UNDER GRAPH = DISTANCE TRAVELLED.

The units of acceleration equal the velocity units per time units. For velocity in m/s and time in seconds the units of acceleration are m/s per s — this is written as m/s^2.

Be careful not to get velocity and distance-time graphs mixed up — always check the axes.

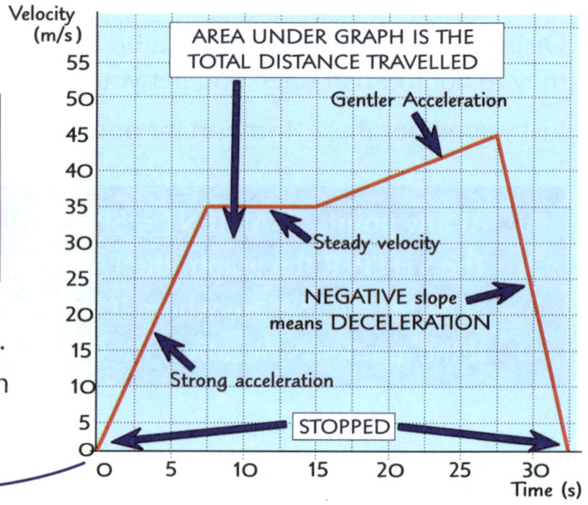

Estimating the Area Under a Curve

GRADE 9

It's easy to find the area under a velocity-time graph if it's made up of straight lines — just split it up into triangles, rectangles and trapeziums and use the area formulas (see p.127).

To estimate the area under a curved graph, divide the area under the graph approximately into trapeziums, then find the area of each trapezium and add them all together.

EXAMPLE

The red graph shows part of Rudolph the super-rabbit's morning run. Estimate the distance he ran during the 24 seconds shown.

1) Divide the area under the graph into trapeziums of equal width.
2) Find the area of each using area = average of parallel sides × distance between:

Area of trap. 1 = $\frac{1}{2} \times (10 + 40) \times 6 = 150$

Area of trap. 2 = $\frac{1}{2} \times (40 + 35) \times 6 = 225$

Area of trap. 3 = $\frac{1}{2} \times (35 + 30) \times 6 = 195$

Area of trap. 4 = $\frac{1}{2} \times (30 + 5) \times 6 = 105$

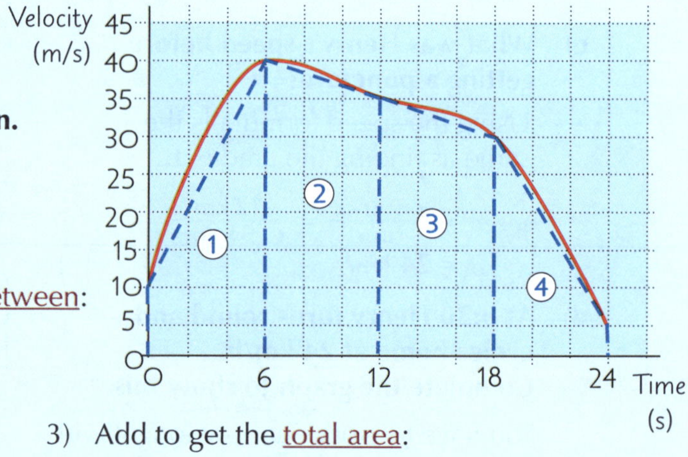

3) Add to get the total area:

Total area = 150 + 225 + 195 + 105 = 675

So Rudolph ran about **675 m** in total.

You could use this to estimate the average speed — just divide the total distance by the time taken.

You can find the average acceleration by finding the gradient between two points on a velocity-time curve, or estimate the acceleration at a specific point by drawing a tangent to the curve (see next page).

The gradient of a velocity-time graph = acceleration

Make sure you're happy with gradients and finding the area underneath a velocity-time graph.

Q1 Calculate the total distance travelled in the velocity-time graph at the top of this page. [3 marks] GRADE 6

Gradients of Real-Life Graphs

Gradients are great — they tell you all sorts of stuff, like 'you're accelerating slowly'.

The **Gradient** of a Graph **Represents** the **Rate**

GRADE 5

No matter what the graph may be, the meaning of the gradient is always simply:

(y-axis UNITS) PER (x-axis UNITS)

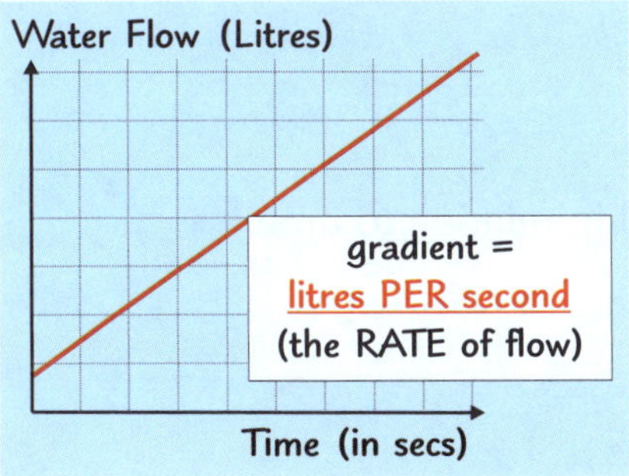

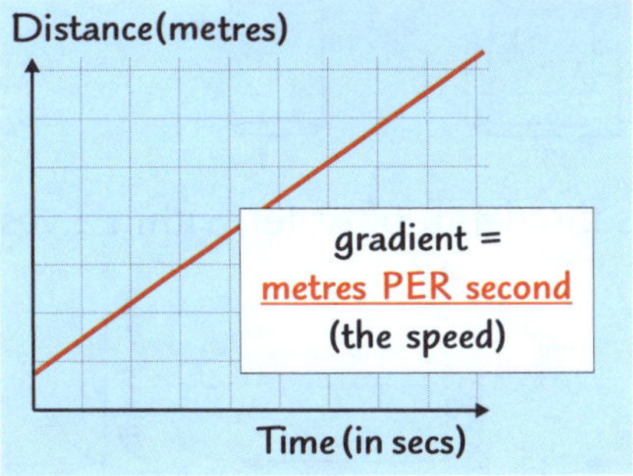

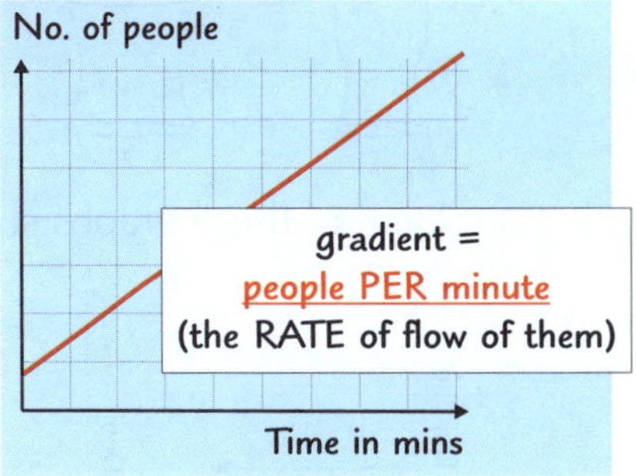

Finding the **Average Gradient**

GRADE 7

You could be asked to find the average gradient between two points on a curve.

EXAMPLE

Vicky is growing a sunflower. She records its height each day and uses this to draw the graph shown. What is the average growth per day between days 40 and 80?

1) Draw a straight line connecting the points.
2) Find the gradient of the straight line.

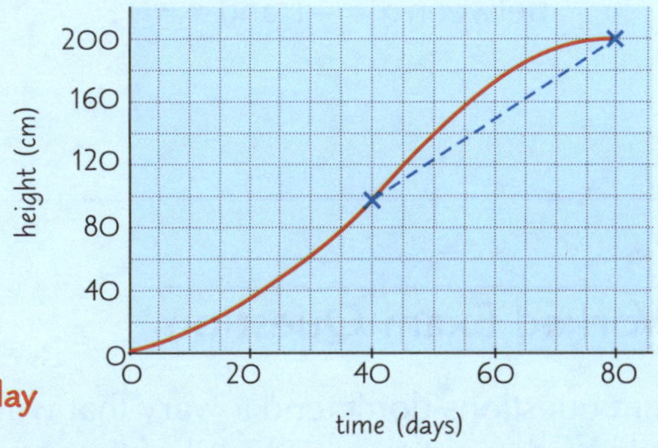

$$\text{Gradient} = \frac{\text{change in } y}{\text{change in } x} = \frac{200 - 100}{80 - 40} = \frac{100}{40} = 2.5 \text{ cm per day}$$

Estimating the **Rate** at a **Given Point**

GRADE 8

To estimate the rate at a single point on a curve, draw a tangent that touches the curve at that point. The gradient of the tangent is the same as the rate at the chosen point.

EXAMPLE

Myra plots a graph to show the distance she travelled during a bike race. Estimate Myra's speed after 40 minutes.

1) Draw a tangent to the curve at 40 minutes.
2) Find the gradient of the straight line.

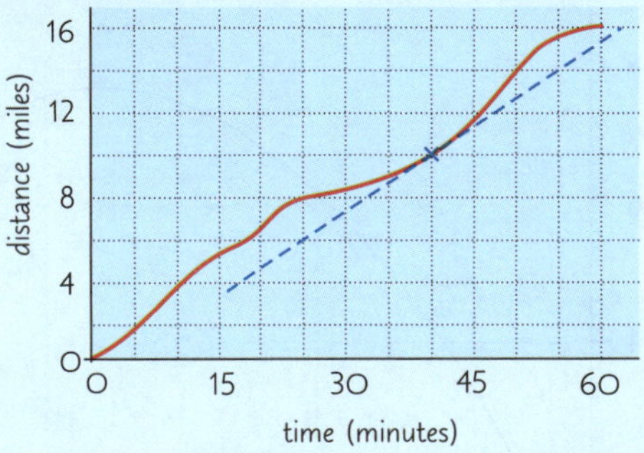

$$\text{Gradient} = \frac{\text{change in } y}{\text{change in } x} = \frac{14 - 10}{55 - 40} = \frac{4}{15} \text{ miles per minute}$$

$$= 16 \text{ miles per hour}$$

The gradient is always y-axis units per x-axis units

Just remember to look at the units and keep a ruler to hand and you'll have no problem with this. Also practise finding the gradient, just to make sure you've got it nailed.

Q1 On the sunflower height graph, estimate the rate of growth on day 20. [2 marks] GRADE 8

Q2 On the cycling graph, calculate the average speed between 25 and 40 minutes. [2 marks] GRADE 7

Q1 Video Solution

Warm-Up and Worked Exam Questions

Here's the last batch of warm-up questions on graphs — they'll get your brain working, ready for those exam practice questions.

Warm-Up Questions

1) Water is poured into each of the vessels below at a constant rate.

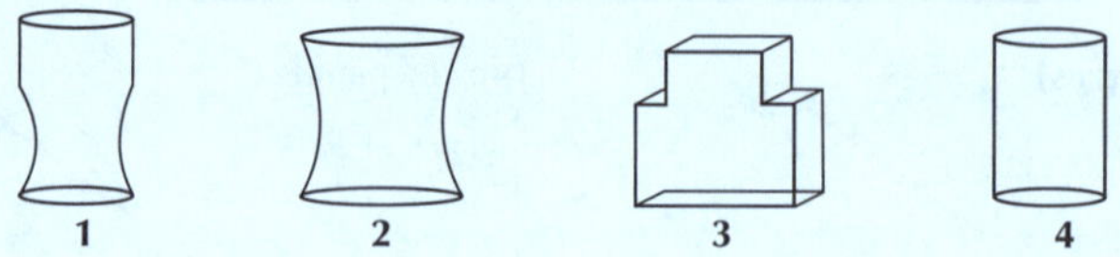

Each of these graphs shows the depth of water within a vessel in relation to time.

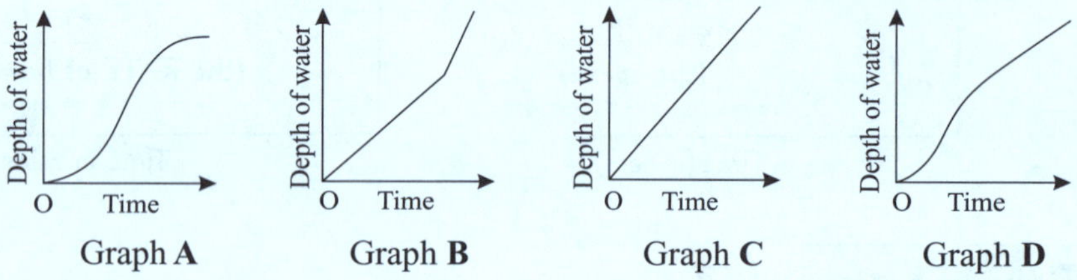

Match each vessel with the correct graph.

2) Find the average gradient of the graph on the right, between $x = -1$ and $x = 1$.

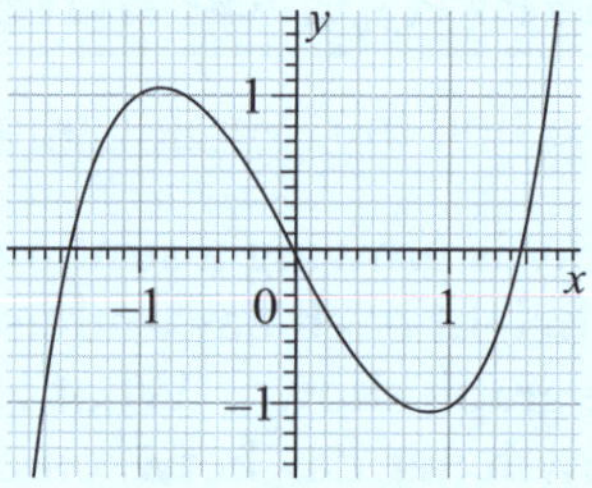

Worked Exam Question

Exam questions don't tend to vary that wildly — the basic format is often pretty similar. You'd be daft not to spend a bit of time learning how to answer a common question, wouldn't you?

1 Rahmiya rolls a ball down a hill and records its velocity. She plots the results on the velocity-time graph shown below. (Grade 9)

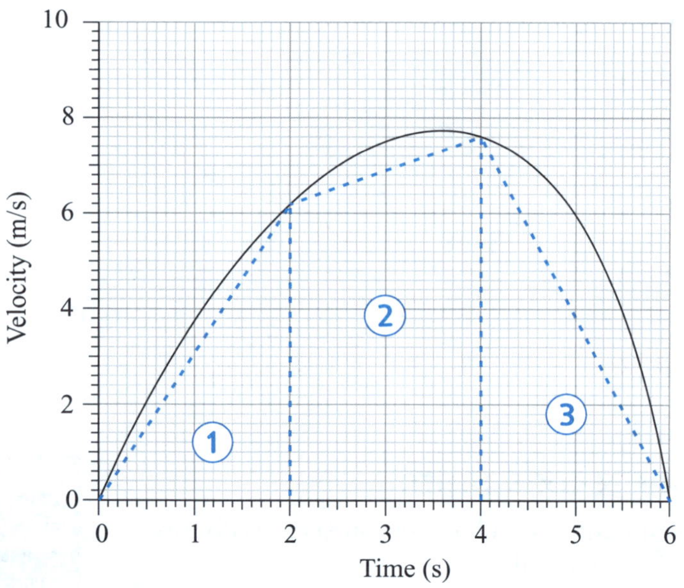

Calculate the average speed of the ball. Give your answer to 1 s.f.

Remember, the area under the graph = total distance travelled.

Area of triangle 1:
$\frac{1}{2} \times 6.2 \times 2 = 6.2$ m

Area of trapezium 2:
$\frac{1}{2} \times (6.2 + 7.6) \times 2 = 13.8$ m

Area of triangle 3:
$\frac{1}{2} \times 7.6 \times 2 = 7.6$ m

Approximate distance travelled
= 6.2 + 13.8 + 7.6 = 27.6 m

Estimate of average speed
= 27.6 ÷ 6 = 5 m/s (1 s.f.)

..........5..... m/s
[4 marks]

Exam Questions

2 An electricity company offers its customers two different price plans. (Grade 3)

Plan **A**:

Monthly tariff of £18, plus 10p for each unit used.

Plan **B**:

No monthly tariff, just pay 40p for each unit used.

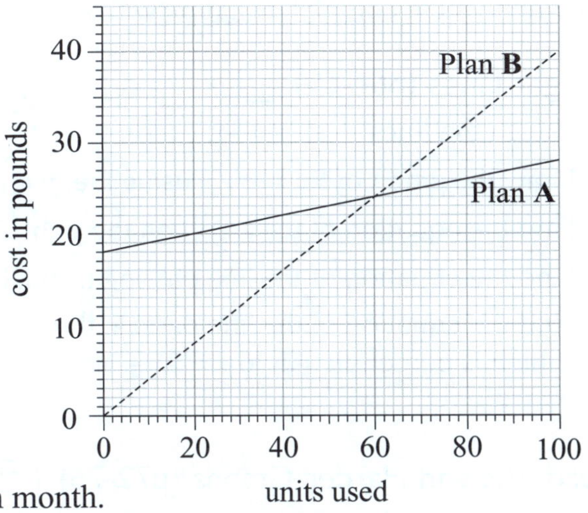

a) Use the graph to find the cost of using 70 units in a month for each plan.

Plan **A** .... Plan **B** ....

[2 marks]

b) Mr Barker uses about 85 units of electricity each month. Which price plan would you advise him to choose? Explain your answer.

....

....

[2 marks]

3 The distance/time graph below shows a 30 km running race between Selby and Tyrone. (Grade 3)

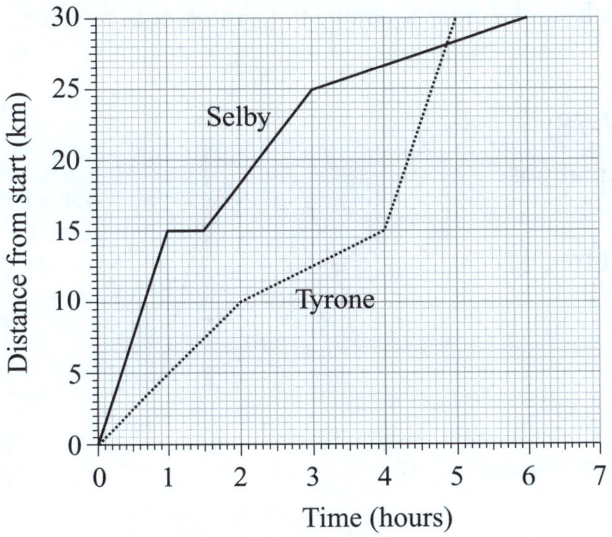

a) During the race Selby stops at a bench to get his breath back. After how many hours did he stop at the bench?

.... hours

[1 mark]

b) Who won the race? How can you tell this from the graph?

....

....

[1 mark]

c) What was Selby's speed between 1.5 and 3 hours into the race? Give your answer to 2 d.p.

.... km/h

[2 marks]

d) During the race, one of the runners injured their leg.
Which runner do you think was injured?
What evidence is there on the graph to support your answer?

....

....

[2 marks]

Revision Questions for Section Three

That wraps up Section Three — time see how much you know.

- Try these questions and tick off each one when you get it right.
- When you're completely happy with a topic, tick that off too.

For even more practice, try the quiz for Section Three — just scan this QR code!

Section Three Quiz

Straight Lines (p64-69)

1) Sketch the lines a) $y = -x$, b) $y = -4$, c) $x = 2$
2) Draw the graph of $5x = 2 + y$ using the '$y = mx + c$' method.
3) Find the equation of the graph on the right.
4) Find the equation of the line passing through (3, –6) and (6, –3).
5) Find the equation of the line passing through (4, 2) which is perpendicular to $y = 2x - 1$.

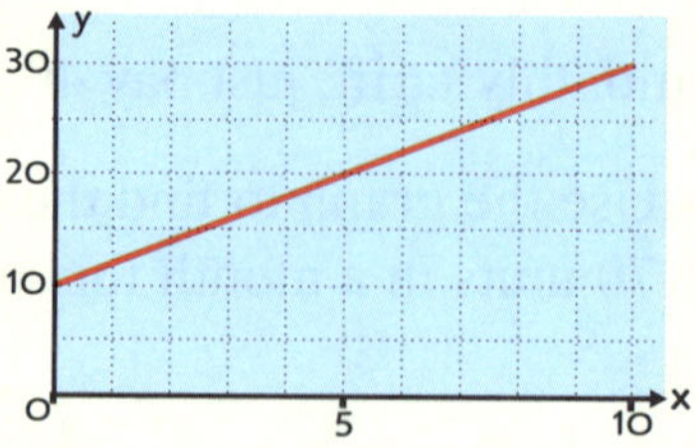

Quadratic and Harder Graphs (p72-76)

6) a) Create and complete a table of values for $-3 \leq x \leq 1$ for the equation $y = x^2 + 3x - 7$
   b) Plot the graph of $y = x^2 + 3x - 7$, labelling the turning point with its exact coordinates.
7) Plot the graph $y = x^2 + 2x - 8$ and use it to estimate the solutions to $-2 = x^2 + 2x - 8$ (to 1 d.p.).
8) Describe in words and with a sketch the forms of these graphs:
   a) $y = ax^3$ b) $xy = a$ c) $y = k^x$ $(k > 1)$ d) $x^2 + y^2 = r^2$
9) The graph of $y = bc^x$ goes through (2, 16) and (3, 128). Given that b and c are positive constants, find their values.
10) Sketch the graph of $\tan x$ for $-360° \leq x \leq 360°$, labelling the points where $\tan x$ intersects the axes.

Solving Equations and Transforming Graphs (p77-78)

11) By plotting their graphs, solve the simultaneous equations $4y - 2x = 32$ and $3y - 12 = 3x$
12) Find the equation of the line you would need to draw on the graph shown on the right to solve $x^2 + 4x = 0$
13) What are the three types of graph transformation you need to learn?

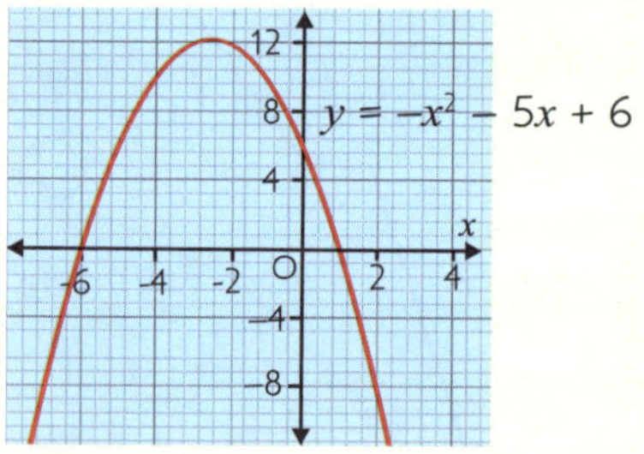

14) Describe how each of the following graphs differs from the graph of $y = x^3 + 1$
   a) $y = (-x)^3 + 1$, b) $y = (x + 2)^3 + 1$, c) $y = (x)^3 + 4$, d) $y = x^3 - 1$

Real-Life Graphs and Gradients (p82-85)

15) Sweets'R'Yum sells chocolate drops. They charge 90p per 100 g for the first kg, then 60p per 100 g after that. Plot a graph to show the cost of buying up to 3 kg of chocolate drops.
16) The graph to the right shows the speed of a sledge on a slope. Find:
   a) an estimate of the total distance travelled by the sledge.
   b) the average acceleration between 6 and 14 seconds.
   c) the acceleration at 12 seconds.

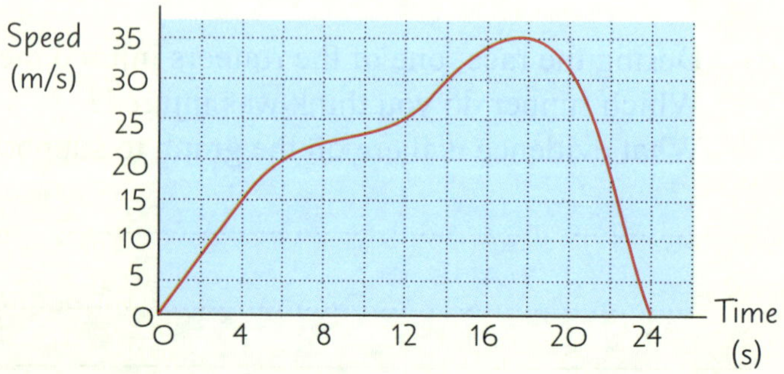

Ratios

Ratios are a pretty important topic — the next three pages should make the whole thing crystal clear.

Writing **Ratios** as **Fractions**

GRADE 3

You can express ratios as fractions in different ways.

For example, if a bag of fruit contains apples and oranges in the ratio 2 : 9, then there are:

- $\frac{2}{9}$ as many apples as oranges, or $\frac{9}{2}$ times as many oranges as apples.
- 2 + 9 = 11 parts in total. So $\frac{2}{11}$ of the bag are apples and $\frac{9}{11}$ of the bag are oranges.

Reducing **Ratios** to their **Simplest Form**

GRADE 3

To reduce a ratio to a simpler form, divide all the numbers in the ratio by the same thing (a bit like simplifying a fraction — see p.6). It's in its simplest form when there's nothing left you can divide by.

EXAMPLE **Write the ratio 15 : 18 in its simplest form.**

For the ratio 15:18, both numbers have a factor of 3, so divide them by 3.

We can't reduce this any further. So the simplest form of 15:18 is **5:6**.

15:18
÷3 ÷3
= 5:6

A **handy trick** for the calculator papers — use the **fraction button**

If you enter a fraction with the [fraction] or [$a\frac{b}{c}$] button, the calculator automatically cancels it down when you press [=]. For the ratio 8 : 12, just enter $\frac{8}{12}$ as a fraction, and it'll simplify to $\frac{2}{3}$. Now you just change it back into a ratio, i.e. 2 : 3.

The More **Awkward Cases**:

GRADE 3

For fractions, multiply by a number that gets rid of both denominators.

1) If the ratio contains **decimals** or **fractions** — **multiply**

EXAMPLE **Simplify the ratio 2.4 : 3.6 as far as possible.**

1) Multiply both sides by 10 to get rid of the decimal parts.
2) Now divide to reduce the ratio to its simplest form.

2.4:3.6
×10 ×10
= 24:36
÷12 ÷12
= **2:3**

2) If the ratio has **mixed units** — convert to the **smaller unit**

EXAMPLE **Reduce the ratio 24 mm : 7.2 cm to its simplest form.**

1) Convert 7.2 cm to millimetres.
2) Simplify the resulting ratio. Once the units on both sides are the same, get rid of them for the final answer.

24 mm : 7.2 cm
= 24 mm : 72 mm
÷24 ÷24
= **1:3**

3) To get to the form **1 : n** or **n : 1** — just **divide**

This form is often the most useful, as it shows the ratio very clearly.

EXAMPLE **Reduce 3 : 56 to the form 1 : n.**

Divide both sides by 3:

3:56
÷3 ÷3
= $1:\frac{56}{3}$ = **$1:18\frac{2}{3}$** (or $1:18.\dot{6}$)

Ratios

*Another page on **ratios** coming up — it's more **interesting** than the first but not as exciting as the next one...*

Scaling Up Ratios (Grade 3)

If you know the ratio between parts and the actual size of one part, you can scale the ratio up to find the other parts.

EXAMPLES

Mortar is made from mixing sand and cement in the ratio 7:2. How many buckets of mortar will be made if 21 buckets of sand are used in the mixture?

You need to multiply by 3 to go from 7 to 21 on the left-hand side (LHS) — so do that to both sides:

sand:cement = 7:2, ×3 → = 21:6

So 21 buckets of sand and 6 buckets of cement are used.

Amount of mortar made = 21 + 6 = 27 buckets

The two parts of a ratio are always in direct proportion (see p.94). So in the example above, sand and cement are in direct proportion, e.g. if the amount of sand doubles, the amount of cement doubles.

Part : Whole Ratios (Grade 3)

You might come across a ratio where the LHS is included in the RHS — these are called part:whole ratios.

EXAMPLE

Mrs Miggins owns tabby cats and ginger cats.
The ratio of tabby cats to the total number of cats is 3:5.

a) What fraction of Mrs Miggins' cats are tabby cats?

The ratio tells you that for every 5 cats, 3 are tabby cats.

$\frac{\text{part}}{\text{whole}} = \frac{3}{5}$

b) What is the ratio of tabby cats to ginger cats?

3 in every 5 cats are tabby, so 2 in every 5 are ginger.

5 − 3 = 2

For every 3 tabby cats there are 2 ginger cats.

tabby:ginger = 3:2

c) Mrs Miggins has 12 tabby cats. How many ginger cats does she have?

Scale up the ratio from part b) to find the number of ginger cats.

tabby:ginger = 3:2, ×4 → = 12:8

There are 8 ginger cats

Proportional Division (Grade 4)

In a proportional division question a TOTAL AMOUNT is split into parts in a certain ratio. The key word here is PARTS — concentrate on 'parts' and it all becomes quite painless:

EXAMPLE

Jess, Mo and Greg share £9100 in the ratio 2:4:7. How much does Mo get?

1) ADD UP THE PARTS:
The ratio 2:4:7 means there will be a total of 13 parts: 2 + 4 + 7 = 13 parts

2) DIVIDE TO FIND ONE "PART":
Just divide the total amount by the number of parts: £9100 ÷ 13 = £700 (= 1 part)

3) MULTIPLY TO FIND THE AMOUNTS:
We want to know Mo's share, which is 4 parts: 4 parts = 4 × £700 = £2800

Ratios

*If you were worried I was running out of **great stuff** to say about ratios then worry no more...*

Using the **Difference Between Two Parts**

GRADE 4

Sometimes questions give you the difference between two parts instead of the total amount.

EXAMPLE

A baguette is cut into 3 pieces in the ratio 1 : 2 : 5. The first piece is 28 cm smaller than the third piece. How long is the second piece?

1) Work out how many parts 28 cm makes up. 28 cm = 3rd piece – 1st piece
= 5 parts – 1 part = 4 parts

2) Divide to find one part. 28 cm ÷ 4 = 7 cm

3) Multiply to find the length of the 2nd piece. 2nd piece = 2 parts = 2 × 7 cm = 14 cm

Changing Ratios

GRADE 8

You'll need to know how to deal with all sorts of questions where the ratio changes.
Have a look at this example to see how to handle them.

EXAMPLE

The ratio of male to female pupils going on a skiing trip is 5 : 3.
Four male teachers and nine female teachers are also going on the trip.
The ratio of males to females going on the trip is 4 : 3 (including teachers).
How many female pupils are going on the trip?

1) Write the ratios as equations.

Let m be the number of male pupils and f be the number of female pupils.

$m:f = 5:3$

$(m + 4):(f + 9) = 4:3$

2) Turn the ratios into fractions (see p.89).

$\frac{m}{f} = \frac{5}{3}$ and $\frac{m+4}{f+9} = \frac{4}{3}$

3) Solve the two equations simultaneously.

$3m = 5f$ and $3m + 12 = 4f + 36$

$3m - 4f = 24$

$-\ 3m - 5f = 0$

$f = 24$

24 female pupils are going on the trip.

See pages 56-57 for more on simultaneous equations.

You need to know how to simplify all kinds of ratios

If you're stuck, it's often really helpful to think about what one 'part' is and take it from there...

Q1 Simplify: a) 25 : 35 b) 3.4 : 5.1 c) $\frac{9}{4} : \frac{15}{2}$ [4 marks] GRADE 3

Q2 Orange squash is made of water and concentrate in the ratio 11 : 2.
a) What fraction of the squash is concentrate? [1 mark]
b) How many litres of water are needed to make 1.95 litres of orange squash? [2 marks] GRADE 3

Q3 The ages of Ben, Graham and Pam are in the ratio 3 : 7 : 8.
Pam is 25 years older than Ben. How old is Graham? [2 marks] GRADE 5

Q4 A bag contains red and blue balls. If two of each colour ball are removed from the bag the ratio of red to blue balls is 5 : 7. If seven of each colour ball are added to the original bag, the ratio of red to blue balls is 4 : 5.
How many red and blue balls are in the original bag? [6 marks] GRADE 8

Q4 Video Solution

Warm-Up and Worked Exam Questions

Make the most of the help on this page by working through everything carefully.

Warm-Up Questions

1) Write these ratios in their simplest forms:

   a) 4 : 8 b) 12 : 27 c) 1.2 : 5.4 d) $\frac{8}{3} : \frac{7}{6}$ e) 0.5 litres : 400 ml

2) Reduce 5 : 22 to the form 1 : n.
3) A recipe uses flour and sugar in the ratio 3 : 2.
   How much flour do you need if you're using 300 g of sugar?
4) Anja collects mugs. The ratio of red mugs to the total number of mugs is 6 : 15.
   Given that Anja has 50 mugs, how many of them are red?
5) Divide 180 in the ratio 3 : 4 : 5.
6) The heights of Niamh, Owen and Maya are in the ratio 5 : 9 : 7.
   Maya is 44 cm taller than Niamh. How tall is Owen?

Worked Exam Questions

I've gone through these questions and filled in the answers like you'll do in the exam, so take a look.

1 Mr Appleseed's Supercompost is made by mixing soil, compost and grit in the ratio 4 : 3 : 1.
Soil costs £8 per 40 kg, compost costs £15 per 25 kg and grit costs £12 per 15 kg.
How much profit will be made if 16 kg of Mr Appleseed's Supercompost is sold for £10? (Grade 5)

16 kg of Mr Appleseed's Supercompost contains:

$\frac{4}{8} \times 16 = 8$ kg of soil. $\frac{3}{8} \times 16 = 6$ kg of compost. $\frac{1}{8} \times 16 = 2$ kg of grit.

Soil costs £8 ÷ 40 = £0.20 per kg.

Compost costs £15 ÷ 25 = £0.60 per kg.

Grit costs £12 ÷ 15 = £0.80 per kg.

16 kg of Supercompost costs: $(8 \times 0.2) + (6 \times 0.6) + (2 \times 0.8) =$ £6.80

Profit: £10 – £6.80 = £3.20

£ ......3.20......

[5 marks]

2 In the morning a baker makes x muffins and y pastries.
After selling 5 muffins and 3 pastries, the ratio of muffins to pastries is 5 : 8. (Grade 6)
She then makes 10 more of each item and the ratio becomes 5 : 7.
Find the values of x and y.

$x - 5 : y - 3 = 5 : 8$ and $x + 5 : y + 7 = 5 : 7$

$\frac{x-5}{y-3} = \frac{5}{8}$ and $\frac{x+5}{y+7} = \frac{5}{7}$

$8(x - 5) = 5(y - 3)$ and $7(x + 5) = 5(y + 7)$

Turn the two ratios into simultaneous equations, then solve.

Expand and simplify to give:

$8x - 5y = 25$ [1] and $7x - 5y = 0$ [2]

[1] – [2]: $x = 25$

Sub $x = 25$ into [1]: $(8 \times 25) - 5y = 25$

$5y = 175$, so $y = 35$

$x =$ ......25......

$y =$ ......35......

[5 marks]

Exam Questions

3 Hannah is making some green paint to paint her kitchen wall.
She makes it by mixing together $3\frac{3}{4}$ tins of yellow paint and $1\frac{1}{2}$ tins of blue paint. (Grade 4)
The tins are all the same size.

a) Express this ratio in its simplest form.

....
[2 marks]

b) How much of each paint will Hannah need to make 2800 ml of green paint?

yellow paint .... ml

blue paint .... ml
[2 marks]

4 Simone, Ariana, Nasir and Catrin shared £660.
Nasir got four times as much money as Catrin, Simone got twice as much money as Nasir, and Ariana got a quarter as much money as Simone. (Grade 5)

How much money did Simone get?

£ ....
[3 marks]

5 Fabio has a large jar containing only black and green olives.
The probability of randomly choosing a black olive from the jar is $\frac{5}{16}$. (Grade 8)
After eating 1 green and 3 black olives the probability of choosing a black olive is $\frac{3}{10}$.

How many black and green olives were originally in the jar?

Start by finding the ratios of black to green olives before and after he eats some — careful though, the original ratio of black:green isn't 5:16.

Black olives: ....

Green olives: ....
[6 marks]

Direct and Inverse Proportion

*There can sometimes be a lot of **information** packed into proportion questions, but the **method** of solving them always stays the same — have a look at this page and see what you think.*

Direct Proportion

1) Two quantities, A and B, are in direct proportion (or just in proportion) if increasing one increases the other one proportionally. So if quantity A is doubled (or trebled, halved, etc.), so is quantity B.
2) Remember this golden rule for direct proportion questions:

DIVIDE for ONE, then TIMES for ALL

EXAMPLE

Midori pays £3.60 per 400 g of cheese.
She uses 220 g of cheese to make 4 cheese pasties.
How much would the cheese cost if she wanted to make 50 cheese pasties?

In 1 pasty there is: 220 g ÷ 4 = 55 g of cheese
So in 50 pasties there is: 55 g × 50 = 2750 g of cheese

1 g of cheese would cost: £3.60 ÷ 400 = 0.9p
So 2750 g of cheese would cost: 0.9 × 2750 = 2475p = £24.75

There will often be lots of stages to direct proportion questions — keep track of what you've worked out at each stage.

Inverse Proportion

1) Two quantities, C and D, are in inverse proportion if increasing one quantity causes the other quantity to decrease proportionally. So if quantity C is doubled (or tripled, halved, etc.), quantity D is halved (or divided by 3, doubled etc.).
2) The rule for finding inverse proportions is:

TIMES for ONE, then DIVIDE for ALL

EXAMPLE

4 bakers can decorate 100 cakes in 5 hours.

a) How long would it take 10 bakers to decorate the same number of cakes?

100 cakes will take 1 baker: 5 × 4 = 20 hours

So 100 cakes will take 10 bakers: 20 ÷ 10 = 2 hours for 10 bakers

b) How long would it take 11 bakers to decorate 220 cakes?

100 cakes will take 1 baker: 20 hours

1 cake will take 1 baker: 20 ÷ 100 = 0.2 hours

220 cakes will take 1 baker: 0.2 × 220 = 44 hours
220 cakes will take 11 bakers: 44 ÷ 11 = 4 hours

The number of bakers is inversely proportional to the number of hours — but the number of cakes is directly proportional to the number of hours.

Direct proportion means 'as one thing increases, so does the other'

With inverse proportion, as one increases, the other decreases. Don't get them muddled up.

Q1 It takes 2 carpenters 4 hours to make 3 bookcases.
How long would it take 5 carpenters to make 10 bookcases? [4 marks]

Q2 It costs £43.20 for 8 people to go on a rollercoaster 6 times.
How much will it cost for 15 people to go on a rollercoaster 5 times? [4 marks]

Q2 Video Solution

Direct and Inverse Proportion

***Algebraic proportion** questions normally involve two variables (often x and y) which are **linked** in some way.*

Types of Proportion

GRADE 7

$\propto$ means 'is proportional to'.

1) The simple proportions are 'y is proportional to x' ($y \propto x$) and 'y is inversely proportional to x' ($y \propto \frac{1}{x}$).
2) You can always turn a proportion statement into an equation by replacing '$\propto$' with '$=$ k' like this:

	Proportionality	Equation
'y is proportional to x'	$y \propto x$	$y = \text{k}x$
'y is inversely proportional to x'	$y \propto \frac{1}{x}$	$y = \frac{\text{k}}{x}$

k is just some constant (unknown number)

3) Trickier proportions involve y varying proportionally or inversely to some function of x, e.g. x^2, x^3, $\sqrt{x}$ etc.

	Proportionality	Equation
'y is proportional to the square of x'	$y \propto x^2$	$y = \text{k}x^2$
't is proportional to the square root of h'	$t \propto \sqrt{h}$	$t = \text{k}\sqrt{h}$
'V is inversely proportional to r cubed'	$V \propto \frac{1}{r^3}$	$V = \frac{\text{k}}{r^3}$

4) Once you've written the proportion statement as an equation you can easily graph it.

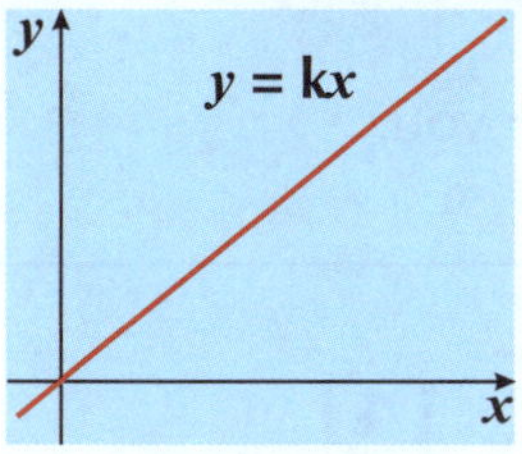

y is proportional to x

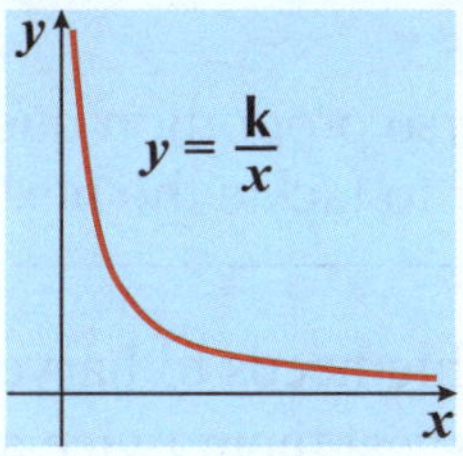

y is inversely proportional to x

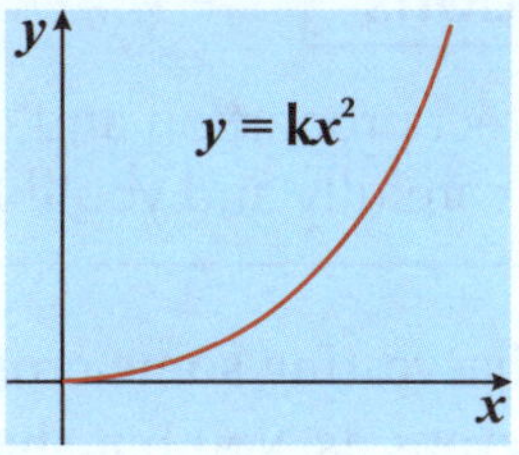

y is proportional to x^2

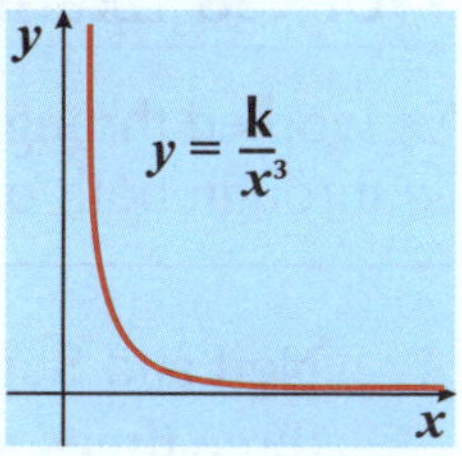

y is inversely proportional to x^3

Handling **Algebra Questions** on Proportion

GRADE 7

1) Write the sentence as a proportionality and replace '$\propto$' with '$=$ k' to make an equation (as above).
2) Find a pair of values (x and y) somewhere in the question — substitute them into the equation to find k.
3) Put the value of k into the equation and it's now ready to use, e.g. $y = 3x^2$.
4) Inevitably, they'll ask you to find y, having given you a value for x (or vice versa).

EXAMPLE **G is inversely proportional to the square root of H. When $G = 2$, $H = 16$. Find an equation for G in terms of H, and use it to work out the value of G when $H = 36$.**

1) Convert to a proportionality and replace $\propto$ with '$=$ k' to form an equation. $G \propto \frac{1}{\sqrt{H}}$ $\quad G = \frac{k}{\sqrt{H}}$
2) Use the values of G and H (2 and 16) to find k. $2 = \frac{k}{\sqrt{16}} = \frac{k}{4} \Rightarrow k = 8$
3) Put the value of k back into the equation. $G = \frac{8}{\sqrt{H}}$
4) Use your equation to find the value of G. When $H = 36$, $G = \frac{8}{\sqrt{36}} = \frac{8}{6} = \frac{4}{3}$

This is the equation for G in terms of H.

Remember that $\propto$ means 'proportional to'

Replace the $\propto$ sign with '$=$ k' and substitute in the values given in the question to find the value of k.

Q1 An object is moving with a velocity that changes proportionally with time. After 5 seconds its velocity is 105 m/s. How fast will it be travelling after 13 seconds? [3 marks] GRADE 7

Q2 P is inversely proportional to the square of Q (P, $Q > 0$). When $P = 3$, $Q = 4$. Find an equation for P in terms of Q and find the exact value of Q when $P = 8$. [4 marks] GRADE 7

Warm-Up and Worked Exam Questions

Question pages take up a large proportion of this section, but that's just because we think they'll be really useful. Give these warm-up questions a shot when you're ready.

Warm-Up Questions

1) It costs £840 to feed 2 elephants for 7 days.
   How much will it cost to feed 6 elephants for 3 days?
2) It takes 5 mechanics 9 hours to build 2 sports cars.
   How long would it take 3 mechanics to build 4 sports cars?
3) The electrical current through a metal wire is proportional to the voltage. When the current is 0.4 A the voltage is 4 V. What would the voltage be when the current is 1.2 A?
4) Write each of the following as an equation:
   a) A is proportional to the square of r b) $D \propto \frac{1}{R}$
   c) H is inversely proportional to the cube of D d) $V \propto S^3$

Worked Exam Questions

I've worked through one direct proportion and one inverse proportion question for you. Go through both questions carefully and you'll be ready to tackle the next page.

1 Neil and Sophie are harvesting some crops. Sophie needs to harvest three times as many crops as Neil but she can harvest them twice as quickly. GRADE 4

Neil takes 3.5 hours to harvest his crops. How long does Sophie take to harvest her crops?

To harvest the same amount as Neil, Sophie will take:
3.5 hours ÷ 2 = 1.75 hours
Sophie needs to harvest three times as much so it will take her:
1.75 × 3 = 5.25 hours

..........5.25.......... hours
[3 marks]

2 The gravitational force, f, between two objects is inversely proportional to the square of the distance, d, between them. When $d = 100$, $f = 20$. GRADE 7
Write an equation connecting f and d and use it to find the value of f when $d = 800$.

$f \propto \frac{1}{d^2}$, so $f = \frac{k}{d^2}$
When $d = 100$ and $f = 20$, $20 = \frac{k}{100^2}$,
so $k = 20 \times 100^2 = 200\,000$
and $f = \frac{200\,000}{d^2}$
When $d = 800$, $f = \frac{200\,000}{800^2} = 0.3125$

$f =$ ....0.3125..........
[3 marks]

Exam Questions

3 Elijah runs a go-kart track. It takes 12 litres of petrol to race 8 go-karts for 20 minutes. Petrol costs £1.37 per litre.

a) 6 go-karts used 18 litres of petrol. How many minutes did they race for?

.... minutes
[4 marks]

b) How much does the petrol cost to run 8 go-karts for 45 minutes?

£ ....
[3 marks]

4 Sketch the following proportions on the axes below them.

y is proportional to x

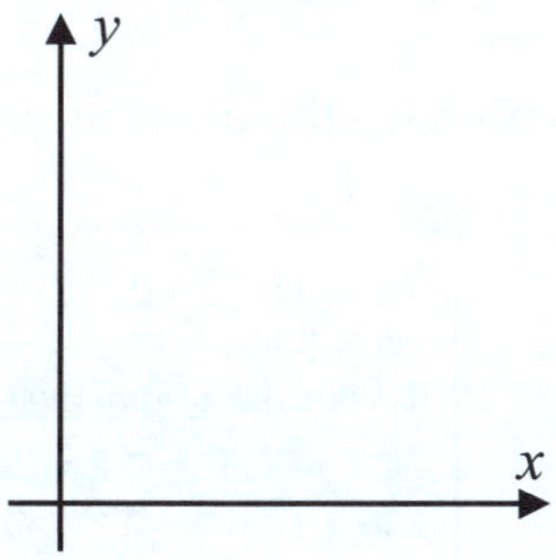

$y \propto \frac{1}{x}$

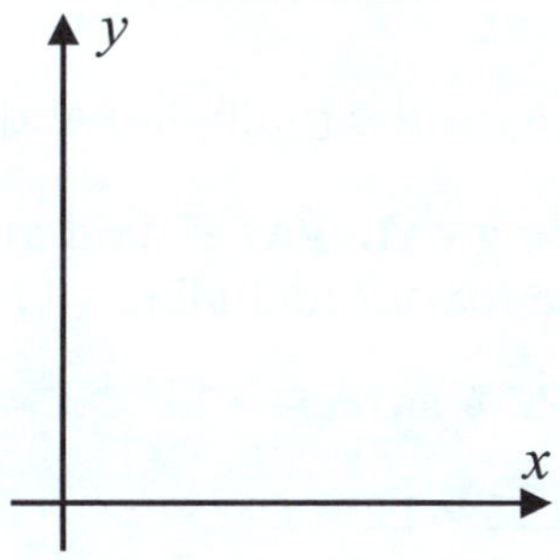

$y = kx^3$

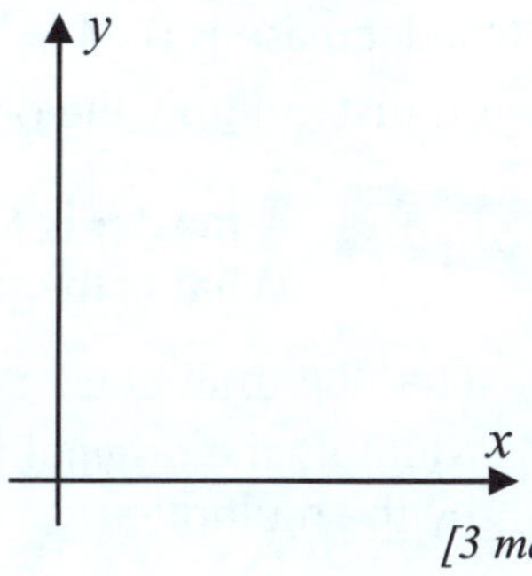

[3 marks]

5 Round a bend on a railway track the height difference (h mm) between the outer and inner rails must vary in direct proportion to the square of the maximum permitted speed (S km/h).

a) When $S = 50$, $h = 35$. Calculate h when $S = 40$.

h = ...
[3 marks]

b) The maximum speed on a bend is to be increased by 30%. What will be the percentage increase in the height difference between the outer and inner rails?

There's more stuff on percentages on the next few pages.

.... %
[4 marks]

Percentages

*'Per cent' means 'out of 100' — remember this and you'll be able to convert percentages into fractions and decimals (p.9). Then you're ready to tackle the first three **simple types** of percentage question.*

Three **Simple** Question Types

Type 1 — "Find x% of y"

GRADE 3

Turn the percentage into a decimal/fraction, then multiply. E.g. 15% of £46 = $\frac{15}{100}$ × £46 = £6.90.

EXAMPLE

A shopkeeper had a box of 140 chocolate bars. He sold 60% of the chocolate bars for 62p each and he sold the other 40% at 2 for £1. How much did the box of chocolate bars sell for in total?

1) Find 60% and 40% of 140:
   60% of 140 bars = 0.6 × 140 = 84 bars
   40% of 140 bars = 0.4 × 140 = 56 bars

2) So he sold 84 bars for 62p each and 56 bars at 2 for £1.
   Total sales = (84 × 0.62) + (56 ÷ 2) = £80.08

Type 2 — "Find the new amount after a % increase/decrease"

GRADE 3

This time, you first need to find the multiplier — the decimal that represents the percentage change.

E.g. 5% increase is 1.05 (= 1 + 0.05)
26% decrease is 0.74 (= 1 – 0.26)

A % increase has a multiplier greater than 1, a % decrease has a multiplier less than 1.

Then you just multiply the original value by the multiplier and voilà — you have the answer.

EXAMPLE

A toaster is £38 excluding VAT. VAT is paid at 20%. What is the price of the toaster including VAT?

1) Find the multiplier:
   20% increase = 1 + 0.2 = 1.2

2) Multiply the original value by the multiplier:
   £38 × 1.2 = £45.60

If you prefer, you can work out the percentage, then add or subtract it from the original value:
20% of £38 = 0.2 × 38 = £7.60
£38 + £7.60 = £45.60

Type 3 — "Express x as a percentage of y"

GRADE 3

Divide x by y, then multiply by 100. E.g. 209 as a percentage of 400 = $\frac{209}{400}$ × 100 = 52.25%.
N.B. if x is bigger than y you'll get a percentage that's bigger than 100.

EXAMPLE

There are 480 pupils in a school. 55% of them are girls and 59 of the girls have blonde hair. What percentage of the girls have blonde hair?

1) Find the number of girls in the school:
   55% = 55 ÷ 100 = 0.55
   55% of 480 = 0.55 × 480 = 264 girls

2) Divide the number of blonde-haired girls by the number of girls and multiply by 100.
   $\frac{59}{264}$ × 100 = 22.348...% = 22.3% (1 d.p.)

There are loads of different types of percentage question

Start by learning these three types, then try this Practice Question:

Q1 A cricket team scored 250 runs in their first innings. One player scored 30% of these runs. In the second innings, the same player scored 105 runs. Express his second innings score as a percentage of his first innings score. [3 marks] GRADE 3

Percentages

*Watch out for these **trickier types** of percentage question — they'll often include lots of real-life context. Just make sure you know the **proper method** for each of them and you'll be fine.*

Two **Trickier** Question Types

Type 1 — Finding the percentage change

GRADE 4

1) This is the formula for giving a change in value as a percentage — LEARN IT, AND USE IT:

$$\text{Percentage 'Change'} = \frac{\text{'CHANGE'}}{\text{ORIGINAL}} \times 100$$

2) You end up with a percentage rather than an amount, as you did for Type 3 on the previous page.
3) Typical questions will ask 'Find the percentage increase/profit/error' or 'Calculate the percentage decrease/loss/discount', etc.

EXAMPLE

A trader buys 6 watches at £25 each. He scratches one of them, so he sells that one for £11. He sells the other five for £38 each. Find his profit as a percentage.

1) Here the 'change' is profit, so the formula looks like this: $\text{percentage profit} = \frac{\text{profit}}{\text{original}} \times 100$
2) Work out the actual profit (amount made – amount spent): profit = (£38 × 5) + £11 – (6 × £25) = £51
3) Calculate the percentage profit: $\text{percentage profit} = \frac{51}{6 \times 25} \times 100 = 34\%$

Type 2 — Finding the original value

GRADE 4

This is the type that most people get wrong — but only because they don't recognise it as this type and don't apply this simple method:

1) Write the amount in the question as a percentage of the original value.
2) Divide to find 1% of the original value.
3) Multiply by 100 to give the original value (= 100%).

EXAMPLE

A house increases in value by 10.5% to £132 600. Find what it was worth before the rise.

Note: The new, not the original value, is given.

1) An increase of 10.5% means £132 600 represents 110.5% of the original value.
2) Divide by 110.5 to find 1% of the original value.
3) Then multiply by 100.

£132 600 = 110.5%

÷110.5

£1200 = 1%

×100

£120 000 = 100%

So the original value was £120 000

If it was a decrease of 10.5%, then you'd put '£132 600 = 89.5%' and divide by 89.5 instead of 110.5.

Think carefully about what each question is asking for...

The methods above are easy to follow but the questions can be a bit tricky. Try these Practice Questions:

Q1 A cereal company has decreased the amount of cereal in a box from 1.2 kg to 900 g. What is the percentage decrease in the amount of cereal per box? [3 marks] GRADE 4

Q2 A shop sells kebabs for a 20% loss. The kebabs sell for £4.88 each. The shop wants to reduce its loss on kebabs to 10%. How much should the shop charge per kebab? [3 marks] GRADE 6

Q2 Video Solution

Percentages

Percentages are almost certainly going to come up in your exam, but they could crop up in lots of ***different topics****. Here are some more* ***examples*** *of where you might see them.*

Simple Interest

GRADE 3

Compound interest is covered on the next page.

Simple interest means a certain percentage of the original amount only is paid at regular intervals (usually once a year). So the amount of interest is the same every time it's paid.

EXAMPLE **Regina invests £380 in an account which pays 3% simple interest each year. How much interest will she earn in 4 years?**

1) Work out the amount of interest earned in one year: $3\% = 3 \div 100 = 0.03$
$3\% \text{ of } £380 = 0.03 \times £380 = £11.40$

2) Multiply by 4 to get the total interest for 4 years: $4 \times £11.40 = £45.60$

Working with Percentages

GRADE 5

1) Sometimes there isn't a set method you can follow to answer percentage questions.
2) You'll have to use what you've learnt on the last couple of pages and do a bit of thinking for yourself.
3) Here are a few examples to get you thinking.

EXAMPLE **80% of the members of a gym are male. 35% of the male members are aged 40 and over. What percentage of gym members are males under 40 years old?**

1) The percentage of male members under 40 is: $100\% - 35\% = 65\%$

2) The percentage of gym members that are male and under 40 is: $65\% \text{ of } 80\% = 0.65 \times 80\%$
$= 52\%$

It's just like finding *x% of y* — but this time the *y* is a percentage too.

EXAMPLE **The side length, x, of a cube is increased by 10%. What is the percentage increase in the volume of the cube?**

GRADE 6

1) Find the volume of the original cube. Original volume $= x^3$

2) Find the volume of the cube after the increase in side length.
10% increase $= 1 + 0.1 = 1.1$
New side length $= 1.1x$
New volume $= (1.1x)^3 = 1.331x^3$

3) Work out the increase in volume. Increase $= 1.331x^3 - x^3 = 0.331x^3$

4) Calculate the percentage increase.
$$\text{percentage increase} = \frac{\text{increase}}{\text{original}} \times 100 = \frac{0.331x^3}{x^3} \times 100 = 33.1\%$$

Percentages are one of the most useful things you'll ever learn

Whenever you open a newspaper, see an advert, watch TV or do a maths exam paper you will see percentages. It's really important you're confident with using them — so practise, practise, practise...

Q1 Faheem invests some money for 5 years in an account at 4% simple interest each year. What is the percentage increase of the investment at the end of the 5 years? [2 marks] GRADE 6

Q2 At a publishing company 60% of the editors are female. 30% of the female editors and 20% of the male editors have a maths degree. What percentage of all the editors have a maths degree? [3 marks] GRADE 5

Compound Growth and Decay

*One more sneaky % type for you... Unlike **simple interest**, in **compound interest** the amount added on **changes** each time — it's a percentage of the **new amount**, rather than the **original amount**.*

The **Formula**

This topic is simple if you LEARN THIS FORMULA. If you don't, it's pretty well impossible:

$$N = N_0 \times (\text{multiplier})^n$$

- N: Amount after n days/hours/years
- N_0: Initial amount
- multiplier: Percentage change multiplier. E.g. 5% increase is 1.05 (= 1 + 0.05), 26% decrease is 0.74 (= 1 – 0.26)
- n: Number of days/hours/years

3 Examples to show you how **EASY** it is:

Compound interest is a popular context for these questions — it means the interest is added on each time, and the next lot of interest is calculated using the new total rather than the original amount.

EXAMPLE

Daniel invests £1000 in a savings account which pays 8% compound interest per annum. How much will there be after 6 years?

Use the formula: Amount = $1000(1.08)^6$ = **£1586.87**

(1000 = initial amount, 1.08 = 8% increase, 6 = 6 years)

'Per annum' just means 'each year'.

Depreciation questions are about things (e.g. cars) which decrease in value over time.

EXAMPLE

Susan has just bought a car for £6500.

a) **If the car depreciates by 9% each year, how much will it be worth in 3 years' time?**

Use the formula: Value = $6500(0.91)^3$ = **£4898.21**

b) **How many complete years will it be before the car is worth less than £3000?**

Use the formula again but this time you know don't know n.

Value = $6500(0.91)^n$

Use trial and error to find how many years it will be before the value drops below £3000.

If $n = 8$, $6500(0.91)^8 = 3056.6414....$

$n = 9$, $6500(0.91)^9 = 2781.5437...$

It will be **9 years** before the car is worth less than £3000.

The compound growth and decay formula can be about population and disease too.

EXAMPLE

The number of bacteria in a sample increases at a rate of 30% each day. After 6 days the number of bacteria is 7500. How many bacteria were there in the original sample?

Put the numbers you know into the formula, then rearrange to find the initial amount, N_0.

$7500 = N_0(1.3)^6$

$N_0 = 7500 \div (1.3)^6 = 1553.82...$

So there were **1554** bacteria originally.

Compound growth — percentages applied again and again

This page is all about the formula, so make sure you learn it... learn it real good. Then, try this:

Q1 Pippa's bank account pays 2.5% compound interest per annum and her balance is £3200. Kyle has the same bank balance but his account pays simple interest at 3% per annum. Who will have the most money after 3 years and how much more will they have? [4 marks]

Q1 Video Solution

Warm-Up and Worked Exam Questions

Have a go at these warm-up questions and see how you get on — the exam questions will be a bit more tricky, so it's important that you can do these first.

Warm-Up Questions

1) Find 15% of £90.
2) A watch is bought for £96 and sold at 135% of its original price. How much did it sell for?
3) Write 36 out of 80 as a percentage.
4) £200 is put in a bank account paying 2% simple interest (per year). Find the total amount in the account after 2 years if no money has been withdrawn.
5) £3000 is invested at 3% compound interest (per year). Work out how much money is in the account at the end of 4 years, to the nearest penny.

Worked Exam Questions

Another couple of worked exam questions for you. Study them well — they might just help.

1 Ian makes and sells lobster pots. He sells them for £32 per pot which is a 60% profit on the cost of the materials. He wants to increase his profit to 88%. (GRADE 5)

How much should Ian start charging per lobster pot?

£32 is a 60% profit so £32 = 160% of cost price

1% of cost price = £32 ÷ 160
= £0.20

He wants an 88% profit = 188% of cost price

188% = £0.20 × 188 = £37.60

£ ......37.60......

[3 marks]

2 The population of fish in a lake is estimated to decrease by 8% every year. (GRADE 5)

a) How many fish will be left after 15 years if the initial population is 2000?

Use the formula: Multiplier = 1 − 0.08 = 0.92

Population after 15 years = $2000 \times (0.92)^{15}$
= 572.59...

You need to round up, as it'll take more than 15 years for the population to fall to 572 fish.

......573......

[2 marks]

b) How many years will it take for the population of fish to be less than $\frac{3}{4}$ of the initial population?

$\frac{3}{4}$ of the initial population = $2000 \times \frac{3}{4} = 1500$

n = 1: $2000 \times 0.92 = 1840$

n = 2: $2000 \times 0.92^2 = 1692.8$

n = 3: $2000 \times 0.92^3 = 1557.376$

n = 4: $2000 \times 0.92^4 = 1432.78592 < 1500$

So the population is less than $\frac{3}{4}$ of the initial population after 4 years.

......4...... years

[2 marks]

Exam Questions

3 The ratio of grapes to cherries in a fruit salad is 2 : 5. (Grade 3)
Circle the correct statement below.

There are 50% more cherries than grapes.

There are 80% more cherries than grapes.

There are 20% as many grapes as cherries.

There are 40% as many grapes as cherries.

[1 mark]

4 A hairdresser recorded some details about her customers one day. (Grade 5)
The ratio of children : adults was 3 : 7.
60% of the children had blond hair and 20% of the adults had blond hair.

What percentage of all the customers had blond hair?

.... %
[4 marks]

5 A conservation company plants pine trees in a forest to increase their number by 16% each year. At the end of each year, a logging company is permitted to cut down up to 75% of the number of new trees planted that year. (Grade 5)

At the start of 2018 there were 5000 pine trees in the forest.

What was the minimum number of pine trees in the forest at the end of 2019?

....
[4 marks]

6 The value of a football player decreases at a rate of 25% each year after the age of 30. At the age of 35 a player was valued at £2 000 000. (Grade 7)

What was the player's value when he was 31 years old?
Give your answer to the nearest £100 000.

£ ....
[3 marks]

Unit Conversions

*A nice easy page for a change — just some **facts** to learn.*

Metric and **Imperial** Units

GRADE 3

COMMON METRIC CONVERSIONS

1 cm = 10 mm	1 tonne = 1000 kg
1 m = 100 cm	1 litre = 1000 ml
1 km = 1000 m	1 litre = 1000 cm^3
1 kg = 1000 g	1 cm^3 = 1 ml

COMMON IMPERIAL CONVERSIONS

1 Yard = 3 Feet	1 Foot = 12 Inches
1 Gallon = 8 Pints	
1 Stone = 14 Pounds	
1 Pound = 16 Ounces	

You only need to remember the metric conversions, but you should be able to use them all.

COMMON METRIC-IMPERIAL CONVERSIONS

1 kg ≈ 2.2 pounds   1 foot ≈ 30 cm   1 gallon ≈ 4.5 litres   1 mile ≈ 1.6 km

Converting **Units**

GRADE 3

Always check your answer looks sensible — if it's not then chances are you divided instead of multiplying or vice versa.

To convert between units, multiply or divide by the conversion factor.

Converting speeds is a bit trickier because speeds are made up of two measures — a distance and a time. You have to convert the distance unit and the time unit separately.

EXAMPLES

1. Convert 10 pounds into kg.

2.2 pounds ≈ 1 kg
So 10 pounds ≈ 10 ÷ 2.2
≈ 4.5 kg

2. A rabbit's top speed is 56 km/h. How fast is this in m/s?

1) First convert from km/h to m/h:
56 km/h = (56 × 1000) m/h = 56 000 m/h

2) Now convert from m/h to m/s:
56 000 m/h = (56 000 ÷ 60 ÷ 60) m/s = 15.6 m/s (1 d.p.)

Converting **Area** and **Volume** Measurements

GRADE 4

Converting areas and volumes from one unit to another is an exam disaster that you have to know how to avoid. 1 m^2 definitely does NOT equal 100 cm^2. Remember this and read on for why.

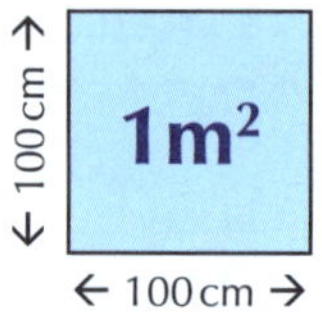

1 m^2 = 100 cm × 100 cm = 10 000 cm^2
1 cm^2 = 10 mm × 10 mm = 100 mm^2

1 m^3 = 100 cm × 100 cm × 100 cm = 1 000 000 cm^3
1 cm^3 = 10 mm × 10 mm × 10 mm = 1000 mm^3

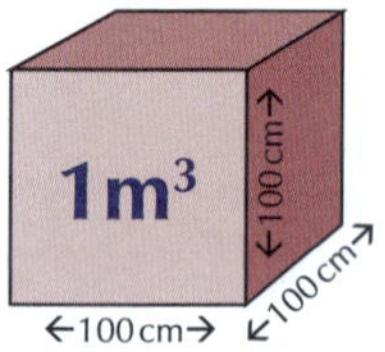

EXAMPLES

1. Convert 9 m^2 to cm^2.

To change area measurements from m^2 to cm^2 multiply by 100 twice.

9 × 100 × 100 = 90 000 cm^2

2. Convert 60 000 mm^3 to cm^3.

To change volume measurements from mm^3 to cm^3 divide by 10 three times.

60 000 ÷ (10 × 10 × 10) = 60 cm^3

There's no way round it — you'll have to learn some conversions

There are a lot of conversions here, and you'll need to remember the metric, area, and volume conversions. Keep covering them up and scribbling them down until you've learned them all.

Q1 Dawn lives 18 km away from work. She drives to work and back 5 days a week. Her car's fuel consumption is 28 mpg (miles per gallon) and petrol costs £1.39 per litre. Calculate the weekly cost of petrol for Dawn travelling to and from work. [5 marks] GRADE 5

Speed, Density and Pressure

Speed, density and pressure. Just a matter of ***learning the formulas****, sticking the* ***numbers*** *in and watching the* ***units****.*

Speed = Distance ÷ Time

Grade 3

Speed is the distance travelled per unit time, e.g. the number of km per hour or metres per second.

$$\text{SPEED} = \frac{\text{DISTANCE}}{\text{TIME}} \qquad \text{TIME} = \frac{\text{DISTANCE}}{\text{SPEED}} \qquad \text{DISTANCE} = \text{SPEED} \times \text{TIME}$$

Formula triangles are a handy tool for remembering formulas like these. The speed one is shown below.

D
S × T

How do you use a formula triangle?

1) Cover up the thing you want to find and write down what's left showing.
2) Now put in the values and calculate — check the units in your answer.

EXAMPLE **A car travels 9 miles at 36 miles per hour. How many minutes does it take?**

Write down the formula, put in the values and calculate:

$$\text{time} = \frac{\text{distance}}{\text{speed}} = \frac{9\text{ miles}}{36\text{ mph}} = 0.25\text{ hours} = 15\text{ minutes}$$

Density = Mass ÷ Volume

Grade 4

Density is the mass per unit volume of a substance. It's usually measured in kg/m^3 or g/cm^3.

$$\text{DENSITY} = \frac{\text{MASS}}{\text{VOLUME}} \qquad \text{VOLUME} = \frac{\text{MASS}}{\text{DENSITY}} \qquad \text{MASS} = \text{DENSITY} \times \text{VOLUME}$$

M
D × V

EXAMPLE **A giant 'Wunda-Choc' bar has a density of 1.3 g/cm³. If the bar's volume is 1800 cm³, what is the mass of the bar in kg?**

Write down the formula, put in the values and calculate:

mass = density × volume
$= 1.3\text{ g/cm}^3 \times 1800\text{ cm}^3 = 2340\text{ g}$
$= 2.34\text{ kg}$

Check your units match. If the density is in g/cm^3, the volume must be in cm^3 and you'll get a mass in g.

Pressure = Force ÷ Area

Grade 4

'N' stands for 'Newtons'.

Pressure is the amount of force acting per unit area. It's usually measured in N/m^2, or pascals (Pa).

$$\text{PRESSURE} = \frac{\text{FORCE}}{\text{AREA}} \qquad \text{AREA} = \frac{\text{FORCE}}{\text{PRESSURE}} \qquad \text{FORCE} = \text{PRESSURE} \times \text{AREA}$$

F
P × A

EXAMPLE **A cylindrical barrel exerts a force of 200 N on the horizontal ground. The radius of the circular face resting on the ground is 0.4 m. Calculate the pressure exerted by the barrel on the ground to 1 d.p.**

Work out the area of the circular face: $\pi \times 0.4^2 = 0.5026...\text{ m}^2$

Write down the pressure formula, put in the values and calculate:

$$\text{pressure} = \frac{\text{force}}{\text{area}} = \frac{200\text{ N}}{0.5026...\text{ m}^2} = 397.8873...\text{ N/m}^2$$
$= 397.9\text{ N/m}^2$ (1 d.p.)

Formula triangles are dead useful

Write down the formula triangles from memory, then use them to generate the formulas.

Q1 A solid lead cone has a vertical height of 60 cm and a base radius of 20 cm. If the density of lead is 11.34 g/cm³, find the mass of the cone in kg to 3 s.f. [3 marks] (Grade 5)
(Hint: you'll need to find the volume of the cone — see p.130.)

Q1 Video Solution

Warm-Up and Worked Exam Questions

Time to check all that lovely revision has sunk in. Try these first to make sure you've learnt the key stuff:

Warm-Up Questions

1) a) Convert 12.7 kg into grams. b) Convert 1430 cm into metres.
2) 22 lbs of apples weighs about how many kilograms?
3) Change 3 m^3 to mm^3.
4) A lump of lead weighing 374 g has a volume of 33 cm^3. What is the approximate density of the lead (to 3 s.f.)?
5) A solid plastic building block measures 5 cm × 4 cm × 6 cm. The density of the plastic is 0.8 g/cm^3. What is the mass of the block?
6) A cheetah runs 100 m in 4 seconds. What is its average speed in km per hour?
7) A cyclist travels for 0.75 hours at a speed of 12 km per hour. What distance does he travel?

Worked Exam Questions

Make sure you really take this stuff in — read it thoroughly and then have a go yourself to check you've understood. You'll kick yourself if this comes up in the exam and you only gave it a quick glance.

1 The cuboid below has three different faces (A, B and C). (Grade 4)
The cuboid has a weight of 40 N.

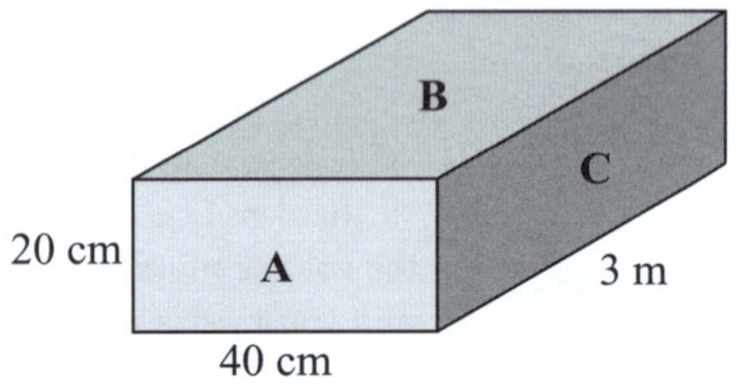

a) Calculate the pressure, in N/m^2, that the cuboid exerts on horizontal ground when the cuboid is resting on face A.

Area of A = 40 cm × 20 cm = 800 cm^2
= 800 ÷ 100 ÷ 100 = 0.08 m^2

Pressure = 40 N ÷ 0.08 m^2
= 500 N/m^2

..........500.......... N/m^2
[3 marks]

b) Three of these cuboids are stacked directly on top of each other and the bottom cuboid is resting on face B. What pressure are they exerting on horizontal ground?

Three cuboids would weigh = 3 × 40N = 120 N

Area of B = 3 m × 0.4 m = 1.2 m^2

Pressure = 120 N ÷ 1.2 m^2 = 100 N/m^2

..........100.......... N/m^2
[3 marks]

2 In 2019 Mo ran a long-distance race and finished with time *t*. (Grade 6)
In 2020 he finished the same race but his time was 10% quicker.
By what percentage did his average speed for the race increase?
Give your answer to 2 decimal places.

s_1 is Mo's speed in 2019, s_2 is Mo's speed in 2020

In 2020 he finished with a time of 0.9t, so $s_1 = \frac{d}{t}$ and $s_2 = \frac{d}{0.9t}$

So, because d is the same each year, $s_1t = 0.9s_2t \rightarrow s_2 = \frac{s_1}{0.9} = 1.11... \times s_1$

So s_2 is 111.11... % of s_1.

His percentage increase was 11.11% (2 d.p.)

..........11.11.......... %
[4 marks]

Exam Questions

3 Adam has been caught speeding by a pair of average speed cameras. The speed limit was 50 mph. The cameras are 2500 m apart.

The time taken for his car to pass between them was 102 seconds.

a) What was Adam's average speed between the cameras? Give your answer to the nearest mph. Take 1 mile as 1.6 km.

.... mph
[3 marks]

b) If Adam had been travelling within the speed limit, what is the minimum time it should have taken him to pass between the cameras? Give your answer to the nearest second.

.... s
[2 marks]

4 An iron cube has side length 4 cm and iron has a density of 7.9 grams per cm^3.

a) Work out the mass of the iron cube.

.... g
[3 marks]

b) A larger iron cube has a mass of 63.2 kg. What is the ratio of the side lengths of the smaller and larger cubes?

....
[4 marks]

5 The cone below has a base diameter of $20x$ cm. When the base of the cone rests on horizontal ground it exerts a pressure of 650 N/m^2.

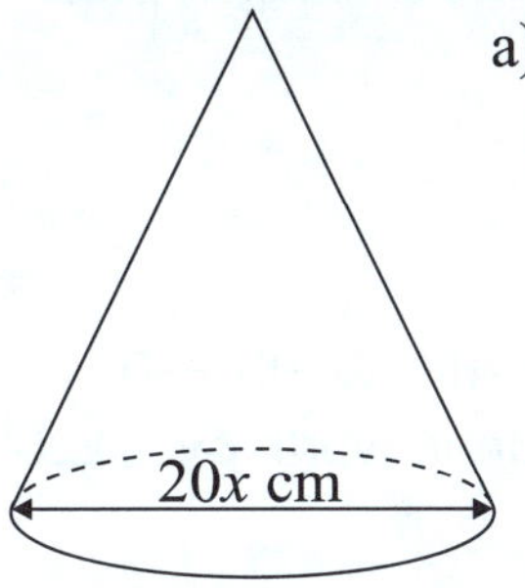

a) Calculate the weight of the cone in terms of x and π.

.... N
[4 marks]

b) The diameter of the cone is halved but the weight is kept the same. What effect will this have on the pressure exerted on the ground?

[2 marks]

Revision Questions for Section Four

Now, let's see what you remember from Section Four...

- Try these questions and tick off each one when you get it right.
- When you're completely happy with a topic, tick that off too.

For even more practice, try the quiz for Section Four — just scan this QR code!

Section Four Quiz

Ratios (p89-91)

1) Pencils and rubbers are in the ratio 13 : 8. How many times more pencils are there than rubbers?
2) Reduce: a) 1.2 : 1.6 to its simplest form b) 49 g : 14 g to the form n : 1
3) Sarah is in charge of ordering stock for a clothes shop. The shop usually sells red scarves and blue scarves in the ratio 5 : 8. Sarah orders 150 red scarves. How many blue scarves should she order?
4) Ryan, Joel and Sam are delivering 800 newspapers. They split the newspapers in the ratio 5 : 8 : 12.
   a) What fraction of the newspapers does Ryan deliver?
   b) How many more newspapers does Sam deliver than Joel?
5) There are 44 oak trees in a forest and the ratio of oak trees to pine trees is 2 : 5. The ratio changes to 9 : 20 when an equal number of each tree are planted. How many of each tree were planted?
6) The ratio of x to y is 4 : 1. If x and y are decreased by 6 they are in the ratio 10 : 1. Find x and y.

Direct and Inverse Proportion (p94-95)

7) 6 gardeners can plant 360 flowers in 3 hours.
   a) How many flowers could 8 gardeners plant in 6 hours?
   b) How many hours would it take for 15 gardeners to plant 1170 flowers?
8) 'y is proportional to the square of x'. a) Write the statement as an equation.
   b) Sketch the graph of this proportion for $x \geq 0$.
9) The pressure a cube exerts on the ground is inversely proportional to the square of its side length. When the side length is 3 cm the pressure is 17 Pa. Find the pressure when the side length is 13 cm.

Percentages (p98-100)

10) If $x = 20$ and $y = 95$: a) Find x% of y. b) Find the new value after x is increased by y%.
    c) Express x as a percentage of y. d) Express y as a percentage of x.
11) What's the formula for finding a change in value as a percentage?
12) An antique wardrobe decreased in value from £800 to £520. What was the percentage decrease?
13) A tree's height has increased by 15% in the last year to 20.24 m. What was its height a year ago?
14) 25% of the items sold by a bakery in one day were pies. 8% of the pies sold were chicken pies. What percentage of the items sold by the bakery were chicken pies?

Compound Growth and Decay (p101)

15) What's the formula for compound growth and decay?
16) Collectable baseball cards increase in value by 7% each year. A particular card is worth £80.
    a) How much will it be worth in 10 years? b) In how many years will it be worth over £200?

Unit Conversions (p104)

17) Convert: a) 5.6 litres to cm^3 b) 8 feet to cm c) 3 m/s to km/h
    d) 12 m^3 to cm^3 e) 1280 mm^2 to cm^2 f) 2.75 cm^3 to mm^3

Speed, Density and Pressure (p105)

18) Find the average speed of a car if it travels 63 miles in an hour and a half.
19) Find the volume of a snowman if its density is 0.4 g/cm^3 and its mass is 5 kg.
20) Find the area of an object in contact with horizontal ground, if the pressure it exerts on the ground is 120 N/m^2 and it pushes on the ground with a force of 1320 N.

Geometry

If you know all these rules thoroughly, you'll at least have a fighting chance of working out problems with lines and angles. If you don't — you've no chance. Sorry to break it to you like that.

5 Simple Rules — that's all

GRADE 3

1) Angles in a triangle add up to 180°.

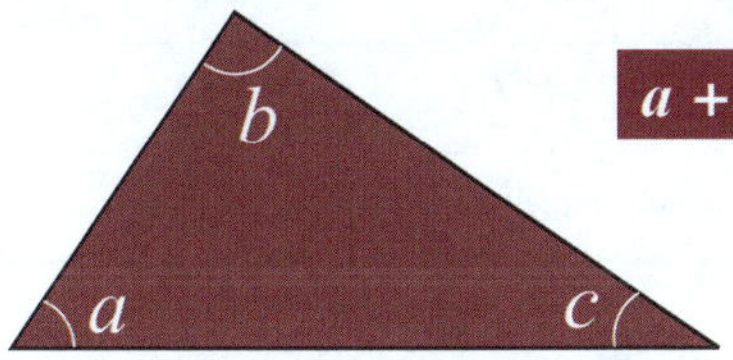

$a + b + c = 180°$

There's a nice proof of this (using parallel lines) on the next page.

2) Angles on a straight line add up to 180°.

$a + b + c = 180°$

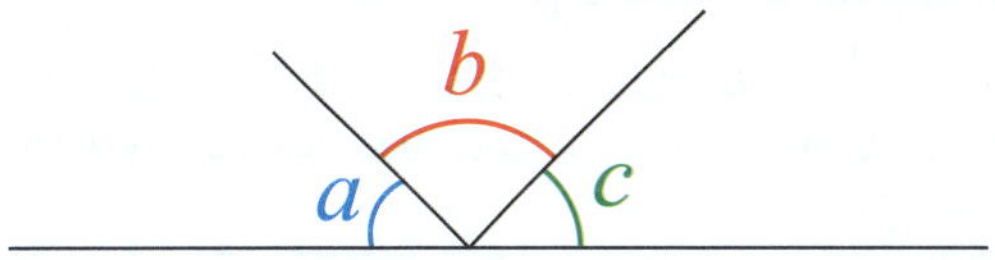

3) Angles in a quadrilateral add up to 360°.

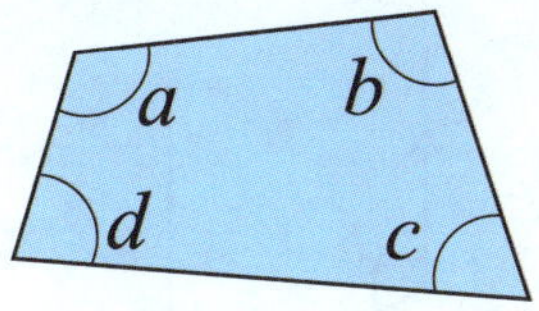

Remember that a quadrilateral is a 4-sided shape.

$a + b + c + d = 360°$

You can see why this is if you split the quadrilateral into two triangles along a diagonal. Each triangle has angles adding up to 180°, so the two together have angles adding up to 180° + 180° = 360°.

4) Angles round a point add up to 360°.

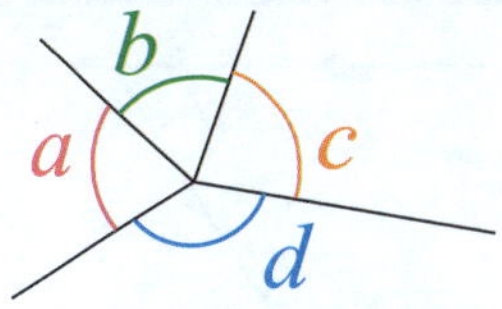

$a + b + c + d = 360°$

5) Isosceles triangles have 2 sides the same and 2 angles the same.

In an isosceles triangle, you only need to know one angle to be able to find the other two.

These dashes indicate two sides the same length.

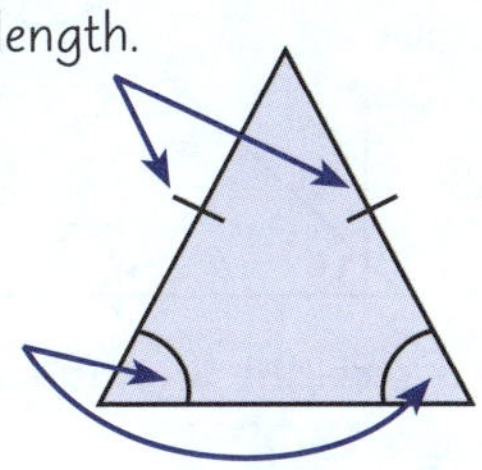

These angles are the same.

EXAMPLE **Find the size of angle x.**

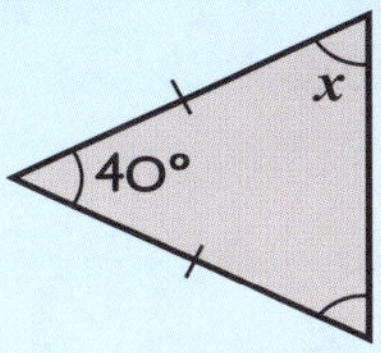

180° – 40° = 140°
The two angles on the right are the same (they're both x) and they must add up to 140°, so $2x$ = 140°, which means x = 70°.

Five simple rules — make sure you LEARN THEM...

Scribble them down again and again until they're ingrained in your brain (or desk). All the basic facts are pretty easy really, but examiners like to combine them in questions to confuse you. There are some examples of these on p.111, but have a go at this one as a warm-up.

Q1 Find the size of the angle marked x. [2 marks] GRADE 3

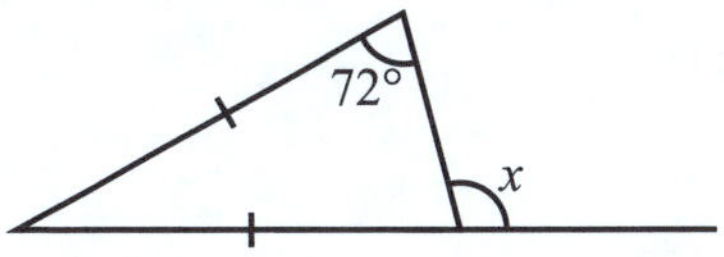

Parallel Lines

***Parallel lines** are quite straightforward really. (They're also quite straight. And parallel.)*
*There are a few **rules** you need to learn — make sure you don't get them mixed up.*

Angles Around **Parallel Lines**

When a line crosses two parallel lines, it forms special sets of angles.

1) The two bunches of angles formed at the points of intersection are the same.
2) There are only actually two different angles involved (labelled a and b here), and they add up to 180° (from rule 2 on the previous page).
3) Vertically opposite angles (ones opposite each other) are equal (in the diagram, a and a are vertically opposite, as are b and b).

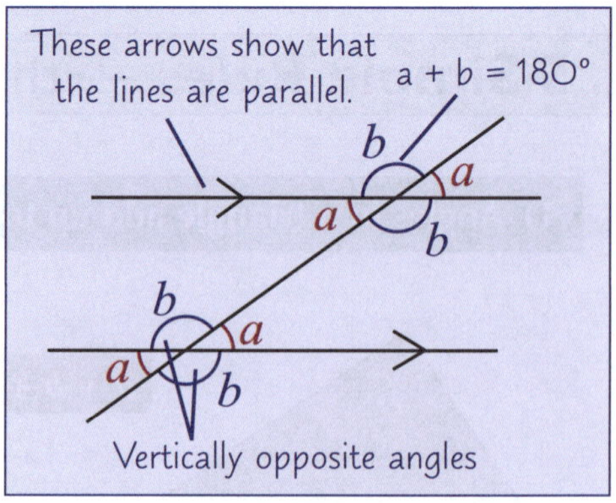

Alternate, Allied and **Corresponding** Angles

The diagram above has some characteristic shapes to look out for — and each shape contains a specific pair of angles. The angle pairs are known as alternate, allied and corresponding angles.

You need to spot the characteristic Z, C, U and F shapes:

ALTERNATE ANGLES

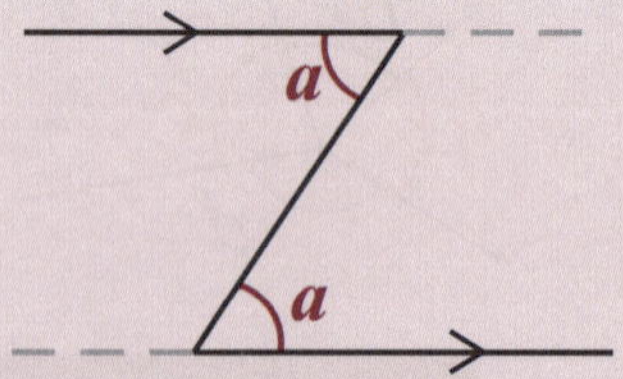

Alternate angles are the same.
They are found in a Z-shape.

ALLIED ANGLES

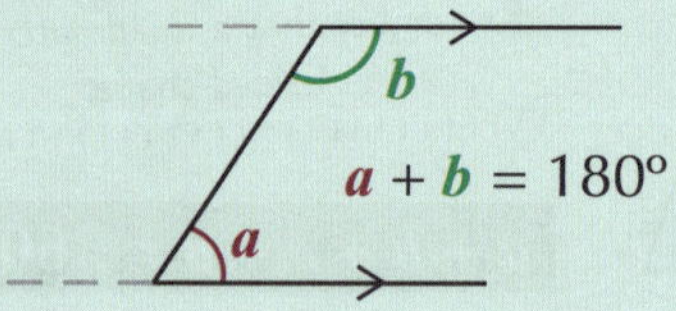

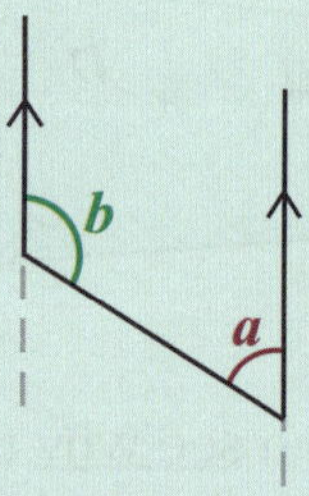

Allied angles add up to 180°.
They are found in a C- or U-shape.

CORRESPONDING ANGLES

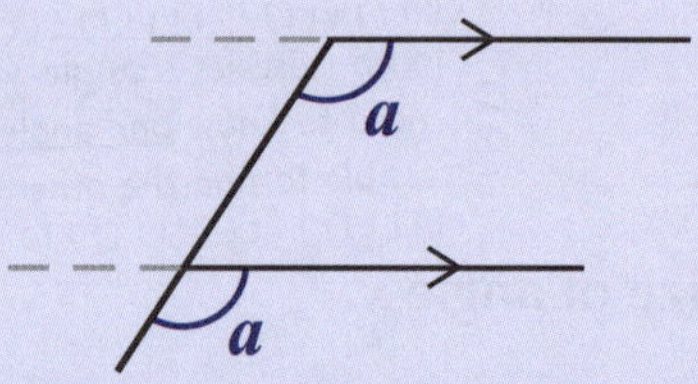

Corresponding angles are the same.
They are found in an F-shape.

It's OK to use the letters Z, C, U and F to help you identify the rule — but you must use the proper names (alternate, allied and corresponding angles) in the exam.

EXAMPLE

Prove that the angles in a triangle add up to 180°.

This is the proof of rule 1 from the previous page.
First, draw a triangle between two parallel lines:

These two angles are the same as they're alternate.

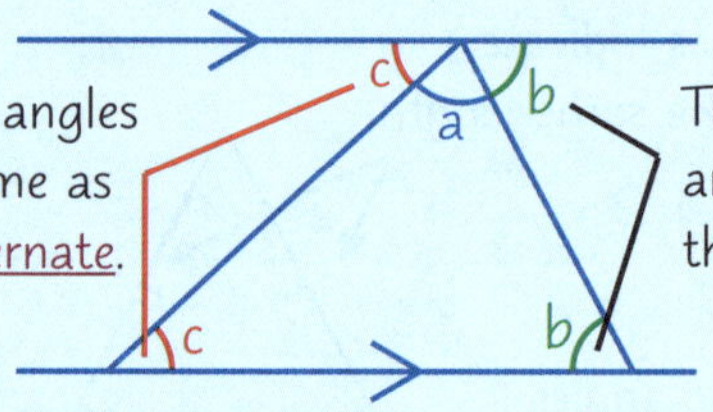

These two angles are the same as they're alternate.

Angles on a straight line add up to 180°, so a + b + c = 180°.

Parallel lines are key things to look out for in geometry

Watch out for hidden parallel lines and Z, C, U and F shapes — extending the lines can make spotting them easier.

Q1 Find the value of x.

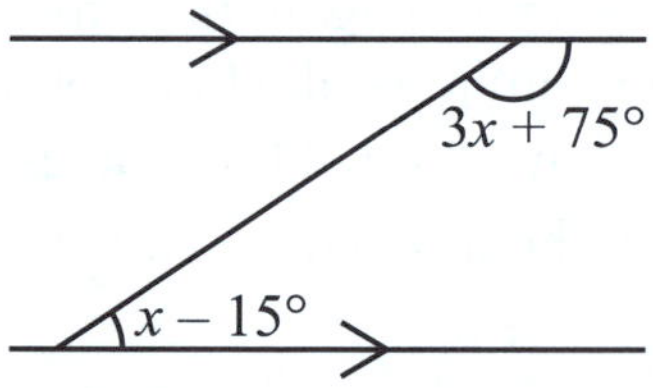

[3 marks]

Geometry Problems

*Once you've learnt **all** of the **geometry rules** on the last two pages you can have a go at using them.*

Try Out **All the Rules** One by One

GRADE 3

Don't concentrate too much on the angle you have been asked to find.
The best method is to find ALL the angles in whatever order they become obvious.

Before we get going, make sure you're familiar with three-letter angle notation, e.g. ∠ABC.
∠ABC, ABC and $A\hat{B}C$ all mean 'the angle formed at B' (it's always the middle letter).
You might even see it written as just $\hat{B}$.

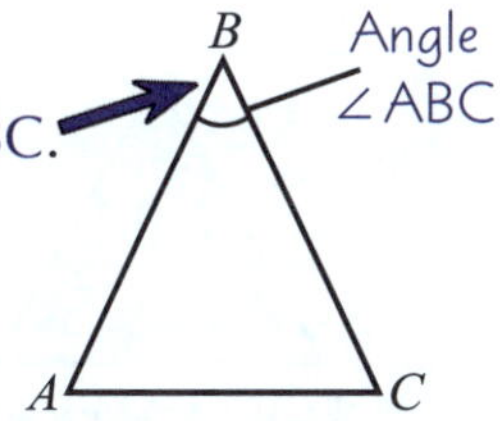

EXAMPLE **Find the size of angles *x* and *y*.**

Write down everything you know (or can work out) about the shape:

Triangle ABD is isosceles,
so ∠BAD = ∠ABD = 76°.
That means ∠ADB = 180° – 76° – 76° = 28°.
∠ADC is a right angle (= 90°),
so angle x = 90° – 28° = **62°**

ABCD is a quadrilateral, so all the angles add up to 360°. 76° + 90° + y + 72° = 360°,
so y = 360° – 76° – 90° – 72° = **122°**

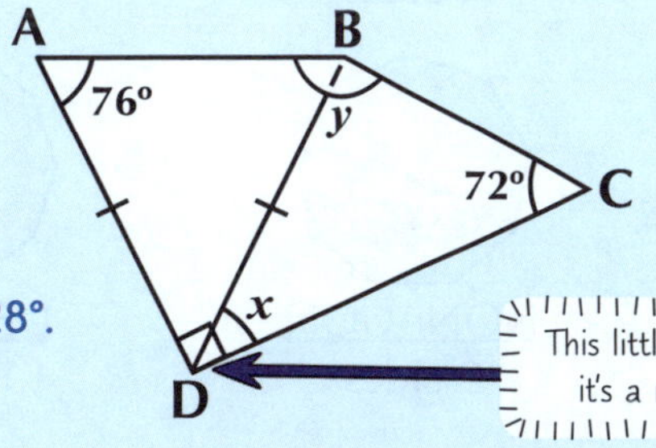

This little square means that it's a right angle (90°).

You could have worked out angle y before angle x.

EXAMPLE **In the diagram below, BDF is a straight line. Find the size of angle BCD.**

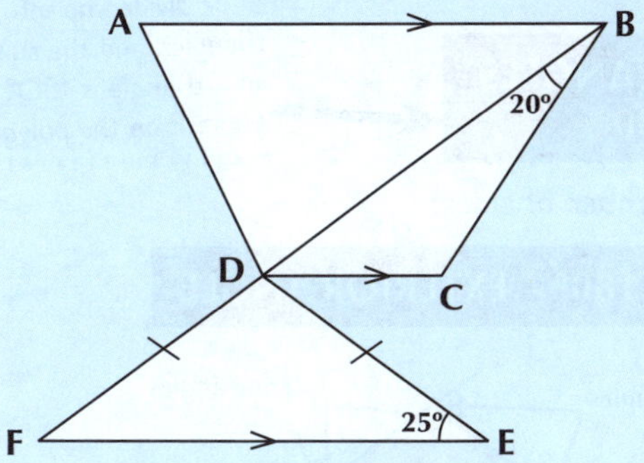

1) Triangle DEF is isosceles, so...
∠DFE = ∠DEF = 25°

2) FE and AB are parallel, so...
∠DFE and ∠ABD are alternate angles.
So ∠ABD = ∠DFE = 25°.

3) ∠ABC = ∠ABD + ∠CBD = 25° + 20° = **45°**

4) DC and AB are parallel, so...
∠BCD and ∠ABC are allied angles.
Allied angles add up to 180°, so
∠BCD + ∠ABC = 180°
∠BCD = 180° – 45° = **135°**.

There's often more than one way of tackling these questions — e.g. you could have found angle BDC using the properties of parallel lines, then used angles in a triangle to find BCD.

The most important rule of all — don't panic

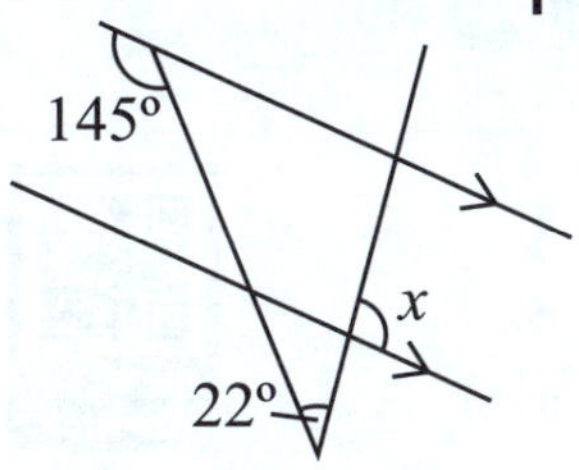

Geometry problems often look a lot worse than they are — don't panic, just write down everything you can work out. Watch out for hidden parallel lines and isosceles triangles — they can help you work out angles.

Q1 Find the size of angle x. [3 marks] GRADE 3

Q1 Video Solution

Polygons

*A **polygon** is a **many-sided shape**, and can be **regular** or **irregular**. A **regular** polygon is one where all the **sides** and **angles** are the **same** (in an **irregular** polygon, the sides and angles are **different**).*

Regular Polygons

Here are the first few regular polygons.

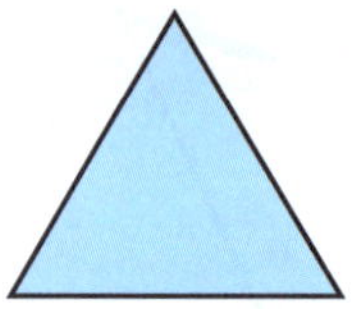

EQUILATERAL TRIANGLE
3 sides

SQUARE
(regular quadrilateral)
4 sides

PENTAGON
5 sides

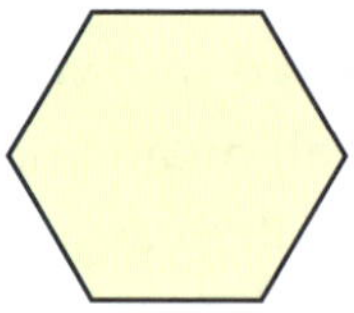

HEXAGON
6 sides

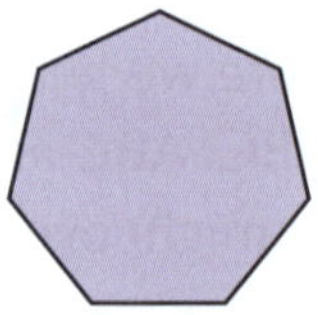

HEPTAGON
7 sides

OCTAGON
8 sides

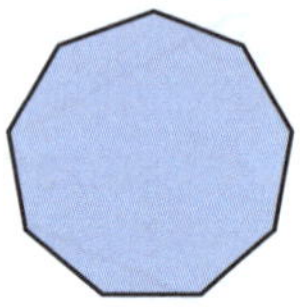

NONAGON
9 sides

DECAGON
10 sides

Regular polygons have the same number of lines of symmetry and the same order of rotational symmetry as the number of sides (rotational symmetry is how many positions you can rotate the shape into so it looks exactly the same).

Interior and **Exterior** Angles

Questions on interior and exterior angles often come up in exams — so you need to know what they are and how to find them. There are a few formulas you need to learn as well.

For ANY POLYGON (regular or irregular):

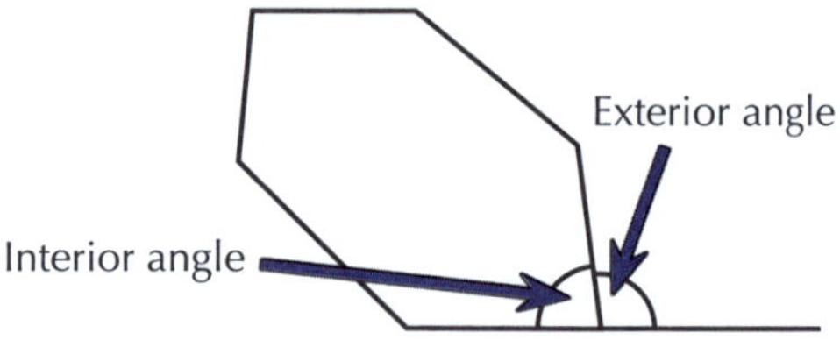

SUM OF EXTERIOR ANGLES = 360°

SUM OF INTERIOR ANGLES = (n – 2) × 180°

(n is the number of sides)

This is because a polygon can be divided up into (n – 2) triangles, and the sum of angles in a triangle is 180°. Try it for yourself on the polygons above.

INTERIOR ANGLE = 180° – EXTERIOR ANGLE

For REGULAR POLYGONS only:

$$\text{EXTERIOR ANGLE} = \frac{360°}{n}$$

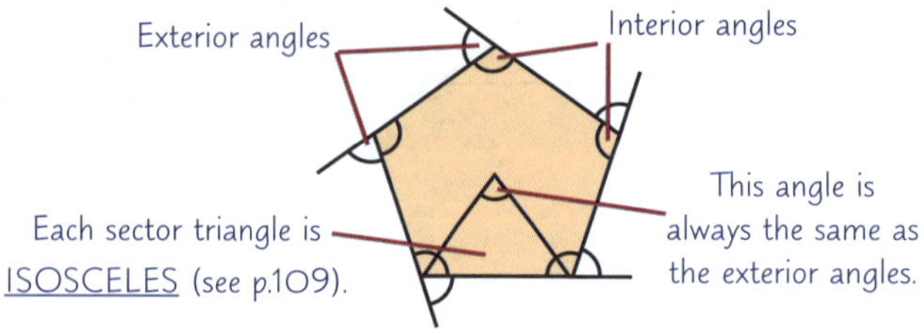

EXAMPLE **The interior angle of a regular polygon is 165°. How many sides does the polygon have?**

First, find the exterior angle of the shape: exterior angle = 180° – 165° = 15°

Use this value to find the number of sides: exterior angle $= \frac{360°}{n}$ so $n = \frac{360°}{\text{exterior angle}} = \frac{360°}{15°} =$ **24 sides**

Four very simple and very important formulas

Learn all the formulas on this page, and which ones go with regular and irregular polygons.

Q1 Find the size of the interior angle of a regular decagon. [2 marks]

Q1 Video Solution

Triangles and Quadrilaterals

*This page is jam-packed with details about **triangles** and **quadrilaterals** — and you need to learn them all.*

Triangles

(Grade 3)

1) EQUILATERAL TRIANGLES

3 equal sides and
3 equal angles of 60°.
3 lines of symmetry,
rotational symmetry order 3.

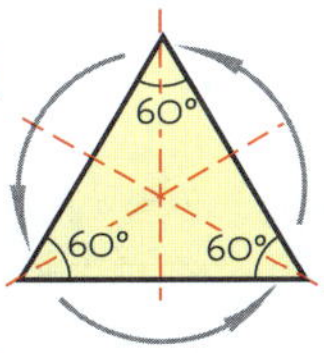

2) RIGHT-ANGLED TRIANGLES

1 right angle (90°).
No lines of symmetry.
No rotational symmetry.

The little square means it's a right angle.

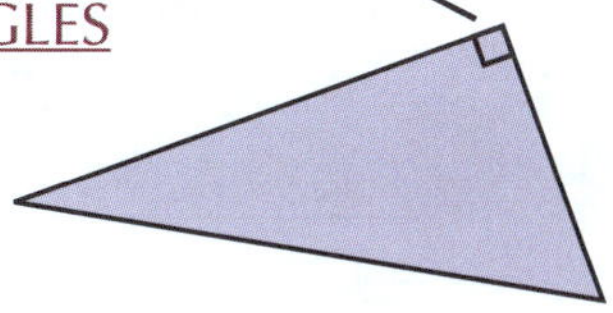

3) ISOSCELES TRIANGLES

2 sides the same.
2 angles the same.
1 line of symmetry.
No rotational symmetry.

These dashes mean that the two sides are the same length.

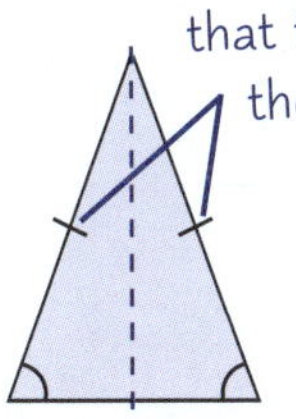

4) SCALENE TRIANGLES

All three sides different.
All three angles different.
No symmetry (pretty obviously).

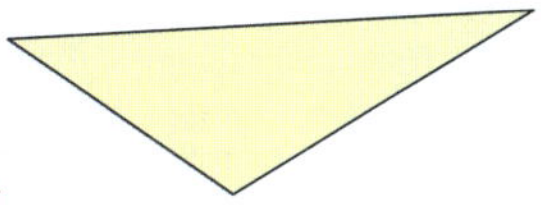

An acute-angled triangle has 3 acute angles, and an obtuse-angled triangle has one obtuse angle.

Quadrilaterals

(Grade 3)

1) SQUARE

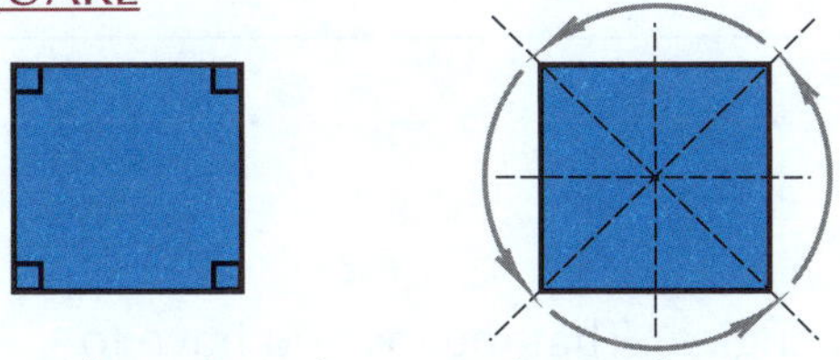

4 equal angles of 90° (right angles).
4 lines of symmetry, rotational symmetry order 4.
Diagonals are the same length and cross at right angles.

2) RECTANGLE

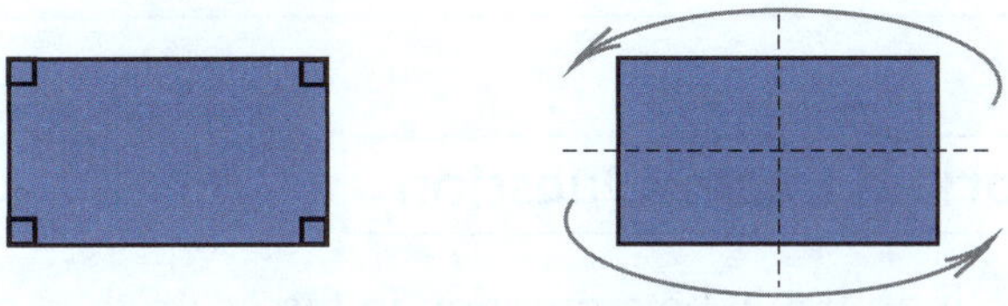

4 equal angles of 90° (right angles).
2 lines of symmetry, rotational symmetry order 2.
Diagonals are the same length.

3) RHOMBUS (A square pushed over)

A rhombus is the same as a diamond.

Matching arrows show parallel sides.

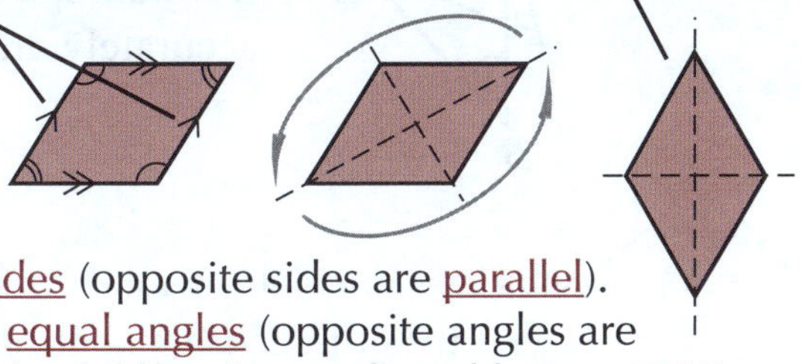

4 equal sides (opposite sides are parallel).
2 pairs of equal angles (opposite angles are equal, and neighbouring angles add up to 180°).
2 lines of symmetry, rotational symmetry order 2.
Diagonals cross at right angles.

4) PARALLELOGRAM (A rectangle pushed over)

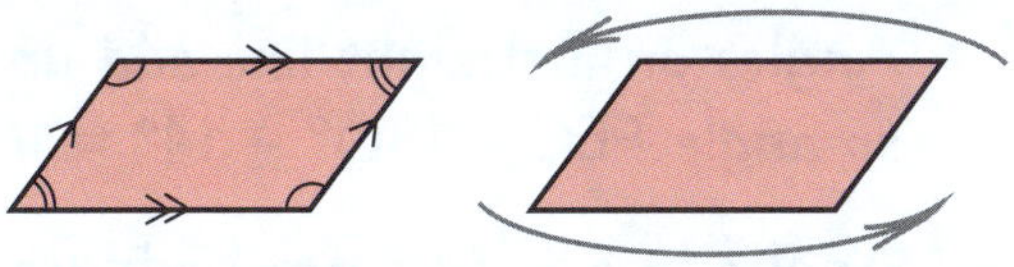

2 pairs of equal sides (each pair are parallel).
2 pairs of equal angles (opposite angles are equal, and neighbouring angles add up to 180°).
NO lines of symmetry, rotational symmetry order 2.

5) TRAPEZIUM

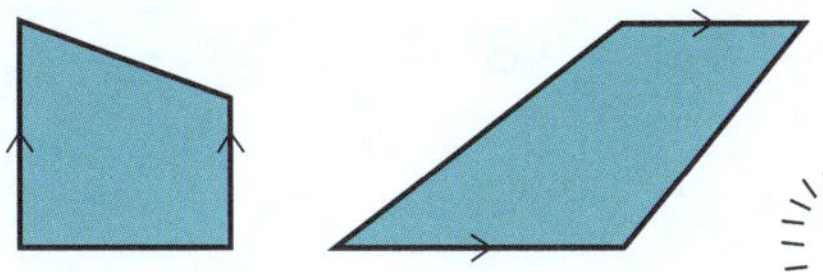

1 pair of parallel sides.
NO lines of symmetry.
No rotational symmetry.

In an isosceles trapezium, the sloping sides are the same length. An isosceles trapezium has 1 line of symmetry.

6) KITE

2 pairs of equal sides.
1 pair of equal angles.
1 line of symmetry.
No rotational symmetry.
Diagonals cross at right angles.

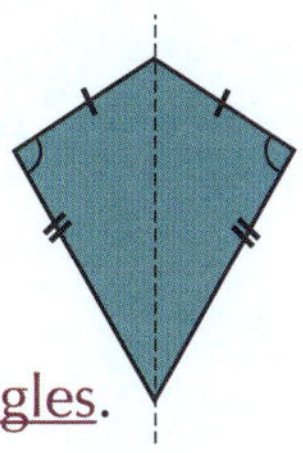

Warm-Up and Worked Exam Questions

Don't just look at those lovely big diagrams — you need to work through the examples in this section one by one to make sure that you've remembered all those rules...

Warm-Up Questions

1) Find the missing angles *a-d* below. State any angle laws used.

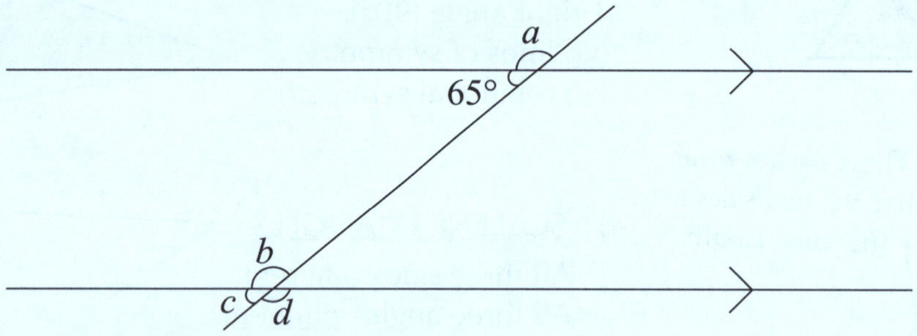

2) How many sides does a hexagon have?
3) Write down the name of the shape to the right:

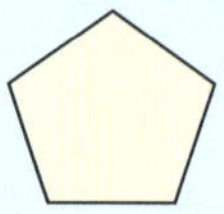

4) A regular polygon has an exterior angle of 24°. How many sides does it have?
5) A quadrilateral has 1 line of symmetry and 2 pairs of equal sides.
   What is the name of the quadrilateral?

Worked Exam Question

There'll probably be a question in the exam that asks you to find angles. That means you have to remember all the different angle rules and practise using them in the right places...

1 *DEF* and *BEC* are straight lines that cross at *E*. *AFB* and *AC* are perpendicular lines. (Grade 3)

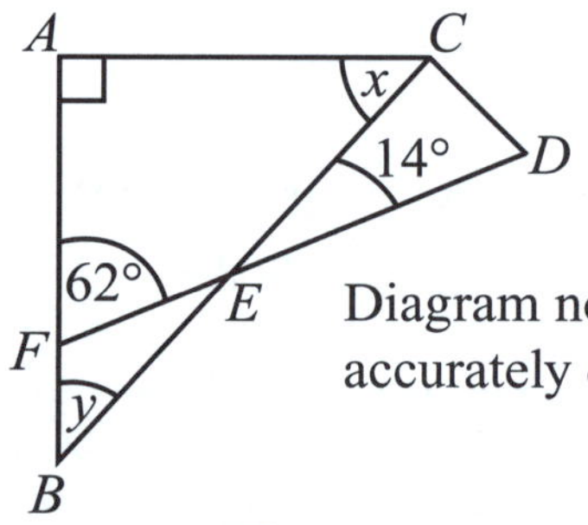

Diagram not accurately drawn

a) Find angle x.
Give a reason for each stage of your working.

Angles on a straight line add up to 180°, so angle FEC = 180° – 14° = 166°

Angles in a quadrilateral add up to 360°, so x = 360° – 90° – 62° – 166° = 42°

Angle x is in quadrilateral ACEF, so find the other missing angle in ACEF, then you can find x.

x = ...42.... °
[2 marks]

b) Hence show that $y = 48°$.

Angles in a triangle add up to 180°, so y = 180° – 90° – 42° = 48°

Angle y and angle x are both in the triangle ABC.

[2 marks]

Exam Questions

2 A triangle is shown in the diagram to the right. (Grade 3)
Prove that $x = y + z$.

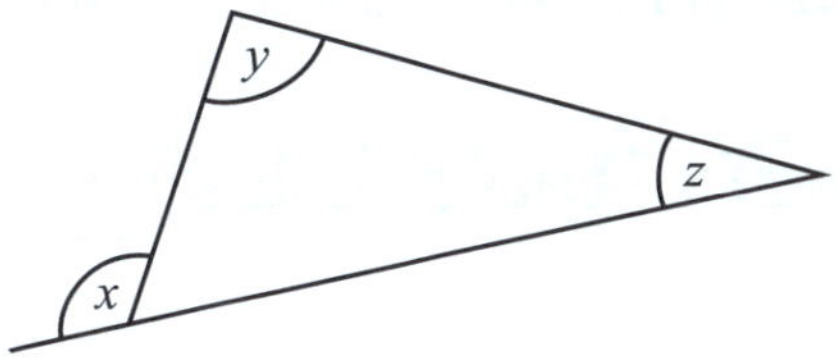

[3 marks]

3 Lines AB and DE are parallel and ABC is a straight line. (Grade 4)
Lines AE, BC and BD are of equal length.

Find an expression for y in terms of x.

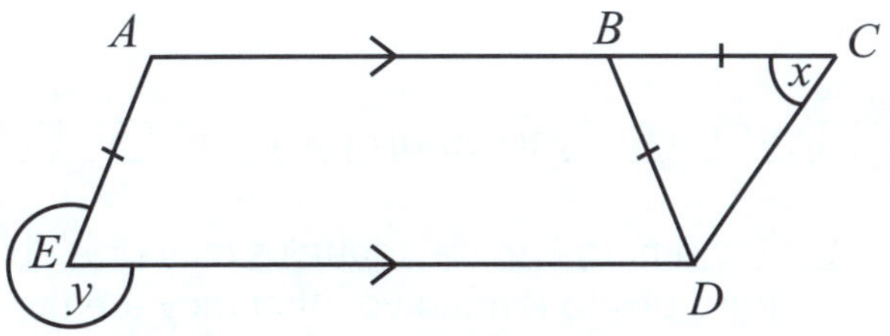

Diagram not accurately drawn

....
[5 marks]

4 Part of a regular polygon is shown below. Each interior angle is 150°. (Grade 4)

Diagram not accurately drawn

Calculate the number of sides of the polygon.

....
[3 marks]

5 The diagram below shows a regular hexagon inside a regular octagon. (Grade 4)
Vertices A and B coincide with vertices I and J respectively.

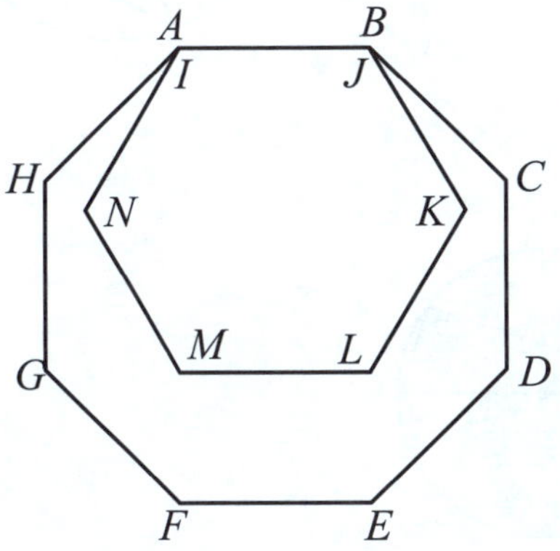

Find the size of angle CBK.

..............°
[2 marks]

Circle Geometry

*It's time to plunge you into the depths of mathematical peril with a 3-page extravaganza on **circle theorems**. There's a lot to learn on these pages, but it's all dead useful stuff.*

9 ~~Simple~~ Rules to Learn

GRADE 8 GRADE

1 A TANGENT and a RADIUS meet at 90°.

A TANGENT is a line that just touches a single point on the circumference of a circle. A tangent always makes an angle of exactly 90° with the radius it meets at this point.

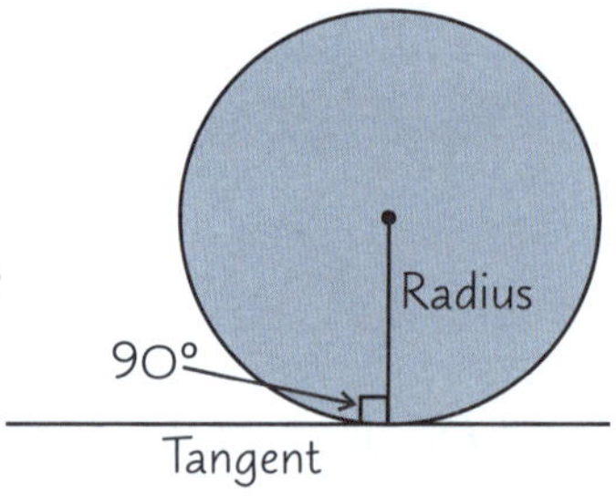

2 TWO RADII form an ISOSCELES TRIANGLE.

Radii is the plural of radius.

Unlike other isosceles triangles they don't have the little tick marks on the sides to remind you that they are the same — the fact that they are both radii is enough to make it an isosceles triangle.

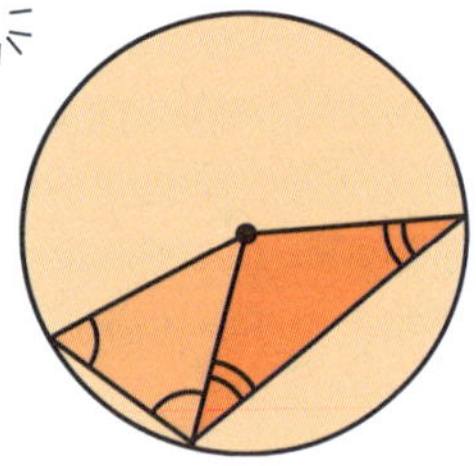

3 The PERPENDICULAR BISECTOR of a CHORD passes through the CENTRE of the circle.

A CHORD is any line drawn across a circle. And no matter where you draw a chord, the line that cuts it exactly in half (at 90°), will go through the centre of the circle.

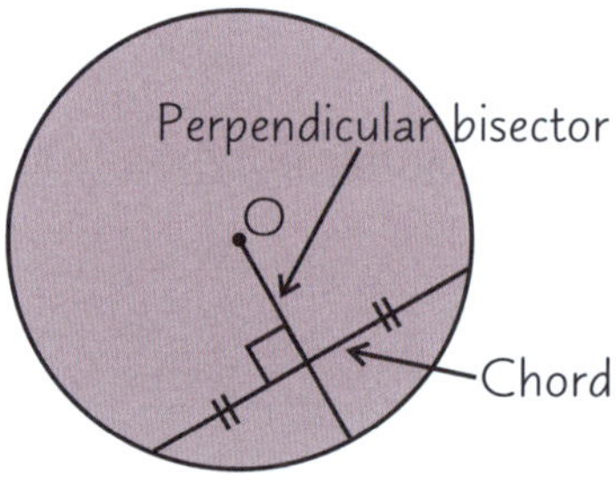

4 The angle at the CENTRE of a circle is TWICE the angle at the CIRCUMFERENCE.

'Angle subtended at' is just a posh way of saying 'angle made at'.

The angle subtended at the centre of a circle is EXACTLY DOUBLE the angle subtended at the circumference of the circle from the same two points (two ends of the same chord).

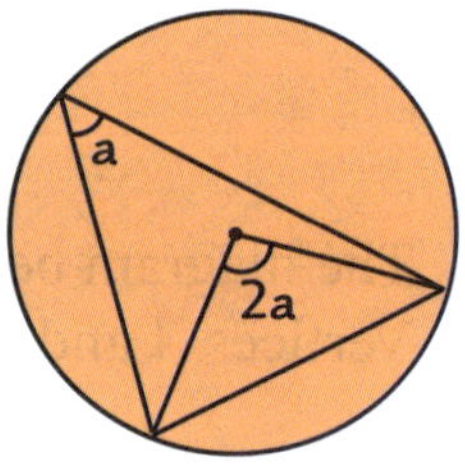

5 The ANGLE in a SEMICIRCLE is 90°.

A triangle drawn from the two ends of a diameter will ALWAYS make an angle of 90° where it hits the circumference of the circle, no matter where it hits.

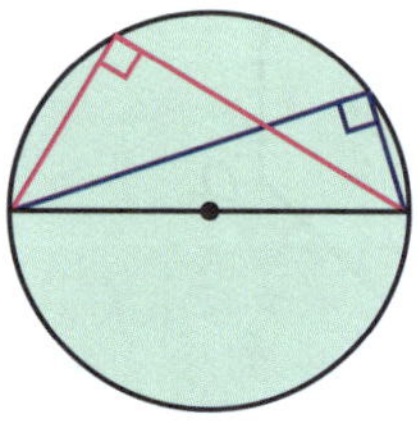

Circle Geometry

6 Angles in the SAME SEGMENT are EQUAL.

All triangles drawn from a chord will have the same angle where they touch the circumference. Also, the two angles on opposite sides of the chord add up to 180°.

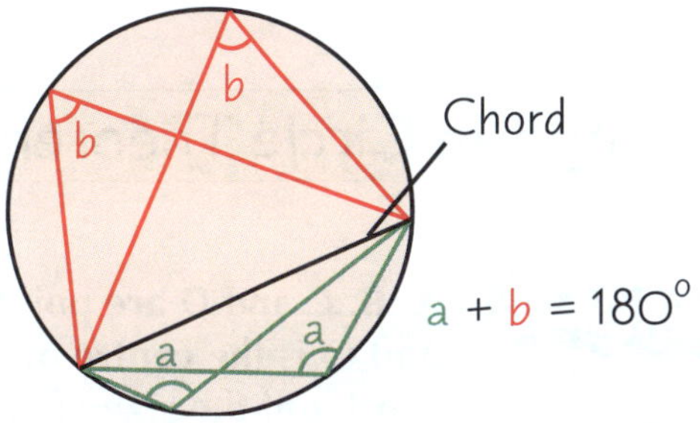

7 OPPOSITE ANGLES in a CYCLIC QUADRILATERAL add up to 180°.

A cyclic quadrilateral is a 4-sided shape with every corner touching the circle. Both pairs of opposite angles add up to 180°.

$a + c = 180°$

$b + d = 180°$

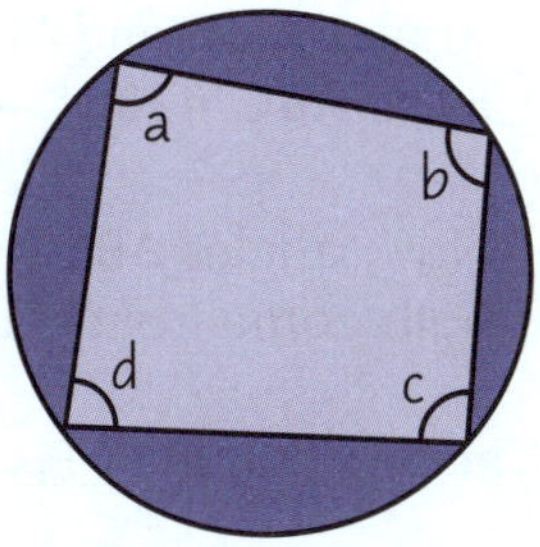

8 TANGENTS from the SAME POINT are the SAME LENGTH.

Two tangents drawn from an outside point are always equal in length, creating two congruent right-angled triangles as shown.

There's more about congruence on p.121.

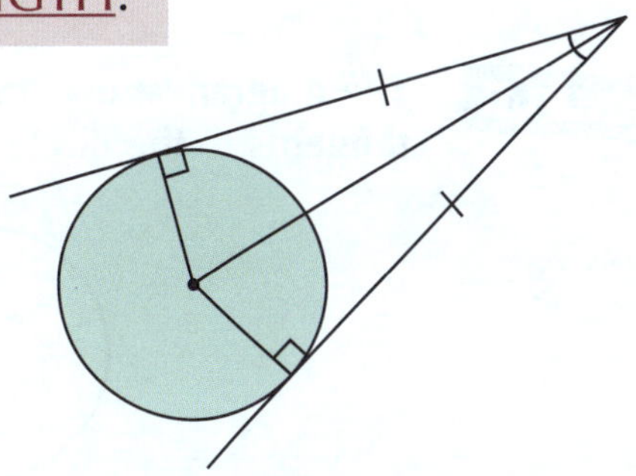

9 The ALTERNATE SEGMENT THEOREM.

The angle between a tangent and a chord is always equal to 'the angle in the opposite segment' (i.e. the angle made at the circumference by two lines drawn from the ends of the chord).

This is probably the hardest rule, so take care.

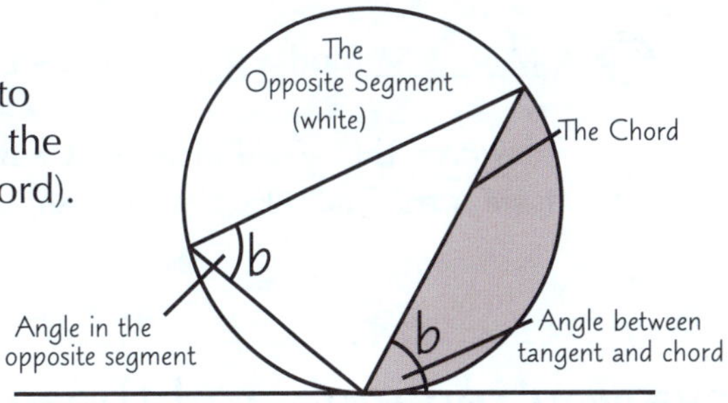

There you have it — nine rules to learn

This page is just more of the same: four more rules to cram into your overcrowded brain. But there's no way round learning this stuff, and once you've learnt all nine, circle geometry becomes much easier.

Circle Geometry

After learning all nine of those rules you might wonder where you'll actually use them. Behold:

Using the Circle Theorems

EXAMPLE **A, B, C and D are points on the circumference of the circle, and O is the centre of the circle. Angle ADC = 109°. Work out the size of angles ABC and AOC.**

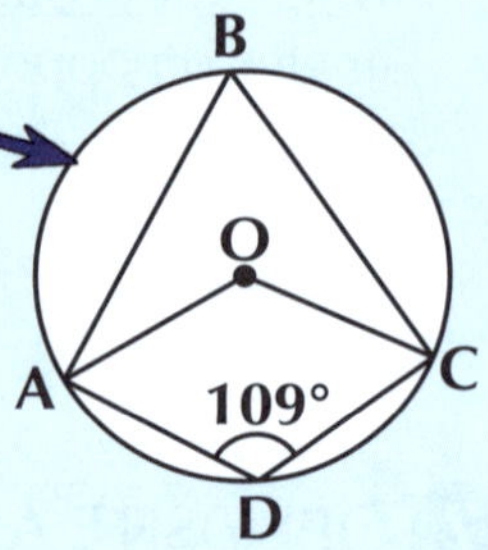

You'll probably have to use more than one rule to solve circle theorem questions — here, ABCD is a cyclic quadrilateral so use rule 7:

7 OPPOSITE ANGLES in a CYCLIC QUADRILATERAL add up to 180°.

Angles ADC and ABC are opposite, so:
angle ABC = 180° – 109° = 71°.

Now, angles ABC (which you've just found) and AOC both come from chord AC, so you can use rule 4:

4 The angle at the CENTRE of a circle is TWICE the angle at the CIRCUMFERENCE.

So angle AOC is double angle ABC, which means:
angle AOC = 71° × 2 = 142°.

EXAMPLE **The diagram shows the triangle ABC, where lines BA and BC are tangents to the circle. Show that line AC is NOT a diameter.**

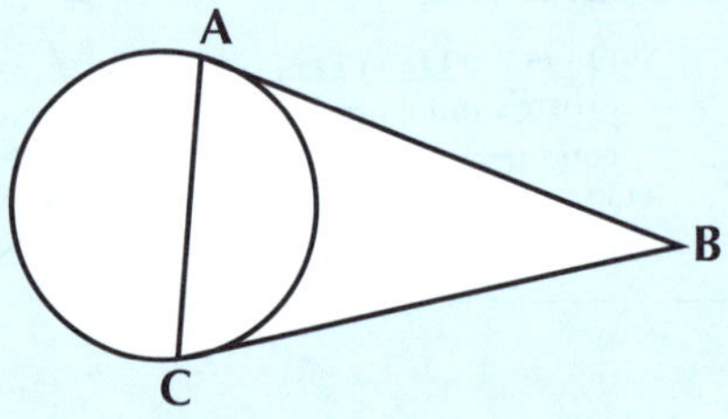

If AC was a diameter passing through the centre, O, then OA and OC would be radii, and angle CAB = angle ACB = 90° by rule 1:

1 A TANGENT and a RADIUS meet at 90°.

However, this would mean that ABC isn't a triangle as you can't have a triangle with two 90° angles, so AC cannot be a diameter.

If angles CAB and ACB were 90°, lines AB and BC would be parallel so would never meet.

Have you remembered those nine rules?

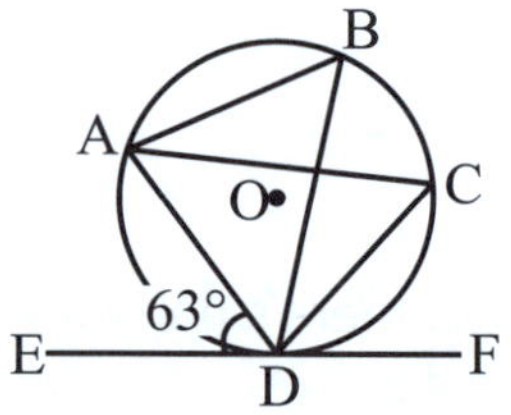

If you find this page isn't making much sense, you need to go back for another look at pages 116 and 117. Sometimes you'll find there's more than one way of finding the angle you want.

Q1 Video Solution

Q1 A, B, C and D are points on the circumference of the circle with centre O. The line EF is a tangent to the circle, and touches the circle at D. Angle ADE is 63°. Find the size of angles ABD and ACD. [2 marks]

Warm-Up and Worked Exam Questions

Time to have a go at using all those lovely circle theorems. The only way this stuff is going to sink in is if you practise answering questions — luckily you've got two pages of them here so you can do just that...

Warm-Up Question

1) *PQR* and *RST* are tangents to the circle.
Find the missing angles *L*, *M* and *N*.

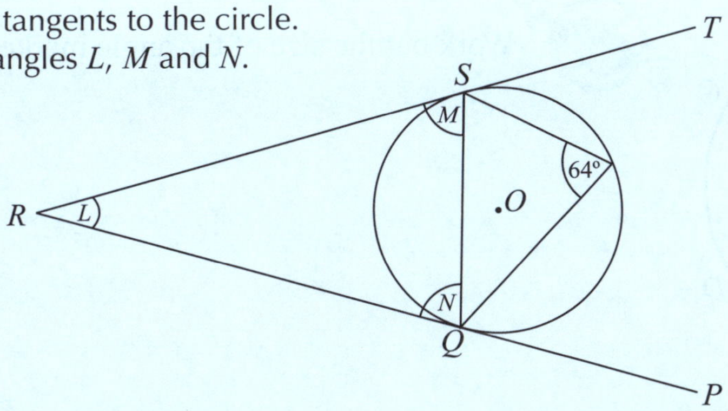

Worked Exam Question

Circle theorem questions can sometimes be a bit overwhelming, and it can be difficult to know where to start. The best approach is to keep finding any angles you can using the circle theorems and the angle rules from pages 109 and 110, until you have enough information to find the angle you want.

1 In the diagram, *O* is the centre of the circle. *A*, *B*, *C* and *D* are points on the circumference of the circle and *DE* and *BE* are tangents. Angle *DEB* is 80°. (GRADE 8)

Not to scale

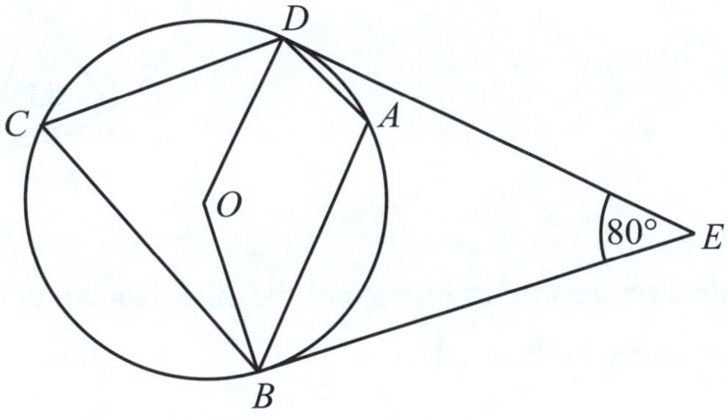

Work out the size of angle *DAB*, giving reasons for each step in your working.

Angles ODE and OBE are both 90° because a tangent always meets a radius at 90°.

Angle DOB = 100° because angles in a quadrilateral add up to 360°.

Angle DCB = 50° because an angle at the centre is twice the angle at the circumference.

Angle DAB = 130° because opposite angles of a cyclic quadrilateral add up to 180°.

..........130.......°
[4 marks]

Exam Questions

2 The diagram below shows a circle with centre O. A, B, C and D are points on the circumference of the circle and AOC is a straight line. (Grade 8)

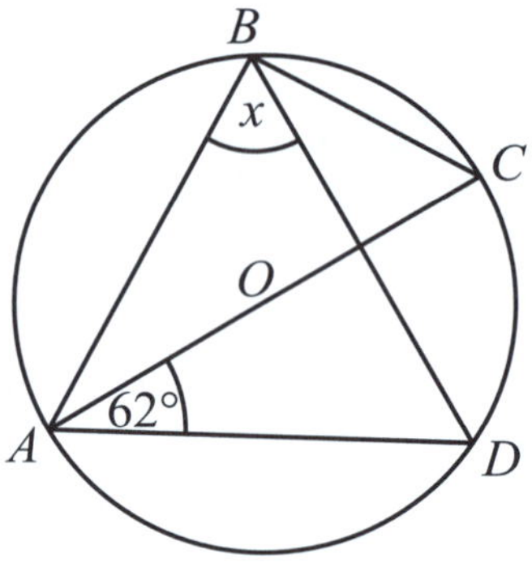

Not to scale

Work out the size of the angle marked x.

$x =$ ....°

[3 marks]

3 Points A, B, C, D and E lie on the circumference of the circle shown in the diagram below. Angle ABE is 37° and angle DCE is 53°. FG is the tangent to the circle at point E. (Grade 9)

Prove that the chord AD passes through the centre of the circle.

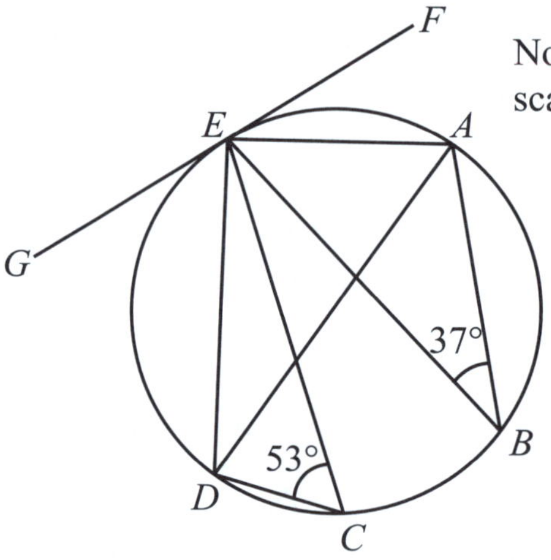

Not to scale

[3 marks]

4 A, B, C and D are points on the circumference of the circle shown below. (Grade 9)

Show that X is not the centre of the circle.

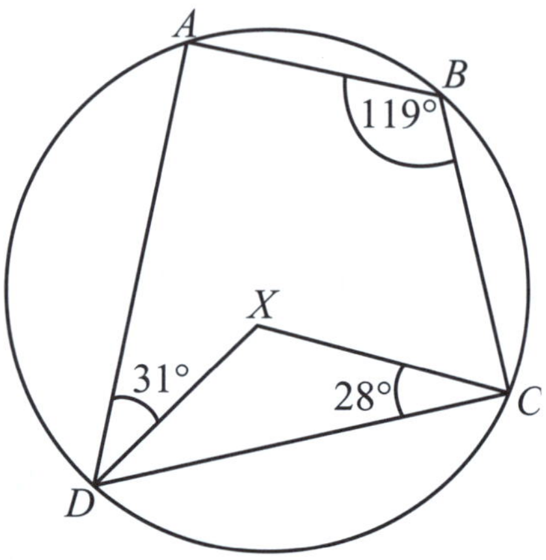

Diagram not accurately drawn

[3 marks]

Congruent Shapes

Congruence *is another ridiculous maths word which sounds really complicated when it's not. If two shapes are congruent, they are simply* ***the same*** *— the* ***same size*** *and the* ***same shape****. That's all it is. They can however be* ***reflected*** *or* ***rotated****.*

CONGRUENT *— same size, same shape*

Proving Triangles are **Congruent**

To prove that two triangles are congruent, you have to show that one of the conditions below holds true:

1) SSS — three sides are the same
2) AAS — two angles and a corresponding side match up
3) SAS — two sides and the angle between them match up
4) RHS — a right angle, the hypotenuse and one other side all match up

The hypotenuse is the longest side of a right-angled triangle — the one opposite the right angle.

SSS

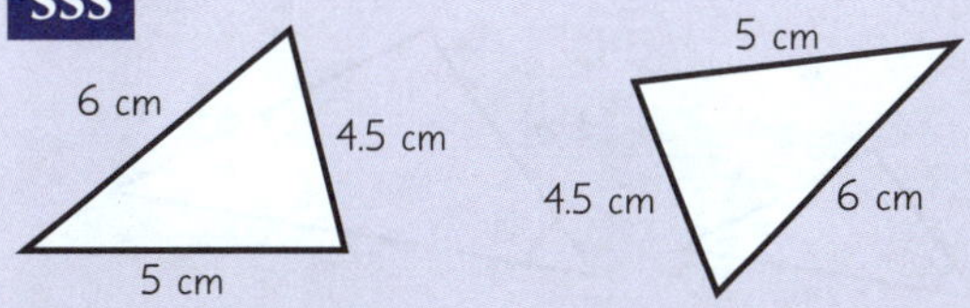

AAS

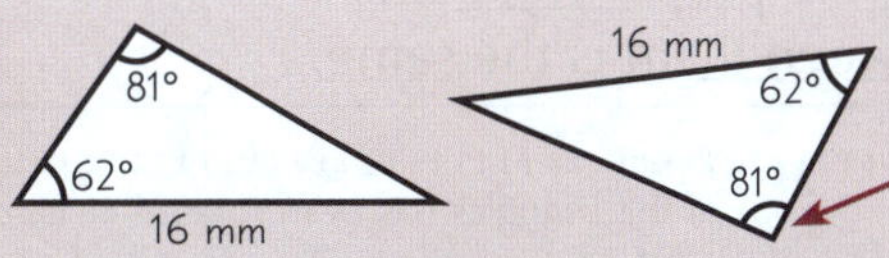

Make sure the sides match up — here, the side is opposite the 81° angle.

SAS

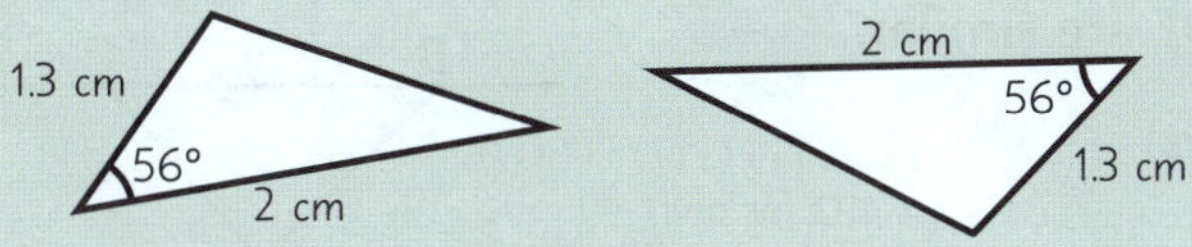

RHS

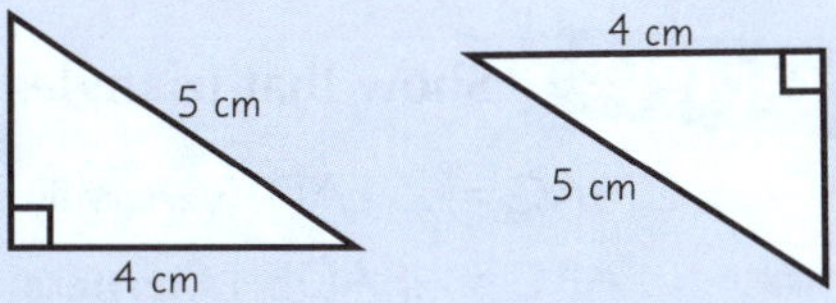

Work Out **All** the **Sides and Angles** You Can Find

The best approach to proving two triangles are congruent is to write down everything you can find out, then see which condition they fit. Watch out for things like parallel lines (p.110) and circle theorems (p.116-117).

EXAMPLE

XY and YZ are tangents to the circle with centre O, and touch the circle at points X and Z respectively. Prove that the triangles OXY and OYZ are congruent.

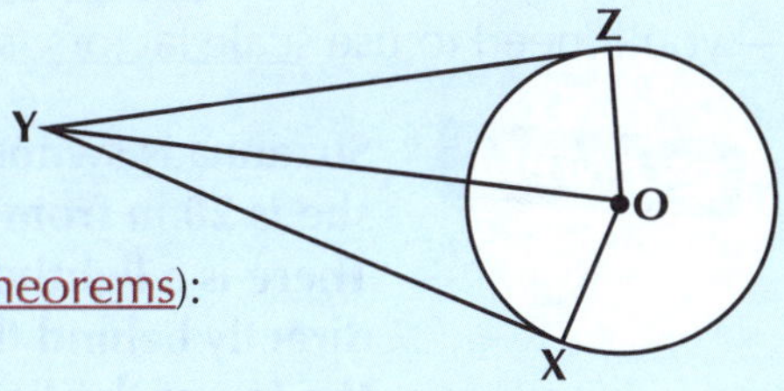

Write down what you know (you're going to have to use circle theorems):

- Sides OX and OZ are the same length (as they're both radii).
- Both triangles have a right angle (OXY and OZY) as a tangent meets a radius at 90°.
- OY is the hypotenuse of each triangle.

So the condition RHS holds, as there is a right angle, the hypotenuse are the same and one other side of each triangle (OX and OZ) are the same.

RHS holds, so OXY and OYZ are congruent triangles.

Congruent just means same size, same shape

Learn all 4 conditions and make sure you know how to use them to prove that triangles are congruent. Then have a go at this Exam Practice Question:

Q1 Prove that triangles ABD and BCD are congruent. [3 marks]

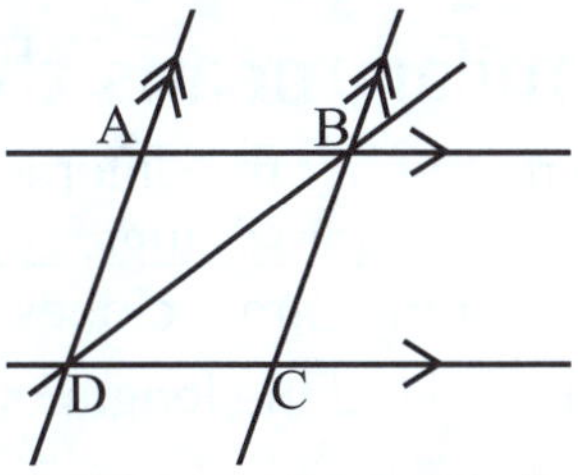

Similar Shapes

Similar *shapes are* ***exactly the same shape****, but can be* ***different sizes*** *(they can also be* ***rotated*** *or* ***reflected****).*

SIMILAR *—*
same shape, ***different size***

Similar Shapes Have the **Same Angles**

Generally, for two shapes to be similar, all the angles must match and the sides must be proportional. But for triangles, there are three special conditions — if any one of these is true, you know they're similar.

Two triangles are similar if:

1) All the angles match up
   i.e. the angles in one triangle are the same as the other.
2) All three sides are proportional
   i.e. if one side is twice as long as the corresponding side in the other triangle, all the sides are twice as long as the corresponding sides.
3) Any two sides are proportional and the angle between them is the same.

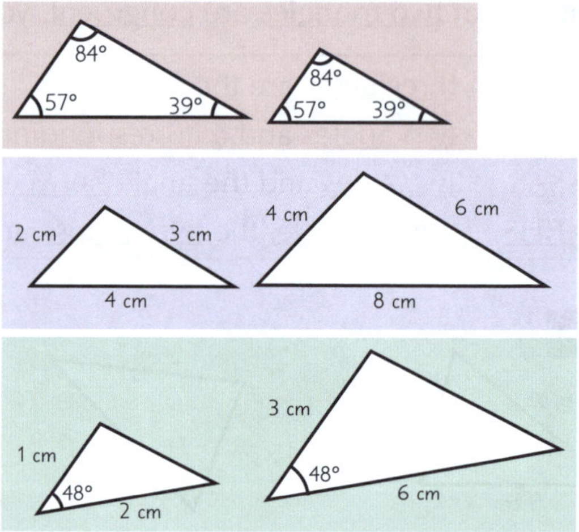

Watch out — if one of the triangles has been rotated or flipped over, it might look as if they're not similar, but don't be fooled.

EXAMPLE **Show that triangles ABC and ADE are similar.**

$\angle BAC = \angle EAD$ (vertically opposite angles)
$\angle ABC = \angle ADE$ (alternate angles)
$\angle BCA = \angle AED$ (alternate angles)

See p.110 for more on angles around parallel lines.

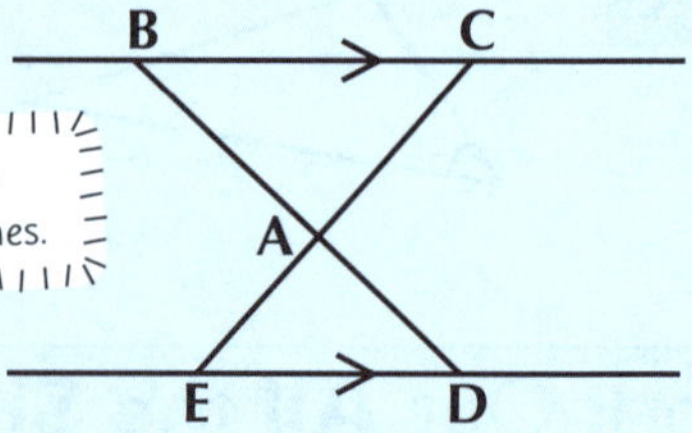

The angles in triangle ABC are the same as the angles in triangle ADE, so the triangles are similar.

Use **Similarity** to Find **Missing Lengths**

You might have to use the properties of similar shapes to find missing distances, lengths etc. — you'll need to use scale factors (see p.124) to find the lengths of missing sides.

EXAMPLE **Suzanna is swimming in the sea. When she is at point B, she is 20 m from a rock that is 8 m tall at its highest point. There is a lighthouse 50 m away from Suzanna that is directly behind the rock. From her perspective, the top of the lighthouse is in line with the top of the rock. How tall is the lighthouse?**

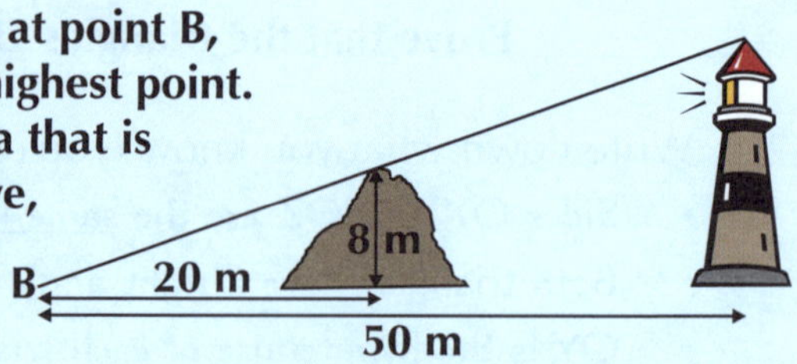

The triangles formed between Suzanna and the rock and Suzanna and the lighthouse are similar, so work out the scale factor: scale factor $= \frac{50}{20} = 2.5$

Now use the scale factor to work out the height of the lighthouse: height $= 8 \times 2.5 = 20$ m

Similar means the same shape but a different size

To remember the difference between similarity and congruence, think 'similar siblings, congruent clones' — siblings are alike but not the same, clones are identical.

Q1 Find the length of DB.

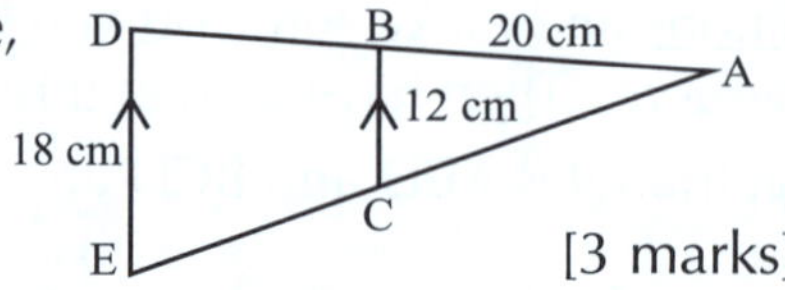

[3 marks]

Q1 Video Solution

The Four Transformations

There are four ***transformations*** *you need to know —* ***translation****,* ***rotation****,* ***reflection*** *and* ***enlargement****.*

1) Translations

In a translation, the amount the shape moves by is given as a vector (see p.155-156) written $\binom{x}{y}$ — where x is the horizontal movement (i.e. to the right) and y is the vertical movement (i.e. up). If the shape moves left and down, x and y will be negative. Shapes are congruent under translation (see p.121).

EXAMPLE

a) Describe the transformation that maps triangle ABC onto A′B′C′.
b) Describe the transformation that maps triangle ABC onto A″B″C″.

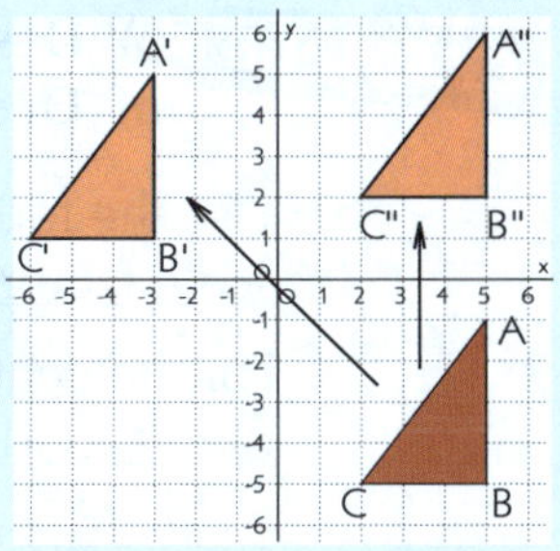

a) To get from A to A′, you need to move 8 units left and 6 units up, so...

The transformation from ABC to A′B′C′ is a translation by the vector $\binom{-8}{6}$.

b) The transformation from ABC to A″B″C″ is a translation by the vector $\binom{0}{7}$.

2) Rotations

To describe a rotation, you must give 3 details:

1) The angle of rotation (usually 90° or 180°).
2) The direction of rotation (clockwise or anticlockwise).
3) The centre of rotation (often, but not always, the origin).

Shapes are congruent under rotation.

For a rotation of 180°, it doesn't matter whether you go clockwise or anticlockwise.

EXAMPLE

a) Describe the transformation that maps triangle ABC onto A′B′C′.
b) Describe the transformation that maps triangle ABC onto A″B″C″.

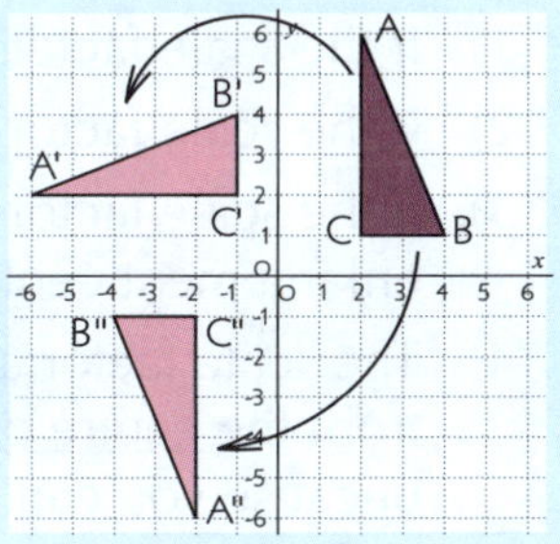

The transformation from ABC to A′B′C′ is a rotation of 90° anticlockwise about the origin.

The transformation from ABC to A″B″C″ is a rotation of 180° clockwise (or anticlockwise) about the origin.

If it helps, you can use tracing paper to help you find the centre of rotation.

3) Reflections

For a reflection, you must give the equation of the mirror line.
Shapes are congruent under reflection as well.

EXAMPLE

a) Describe the transformation that maps shape A onto shape B.
b) Describe the transformation that maps shape A onto shape C.

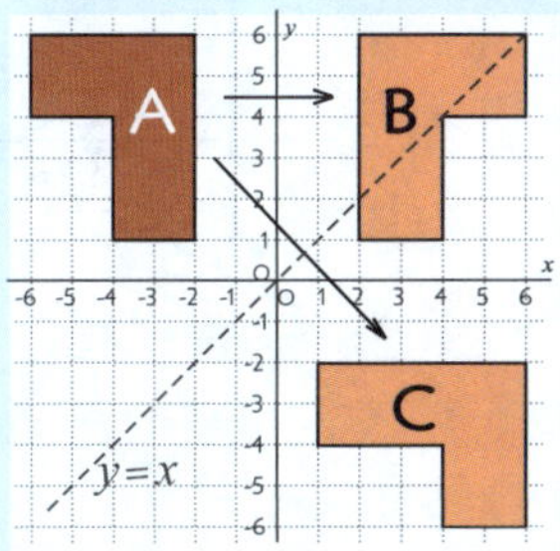

a) The transformation from A to B is a reflection in the y-axis.

b) The transformation from A to C is a reflection in the line $y = x$.

Points are invariant if they remain the same after a transformation — for reflections any point on the mirror line will be invariant.

Learn what details you need to give to describe each transformation

The reason that shapes are congruent under translation, reflection and rotation is because their size and shape don't change, just their position and orientation. Now have a go at this question:

Q1 On a grid, copy shape A above and rotate it 90° clockwise about the point (–1, –1). [2 marks]

The Four Transformations

4) Enlargements

For an enlargement, you must specify:

1) The scale factor. ← $\text{scale factor} = \dfrac{\text{new length}}{\text{old length}}$
2) The centre of enlargement.

Shapes are similar under enlargement — the position and the size change, but the angles and ratios of the sides don't (see p.122).

EXAMPLE

a) Describe the transformation that maps triangle A onto triangle B.
b) Describe the transformation that maps triangle B onto triangle A.

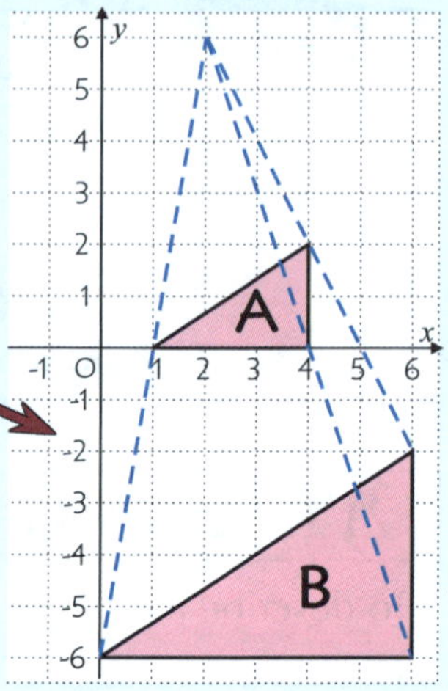

a) Use the formula above to find the scale factor (just choose one side): scale factor $= \frac{6}{3} = 2$

For the centre of enlargement, draw lines that go through corresponding vertices of both shapes and see where they cross.

The transformation from A to B is an enlargement of scale factor 2, centre (2, 6)

b) Use a similar method for the scale factor: scale factor $= \frac{3}{6} = \frac{1}{2}$

and the same centre of enlargement as before, so the transformation from B to A is an enlargement of scale factor $\frac{1}{2}$, centre (2, 6)

Scale Factors — Four Key Facts

1) If the scale factor is bigger than 1 the shape gets bigger.
2) If the scale factor is smaller than 1 (e.g. ½) it gets smaller.
3) If the scale factor is negative then the shape pops out the other side of the enlargement centre. If the scale factor is –1, it's exactly the same as a rotation of 180°.
4) The scale factor also tells you the relative distance of old points and new points from the centre of enlargement — this is very useful for drawing an enlargement, because you can use it to trace out the positions of the new points.

EXAMPLE

Enlarge shape A below by a scale factor of –3, centre (1, 1).
Label the transformed shape B.

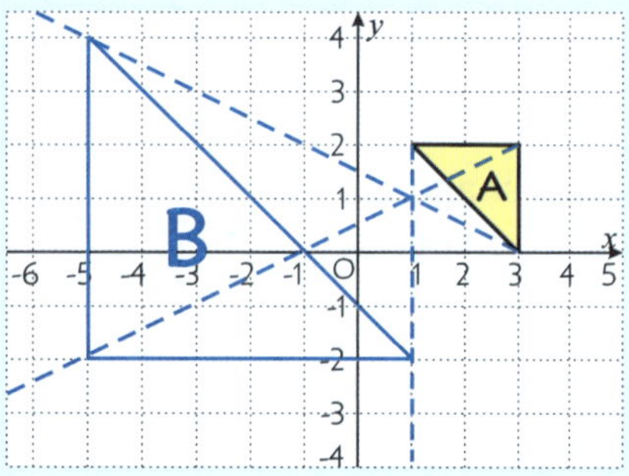

1) First, draw lines going through (1, 1) from each vertex of shape A.
2) Then, multiply the distance from each vertex to the centre of enlargement by 3, and measure this distance coming out the other side of the centre of enlargement. So on shape A, vertex (3, 2) is 2 right and 1 up from (1, 1) — so the corresponding point on shape B will be 6 left and 3 down from (1, 1). Do this for every point.
3) Join the points you've drawn to form shape B.

You'll need to learn the 4 key facts about scale factors

Shapes are congruent under translation, rotation and reflection, and similar under enlargement. If you have to do more than one transformation, do them one at a time — here's some practice.

Q1 On a grid, draw triangle A with vertices (2, 1), (4, 1) and (4, 3). Enlarge it by a scale factor of –2 about point (1, 1), then reflect it in the line $x = 0$. [3 marks]

Q1 Video Solution

Warm-Up and Worked Exam Questions

These warm-up questions cover some of the basics you'll need for the exam — use them to make sure you've learnt all the key information properly before you move on to tackling some exam questions.

Warm-Up Questions

1) From the diagram to the right, pick out:
   a) a pair of congruent shapes
   b) a pair of similar shapes

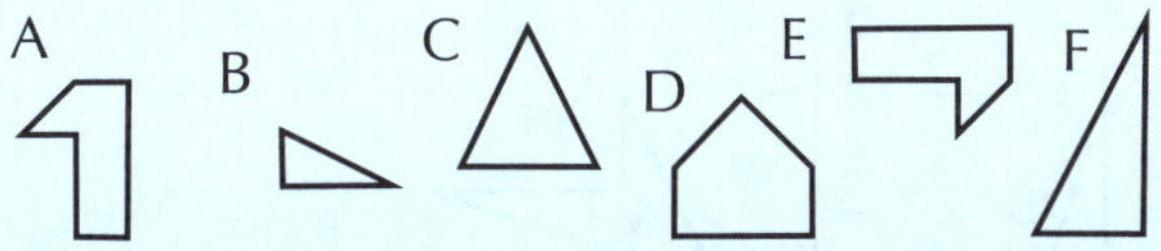

2) Triangles ABC and DEF are similar.
   a) Triangle DEF is an enlargement of triangle ABC. What is the scale factor of the enlargement?
   b) What is the length of DF?

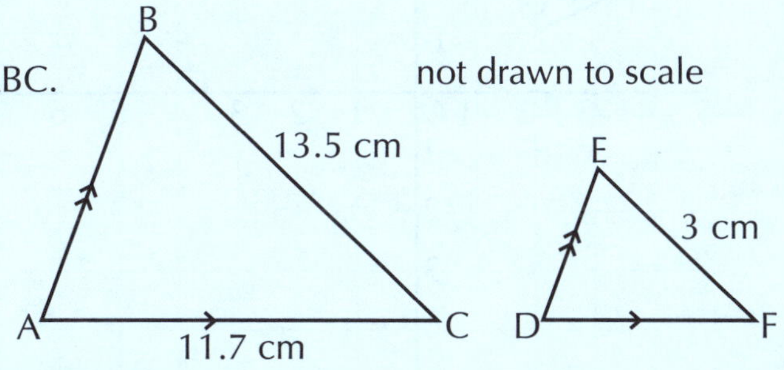

3) What translation would map the point (1, 3) onto (–2, 6)?
4) Point A is reflected in the y-axis to give point B. Given that point A is found at (3, 5), write down the coordinates of point B.

Worked Exam Question

I'm afraid this helpful blue writing won't be there in the exam, so if I were you I'd make the most of it and make sure you fully understand it now.

1 *ABC* is a triangle. *FDEC* is a parallelogram such that *F* is the midpoint of *AC*, *D* is the midpoint of *AB* and *E* is the midpoint of *BC*. (Grade 7)

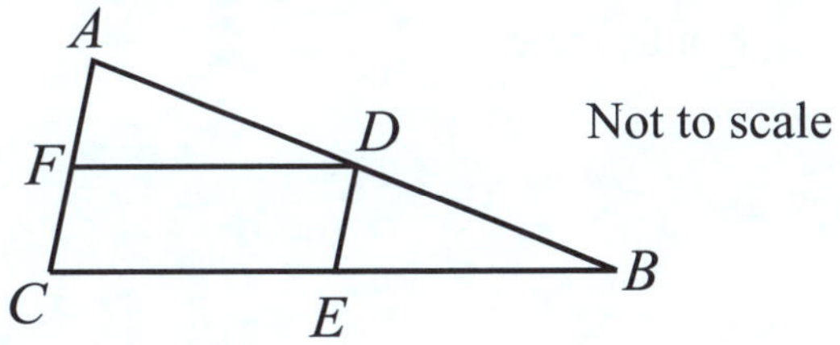

Prove that triangles *AFD* and *DEB* are congruent.

F is the midpoint of AC so AF = FC,
and opposite sides of a parallelogram are equal so DE = FC.
Therefore AF = DE.

E is the midpoint of CB so CE = EB,
and opposite sides of a parallelogram are equal so CE = FD.
Therefore FD = EB.

D is the midpoint of AB, so AD = DB.

Satisfies condition SSS so the triangles are congruent.

[4 marks]

Exam Questions

2 In the diagram below, **B** is an image of **A**. (Grade 3)

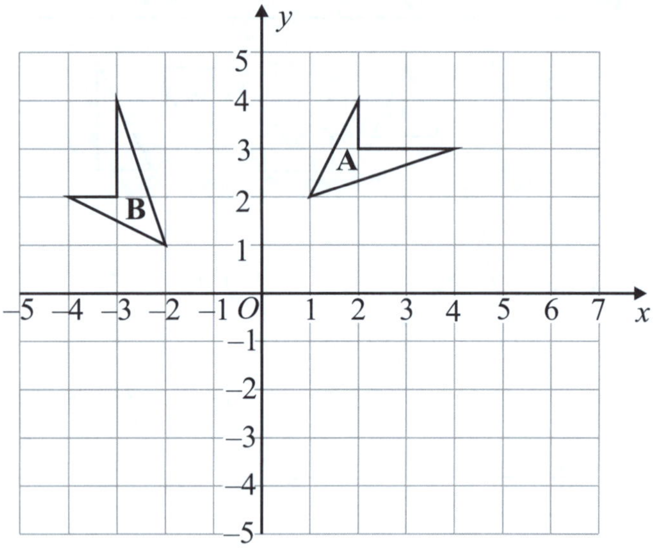

a) Describe fully the single transformation that maps **A** onto **B**.

....

....

....

[3 marks]

b) Translate shape **B** by the vector $\begin{pmatrix}-1\\-4\end{pmatrix}$. Label the image as **C**.

[1 mark]

3 Huan wants to estimate the height of a flagpole in his local park. He finds that if he stands a horizontal distance of 63 m away from the flagpole and holds his index finger up in front of him it exactly covers the flagpole. (Grade 5)

Huan's finger is 8 cm long and he holds it at a horizontal distance of 60 cm away from his body. Use this information to find an estimate for the height of the flagpole.

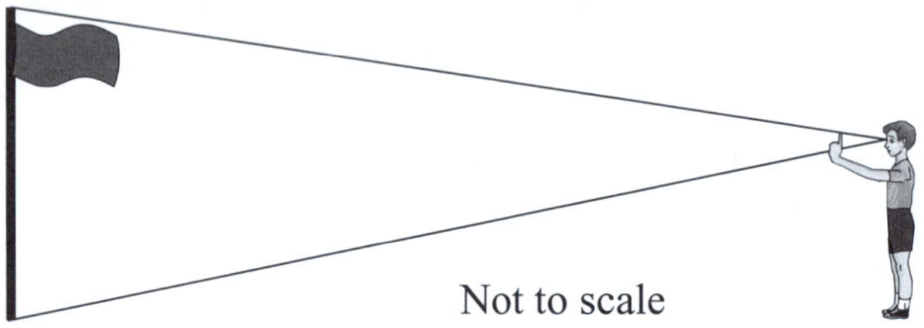

.... m
[3 marks]

4 A triangle has been drawn on the grid below. (Grade 7)

Enlarge the triangle by a scale factor of –2 about the point **C**.

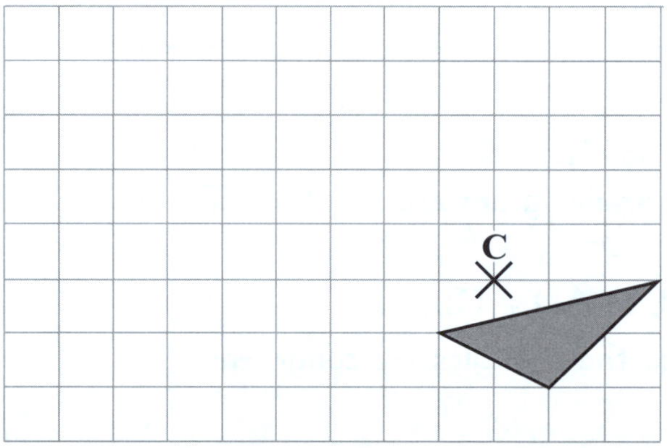

[2 marks]

Area — Triangles and Quadrilaterals

Be warned — there are lots of ***area formulas*** *coming up on the next two pages.*
You should remember the formulas for the area of a ***rectangle*** *(A = l × w) and the area of a* ***square*** *(A = l²).*

Areas of Triangles and Quadrilaterals

Note that in each case the height must be the vertical height, not the sloping height.

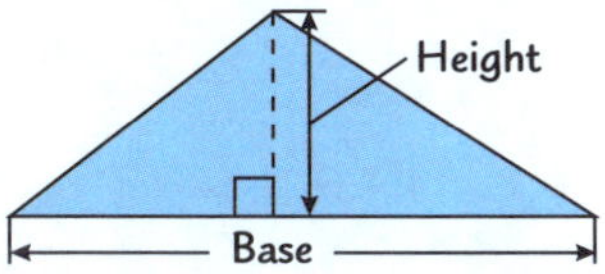

Area of triangle = ½ × base × vertical height

$A = \frac{1}{2} \times b \times h_v$

The alternative formula is:
Area of triangle = ½ ab sin C
This is covered on p.151.

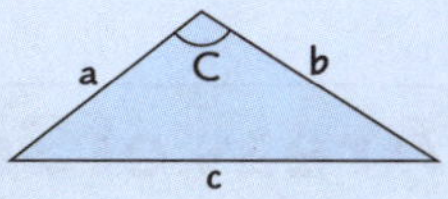

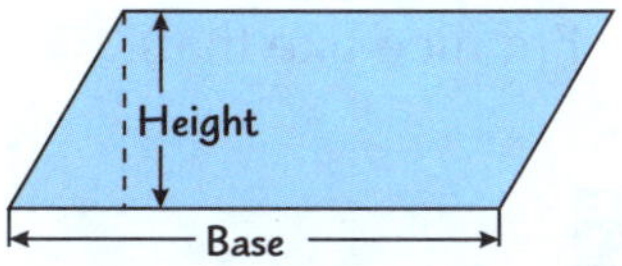

Area of parallelogram = base × vertical height

$A = b \times h_v$

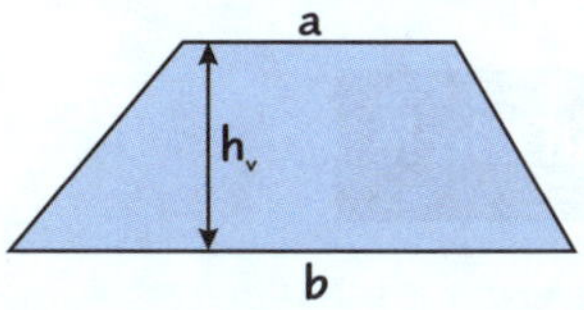

Area of trapezium = average of parallel sides × distance between them (vertical height)

$A = \frac{1}{2}(a + b) \times h_v$

Use the Formulas to Solve Problems

Examiners like to sneak bits of algebra into area and perimeter questions — you'll often have to set up and then solve an equation to find a missing side length or area of a shape.

EXAMPLE

The shape on the right shows a square with sides of length x cm drawn inside an isosceles trapezium. The base of the trapezium is three times as long as one side of the square.

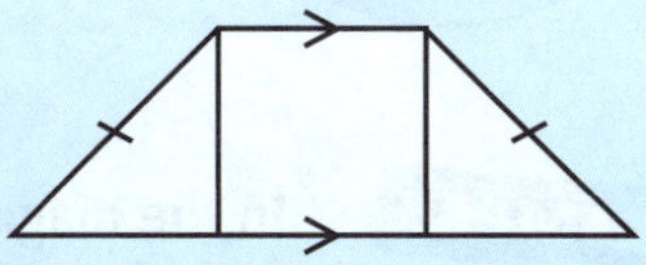

In an isosceles trapezium, the sloping sides are the same length.

a) Find an expression for the area of the trapezium in terms of x.

Top of trapezium = side of square = x cm
Base of trapezium = 3 × side of square = $3x$ cm
Height of trapezium = side of square = x cm
Area of trapezium = $\frac{1}{2}(x + 3x) \times x = 2x^2$ cm²

b) The area of the trapezium is 60.5 cm². Find the side length of the square.

Set your equation from part a) equal to 60.5 and solve to find x:

$2x^2 = 60.5$
$x^2 = 30.25$
$x = 5.5$ cm

You can split tricky shapes into easier ones

If you have a composite shape (a shape made up of different shapes stuck together), split it into triangles and quadrilaterals, work out the area of each bit and add them together.

Q1 The triangle and rectangle shown on the right have the same area. Find the value of x. [2 marks]

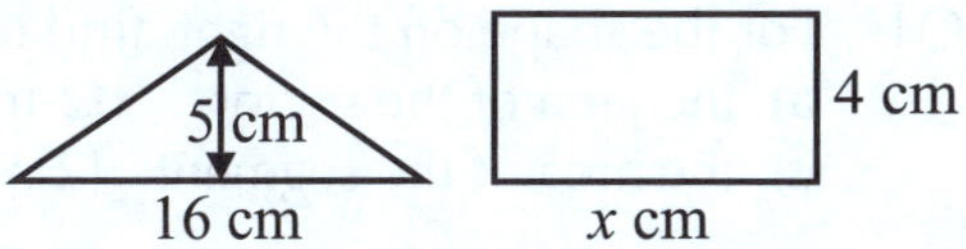

Area — Circles

Yes, I thought I could detect some groaning when you realised that this is another page of formulas.

Area and Circumference of Circles

Grade 3

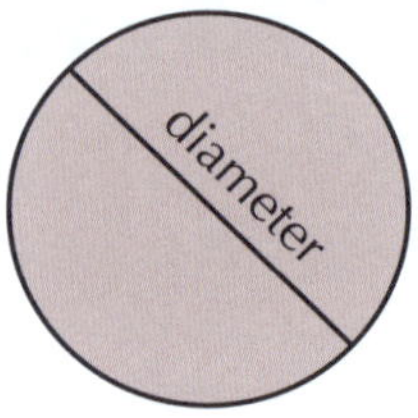

Area of circle = π × (radius)² $A = \pi r^2$

Remember that the radius is half the diameter.

For these formulas, use the π button on your calculator. For non-calculator questions, use π ≈ 3.142.

Circumference = π × diameter
= 2 × π × radius $C = \pi D = 2\pi r$

Areas of Sectors and Segments

Grade 6

These next ones are a bit more tricky — before you try and learn the formulas, make sure you know what a sector, an arc and a segment are (I've helpfully labelled the diagrams below — I'm nice like that).

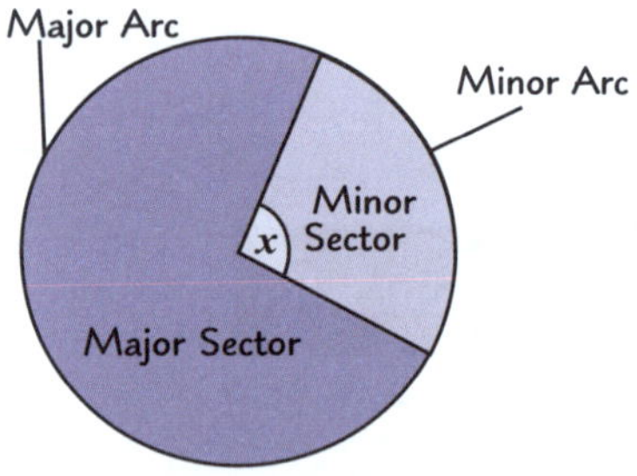

$$\text{Area of Sector} = \frac{x}{360} \times \text{Area of full Circle}$$

$$\text{Length of Arc} = \frac{x}{360} \times \text{Circumference of full Circle}$$

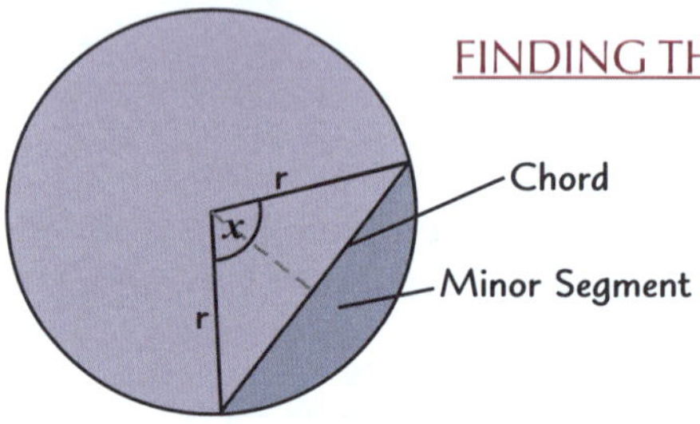

FINDING THE AREA OF A SEGMENT is OK if you know the formulas.

1) Find the area of the sector using the above formula.
2) Find the area of the triangle, then subtract it from the sector's area. You can do this using the '½ ab sin C' formula for the area of the triangle (see previous page), which becomes: $\frac{1}{2}r^2\sin x$.

EXAMPLE **In the diagram on the right, a sector with angle 60° has been cut out of a circle with radius 3 cm. Find the exact area of the shaded shape.**

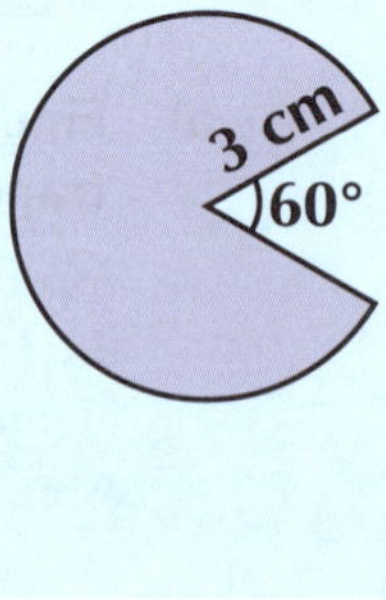

First find the angle of the shaded sector (this is the major sector):

360° – 60° = 300°

Then use the formula to find the area of the shaded sector:

$$\text{area of sector} = \frac{x}{360} \times \pi r^2 = \frac{300}{360} \times \pi \times 3^2$$
$$= \frac{5}{6} \times \pi \times 9 = \frac{15}{2}\pi \text{ cm}^2$$

'Exact area' means leave your answer in terms of π.

Four more lovely formulas for you

One more thing — if you're asked to find the perimeter of a semicircle or quarter circle, don't forget to add on the straight edges too. It's an easy mistake to make, and it'll cost you marks.

Q1 Video Solution

Q1 For the shape on the right, find to 2 decimal places:
a) the area of the sector [2 marks] b) the arc length [2 marks] (Grade 6)
c) the area of the segment [2 marks]

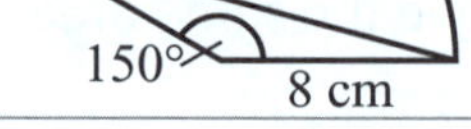

3D Shapes — Surface Area

It's time now to move on to the next ***dimension*** *— yep, that's right,* ***3D shapes****.*

Vertices, Faces and Edges

GRADE 3

There are different parts of 3D shapes you need to be able to spot. These are vertices (corners), faces and edges. You might be asked for the number of vertices, faces and edges in the exam — just count them up, and don't forget the hidden ones.

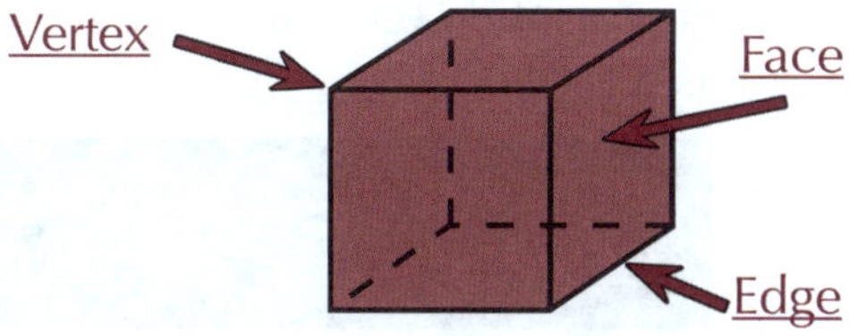

Curved faces are sometimes called surfaces.

Surface Area

GRADE 5

1) SURFACE AREA only applies to 3D objects — it's just the total area of all the faces added together.
2) SURFACE AREA OF SOLID = AREA OF NET (remember that a net is just a 3D shape folded out flat). So if it helps, imagine the net and add up the area of each bit.
3) SPHERES, CONES AND CYLINDERS have surface area formulas that you need to know:

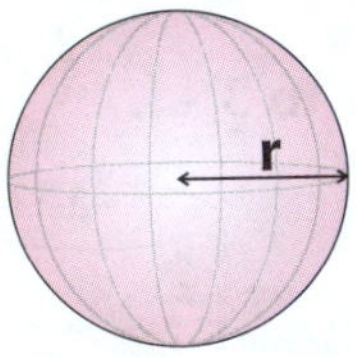

Surface area of a SPHERE = $4\pi r^2$

curved area of cone (l is the slant height)

area of circular base

Surface area of a CONE = $\pi rl + \pi r^2$

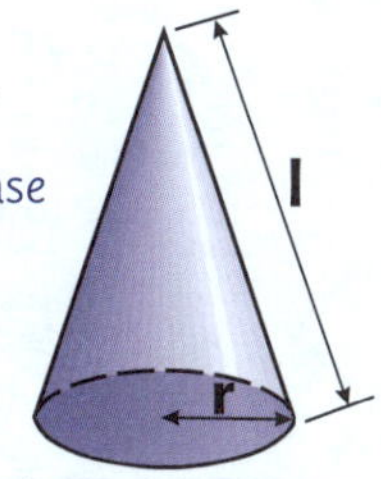

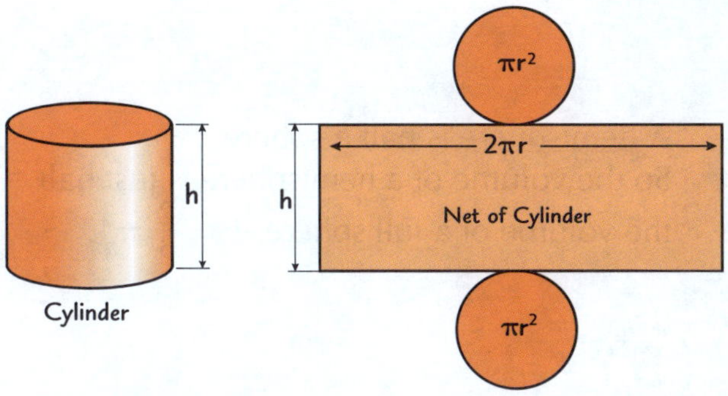

Surface area of a CYLINDER = $2\pi rh + 2\pi r^2$

Note that the length of the rectangle is equal to the circumference of the circular ends.

EXAMPLE **Find the exact surface area of a hemisphere with radius 4 cm.**

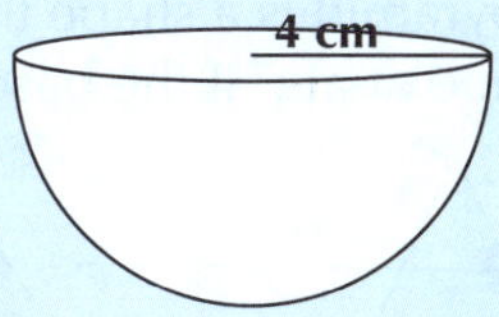

A hemisphere is half a sphere — so the surface area of the curved face is $4\pi r^2 \div 2 = 2\pi r^2 = 2 \times \pi \times 4^2 = 32\pi$ cm².

Don't forget the area of the flat face though — this is just the area of a circle with radius 4 cm: $\pi r^2 = 16\pi$ cm².

So the total surface area is $32\pi + 16\pi = 48\pi$ cm².

You're asked for the exact value, so leave your answer in terms of π.

To find the surface area of a solid just add up the areas of each face

In a net all the faces are folded out flat — which makes it easier to see the shapes you're dealing with. Now have a go at this lovely Practice Question:

Q1 The surface area of a cone with radius 5 cm is 125π cm². Find the slant height, l, of the cone. [3 marks] GRADE 6

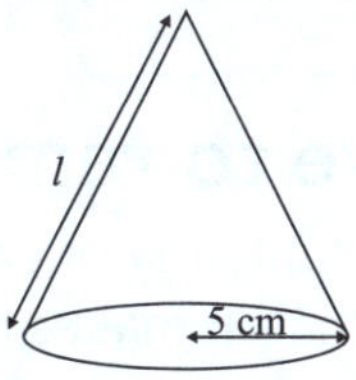

Q1 Video Solution

3D Shapes — Volume

Here are two whole pages on ***volumes*** *of* ***3D shapes****. Remember that the volume of a* ***cuboid*** *is length × width × height (and the volume of a* ***cube*** *is length³).*

Volumes of Prisms

GRADE 3

A PRISM is a solid (3D) object which is the same shape all the way through — i.e. it has a CONSTANT AREA OF CROSS-SECTION.

VOLUME OF PRISM = CROSS-SECTIONAL AREA × LENGTH

V = A × L

Triangular Prism

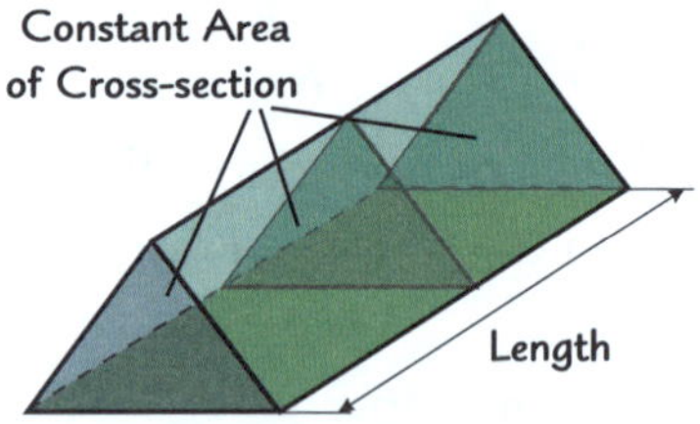

Cylinder

Here, the cross-sectional area is a circle, so the formula for the volume of a cylinder is:

$V = \pi r^2 h$

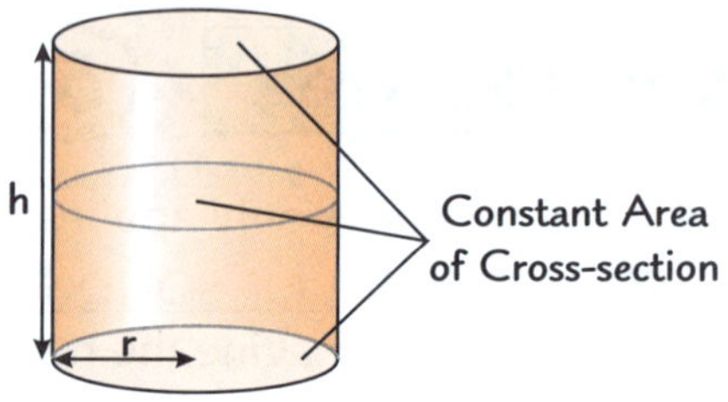

EXAMPLE GRADE 4

Honey comes in cylindrical jars with radius 4.5 cm and height 12 cm. The density of honey is 1.4 g/cm³. Work out the mass of honey in this jar to 3 s.f.

First, work out the volume of the jar — just use the formula above:

$V = \pi r^2 h = \pi \times 4.5^2 \times 12 = 763.4070... \text{ cm}^3$

See p.105 for more on density.

Now use the formula mass = density × volume:

mass of honey = 1.4 × 763.4070... = 1068.7698... = 1070 g (3 s.f.)

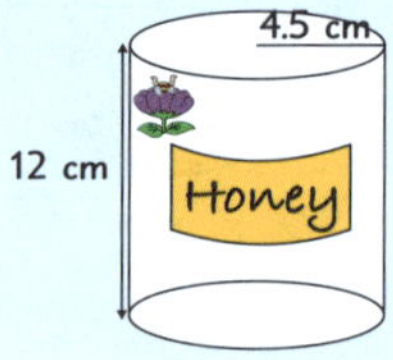

Volumes of Spheres

GRADE 5

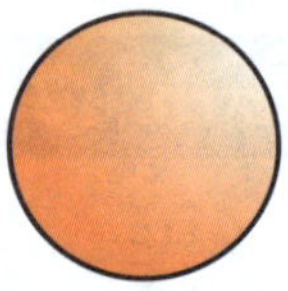

Volume of Sphere = $\frac{4}{3}\pi r^3$

A hemisphere is half a sphere. So the volume of a hemisphere is just half the volume of a full sphere, $V = \frac{2}{3}\pi r^3$.

Volumes of Pyramids and Cones

GRADE 5

A pyramid is a shape that goes from a flat base up to a point at the top. Its base can be any shape at all. If the base is a circle then it's called a cone (rather than a circular pyramid).

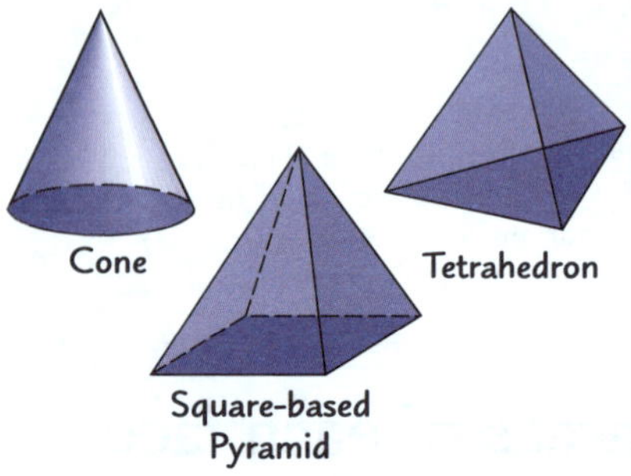

VOLUME OF PYRAMID = $\frac{1}{3}$ × BASE AREA × VERTICAL HEIGHT

VOLUME OF CONE = $\frac{1}{3} \times \pi r^2 \times h_v$

Make sure you use the vertical (perpendicular) height in these formulas — don't get confused with the slant height, which you used to find the surface area of a cone.

You have to remember what a prism is

It's the constant area of cross-section which is important — that's what makes a prism a prism. If you remember that, it makes perfect sense that to get the volume you just multiply that area by the length.

3D Shapes — Volume

Another page on volumes now, but this is a bit of a weird one.
First up, it's volumes of cones with a bit ***chopped off****, then it's on to* ***rates of flow****.*

Volumes of Frustums

GRADE 6

A frustum of a cone is what's left when the top part of a cone is cut off parallel to its circular base.

VOLUME OF FRUSTUM = VOLUME OF ORIGINAL CONE − VOLUME OF REMOVED CONE

$$= \frac{1}{3}\pi R^2 H - \frac{1}{3}\pi r^2 h$$

The bit that's chopped off is a mini cone that's similar to the original cone.

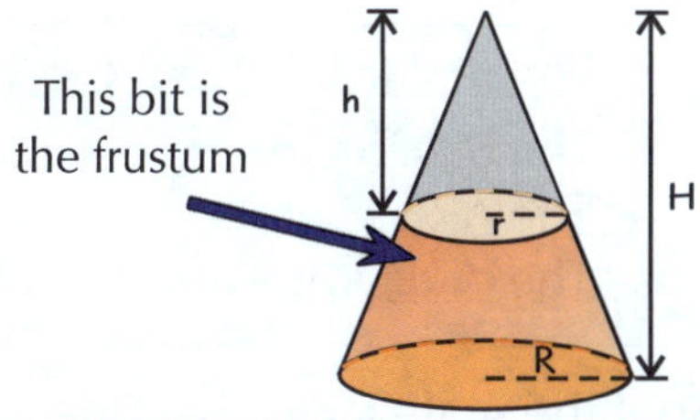

EXAMPLE

A waste paper basket is the shape of a frustum formed by removing a cone of height 10 cm from a cone of height 50 cm and radius 35 cm. Find the volume of the waste paper basket to 3 significant figures.

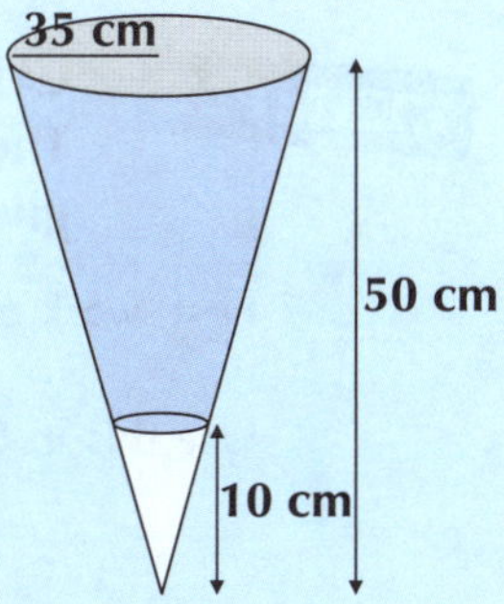

Volume of original cone $= \frac{1}{3}\pi R^2 H = \frac{1}{3} \times \pi \times 35^2 \times 50 = 64140.850...$ cm^3

Radius of removed cone = 35 ÷ 5 = 7 cm (because the cones are similar — the large cone is an enlargement of the small cone with scale factor 5)

Volume of removed cone $= \frac{1}{3}\pi r^2 h = \frac{1}{3} \times \pi \times 7^2 \times 10 = 513.126...$ cm^3

Volume of frustum = 64140.850... − 513.126... = 63627.723... = 63600 cm^3 (3 s.f.)

Rates of Flow

GRADE 8

You need to be really careful with units in rates of flow questions. You might be given the dimensions of a shape in cm or m but the rate of flow in litres (e.g. litres per minute). Remember that 1 litre = 1000 cm^3.

EXAMPLE

A spherical fish tank with a radius of 15 cm is being filled with water at a rate of 4 litres per minute. How long will it take to fill the fish tank $\frac{2}{3}$ full (by volume)? Give your answer in minutes and seconds, to the nearest second.

Find the volume of the fish tank:

$V = \frac{4}{3}\pi r^3 = \frac{4}{3} \times \pi \times 15^3 = 14\,137.166...$ cm^3

So $\frac{2}{3}$ of the fish tank is: $\frac{2}{3} \times 14\,137.166... = 9424.777...$ cm^3

Then convert the rate of flow into cm^3/s:

4 litres per minute = 4000 cm^3/min = 66.666... cm^3/s

So it will take 9424.777... ÷ 66.666... = 141.371... seconds
= 2 minutes and 21 seconds (to the nearest second) to fill the fish tank.

Remember that a frustum is just a cone with the top chopped off

The frustum formula is a bit nasty, but you'll have to learn it. When dealing with rates of flow, keep an eye on the units of rate of flow and volume.

Q1 A cone and a sphere both have radius 9 cm. Their volumes are the same. Find the vertical height, h, of the cone. [4 marks] GRADE 8

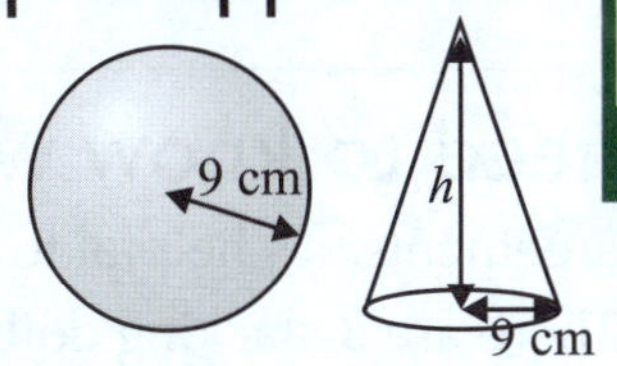

Q1 Video Solution

Q2 A square-based pyramid with base sides of length 60 cm and height 110 cm is being filled with water at a rate of 0.1 litres per second. Does it take longer than 20 minutes to fill? [4 marks] GRADE 8

More Enlargements and Projections

The two topics on this page aren't really related, but you'll need to learn about them both regardless.

How **Enlargement** Affects **Area** and **Volume**

GRADE 7

If a shape is enlarged by a scale factor (see page 124), its area, or surface area and volume (if it's a 3D shape), will change too. However, they don't change by the same value as the scale factor:

For a SCALE FACTOR n:

The SIDES are n times bigger
The AREAS are n^2 times bigger
The VOLUMES are n^3 times bigger

$$n = \frac{\text{new length}}{\text{old length}} \qquad n^2 = \frac{\text{new area}}{\text{old area}} \qquad n^3 = \frac{\text{new volume}}{\text{old volume}}$$

Or... AS RATIOS:

Lengths	a : b
Areas	$a^2 : b^2$
Volumes	$a^3 : b^3$

So if the scale factor is 2, the lengths are 2 times as long, the area is $2^2 = 4$ times as big, and the volume is $2^3 = 8$ times as big. As ratios, these enlargements are 1 : 2 (length), $1^2 : 2^2 = 1 : 4$ (area) and $1^3 : 2^3 = 1 : 8$ (volume).

EXAMPLE

Cylinder A has surface area 6π cm², and cylinder B has surface area 54π cm². The volume of cylinder A is 2π cm³. Find the volume of cylinder B, given that B is an enlargement of A.

First, work out the scale factor, n: $n^2 = \frac{\text{Area B}}{\text{Area A}} = \frac{54\pi}{6\pi} = 9$, so $n = 3$

Use this in the volume formula: $n^3 = \frac{\text{Volume B}}{\text{Volume A}} \Rightarrow 3^3 = \frac{\text{Volume B}}{2\pi}$

$\Rightarrow$ Volume of B $= 2\pi \times 27 = 54\pi$ cm³

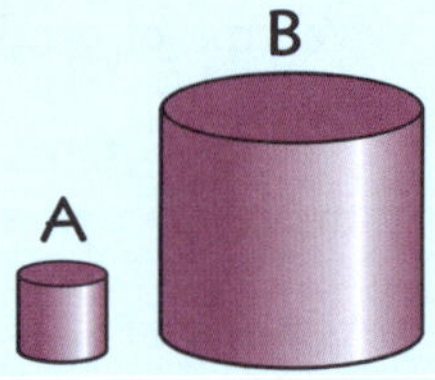

EXAMPLE

The ratio of the surface area of sphere X to the surface area of sphere Y is 16 : 25. Sphere X has a volume of 448 cm³. What is the volume of sphere Y?

Work out the ratio of the volumes: $a^2:b^2 = 16:25$ so $a:b = 4:5$ and $a^3:b^3 = 64:125$

Use the ratio to find the volume of sphere Y: $448 \div 64 = 7$
so volume of sphere Y $= 125 \times 7 = 875$ cm³

Projections Show a **3D Shape** From **Different Viewpoints**

GRADE 3

There are three different types of projection — front elevation, side elevation and plan (elevation is just another word for projection).

1

FRONT ELEVATION — the view you'd see from directly in front (in the direction of the arrow)

2

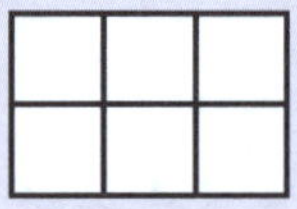

PLAN — the view you'd see from directly above

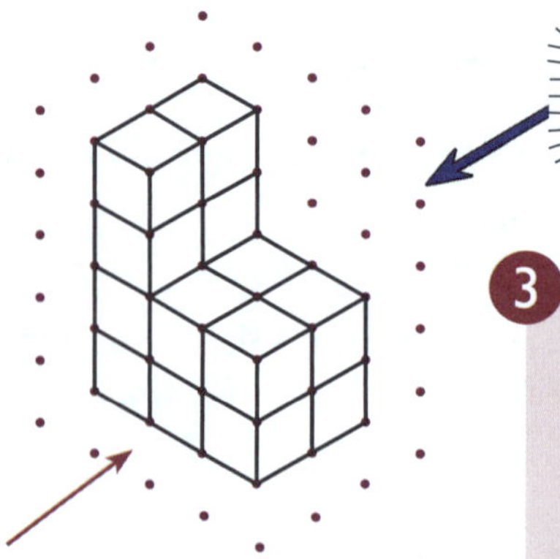

This is an isometric drawing — count the number of dots to find the dimension of the shape.

3

SIDE ELEVATION — the view you'd see from directly to one side

You need to know what the three types of projection are

For enlargements, it's helpful to think about the units — areas are in units², so they are n^2 times bigger.

Q1 There are 3 stacking dolls in a set. The dolls are mathematically similar and have heights of 5 cm, 10 cm and 15 cm. The surface area of the middle doll is 80 cm², and the volume of the largest doll is 216 cm³. Find the surface area and volume of the smallest doll. [4 marks] GRADE 8

Warm-Up and Worked Exam Questions

There are lots of formulas in this section. The best way to see what you know is to practise these questions.

Warm-Up Questions

1) A woodworking template has the shape shown.
   a) Calculate the area of one of the round holes.
   b) Use this to calculate the area of the template.
   c) If the template is 4 mm thick, calculate its volume.

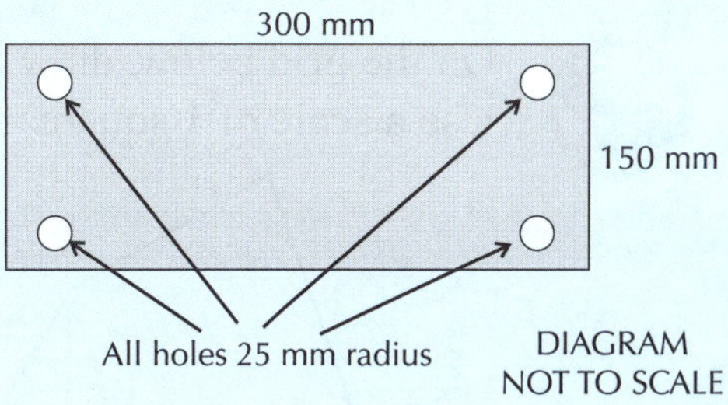

2) Calculate the volume of this triangular prism.

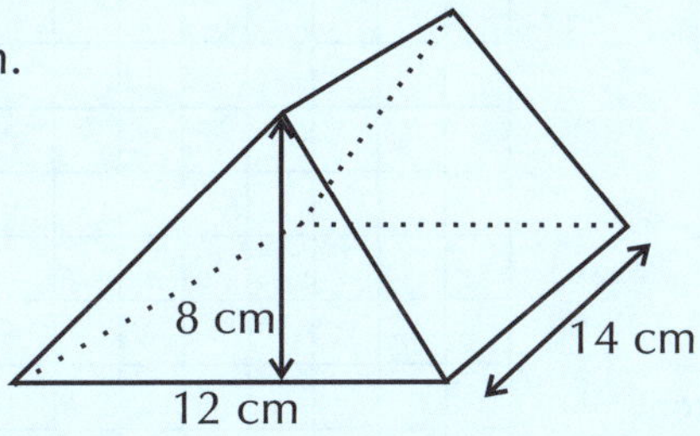

3) Calculate the volume of a sphere with a radius of 7 metres. Give your answer to 3 s.f.

4) From the isometric projection shown on the right, draw:
   a) both side elevations
   b) front elevation
   c) plan view

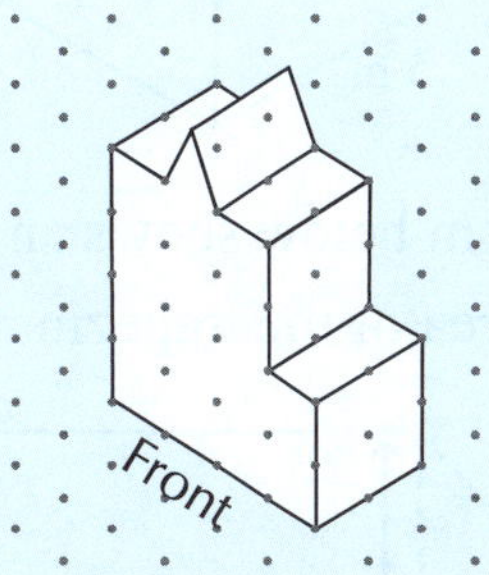

Worked Exam Question

Take a look at this worked exam question — it'll help you to prepare for the real exam:

1 Look at the sector shown in the diagram below. (GRADE 5)

Find the perimeter and the area of the sector. Give your answers to 3 significant figures.

Diagram not accurately drawn

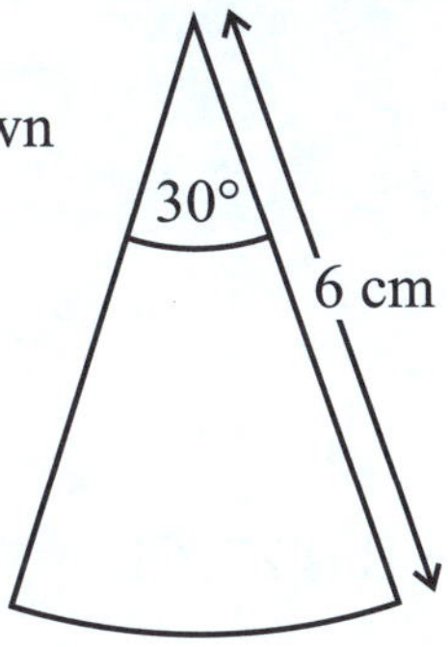

Circumference of full circle = $2 \times \pi \times 6$
$= 12\pi$ cm

Length of arc = $(30 \div 360) \times$ circumference of circle
$= (30 \div 360) \times 12\pi = \pi$ cm

Perimeter of sector = $\pi + 6 + 6 = 15.1$ cm (3 s.f.)

Area of full circle = $\pi \times 6^2 = 36\pi$ cm^2

Area of sector = $(30 \div 360) \times$ area of circle
$= (30 \div 360) \times 36\pi = 3\pi$ cm^2
$= 9.42$ cm^2 (3 s.f.)

Perimeter = ...15.1....... cm

Area = ...9.42....... cm^2

[5 marks]

Exam Questions

2 The diagram shows a house made of a 5 m × 5 m × 6 m cuboid and a triangular roof of width 4 m, length 5 m and vertical height 4 m.

On the grid below, draw the front elevation of the house.
Use a scale of 1 square = 1 m.

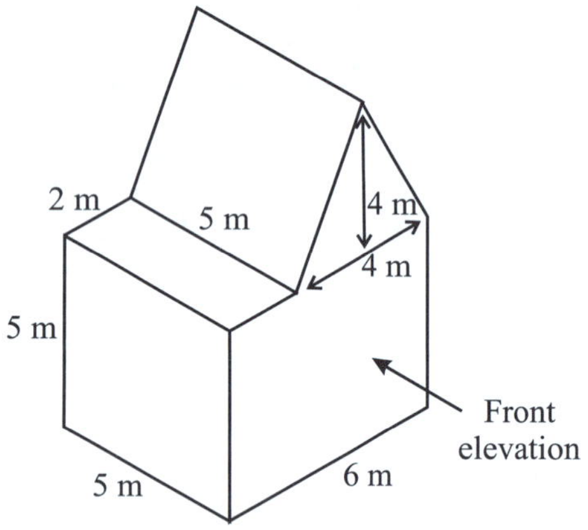

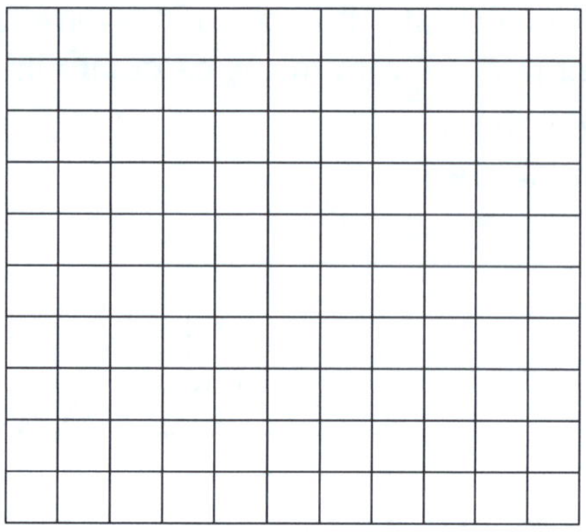

[2 marks]

3 The diagram below shows an isosceles trapezium.
Find the area of the trapezium.

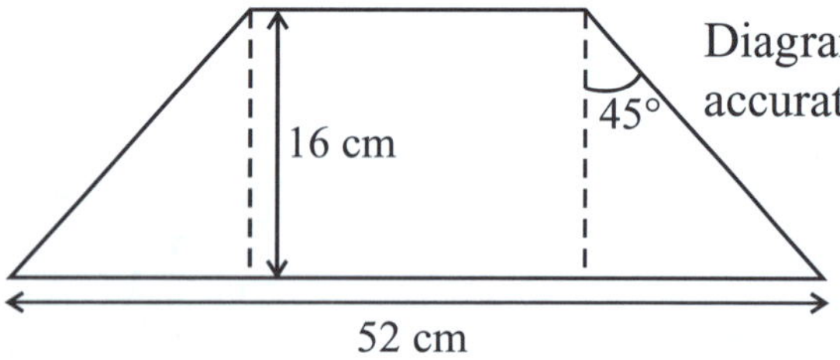

.... cm^2
[2 marks]

4 The diagram below shows a wooden spinning top made from a hemisphere and a cone.

The hemisphere has a diameter of 14 cm. The slanting length of the cone is 12 cm and the radius of its base is 2 cm.

Work out the total surface area of the spinning top.
Give your answer to 3 significant figures.

.... cm^2
[4 marks]

5 Cylinder B is an enlargement of cylinder A.
The ratio of the volume of cylinder A to the volume of cylinder B is 27 : 64.
The surface area of cylinder A is 81π cm^2.

Find the surface area of cylinder B.

.... cm^2
[3 marks]

Exam Questions

6 **A**, **B** and **C** are three solid cones which are mathematically similar.
The surface area of each cone is given below. The perpendicular height of **A** is 4 cm.
The volume of **C** is 135π cm^3.

Not to scale

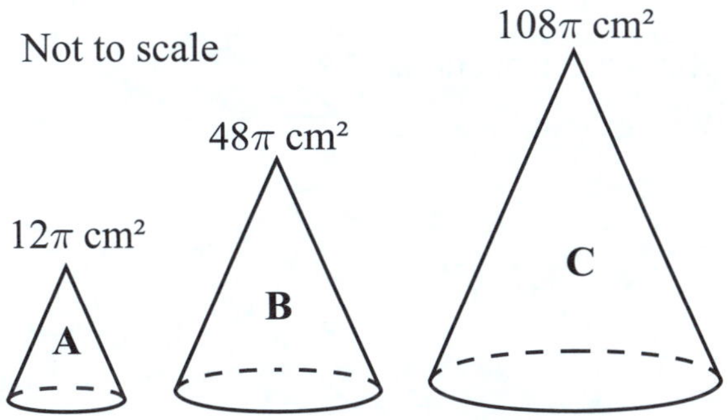

a) Calculate the volume of **A**.

.... cm^3
[4 marks]

b) Calculate the perpendicular height of **B**.

.... cm
[3 marks]

7 The diagram shows how two identical solid spheres fit **exactly** inside a cuboid box.
Find the percentage of the volume of the box which is occupied by the spheres.
Give your answer to 1 d.p.

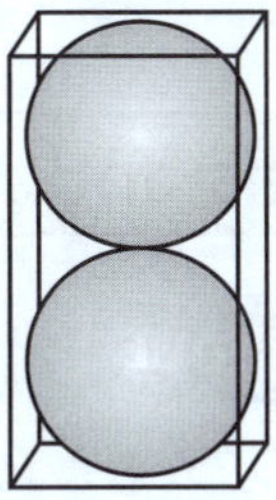

...........................%
[4 marks]

8 Water is flowing through the cylindrical pipe shown in the diagram below.
The radius of the pipe is 0.2 m, and the water comes halfway up the pipe.

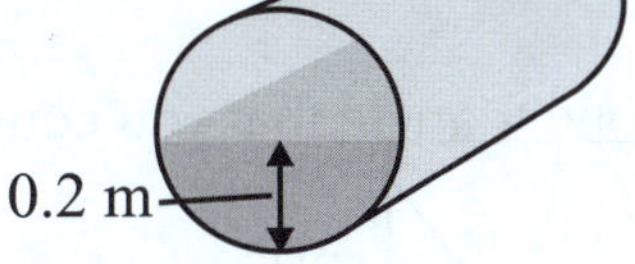

a) Find the cross-sectional area of the water in the pipe.

.... m^2
[1 mark]

The water is flowing at a rate of 2520 litres per minute.

b) Find the speed of the water in m/s to 3 s.f.

Convert the rate of flow to m^3/s, then use speed = rate of flow ÷ cross-sectional area of water.

.... m/s
[4 marks]

Triangle Construction

How you construct a triangle depends on what ***info you're given*** *about the triangle...*

Three sides — use a Ruler and Compasses

GRADE 3

EXAMPLE **Construct the triangle ABC where AB = 6 cm, BC = 4 cm, AC = 5 cm.**

First, sketch and label a triangle so you know roughly what's needed. It doesn't matter which line you make the base line.

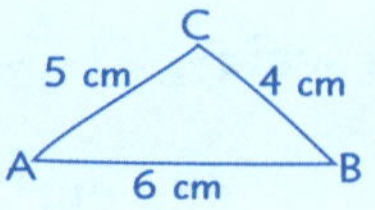

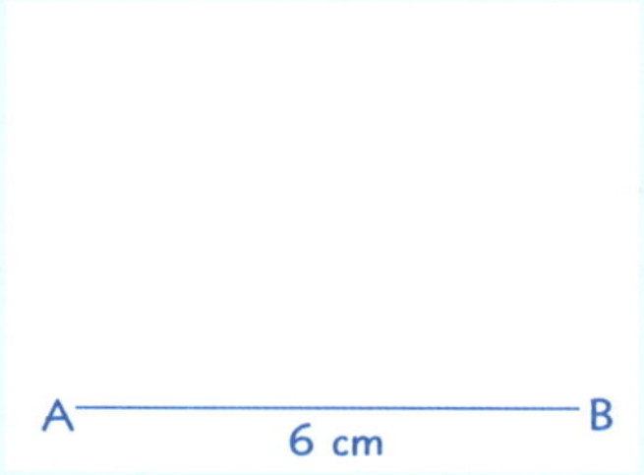

Draw the base line accurately. Label the ends A and B.

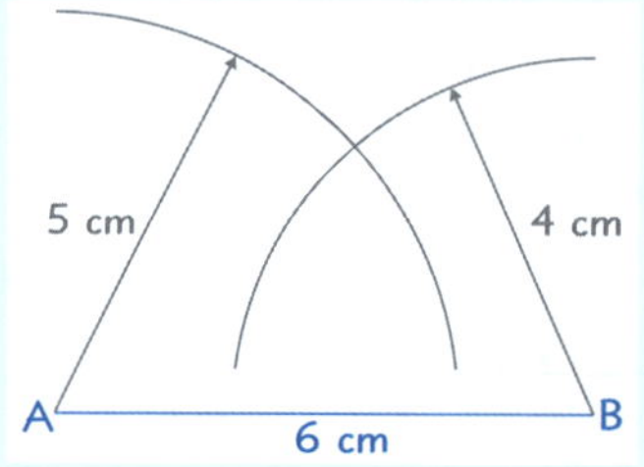

For AC, set the compasses to 5 cm, put the point at A and draw an arc. For BC, set the compasses to 4 cm, put the point at B and draw an arc.

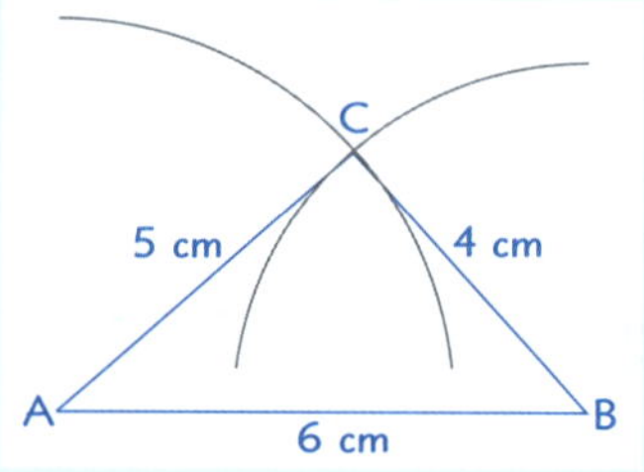

Where the arcs cross is point C. Now you can finish your triangle.

Sides and Angles — use a Ruler and Protractor

GRADE 3

EXAMPLE **Construct triangle DEF. DE = 5 cm, DF = 3 cm, and angle EDF = 40°.**

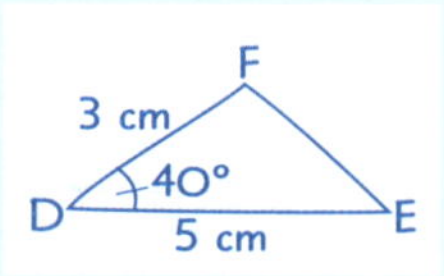

Roughly sketch and label the triangle.

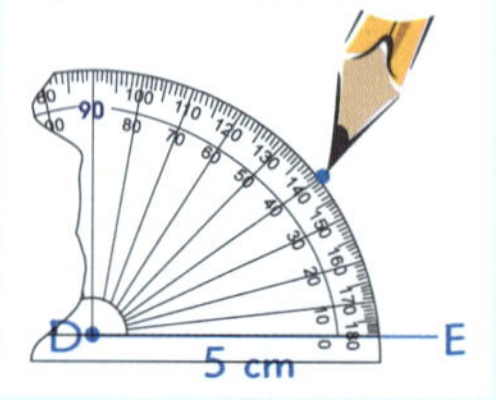

Draw the base line accurately. Then draw angle EDF (the angle at D) — place the centre of the protractor over D, measure 40° and put a dot.

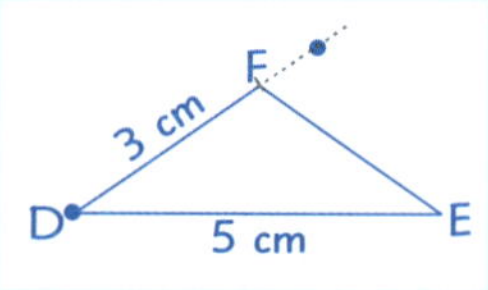

Measure 3 cm towards the dot and label it F. Join up D and F. Now you've drawn the two sides and the angle. Just join up F and E to complete the triangle.

If you're given 3 pieces of information about a triangle, there's usually only one triangle that you could draw.

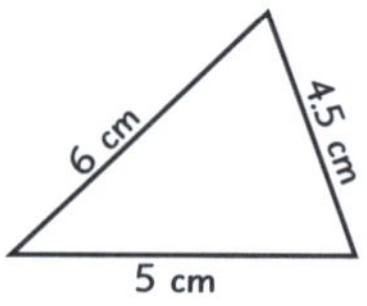

SSS — 3 sides

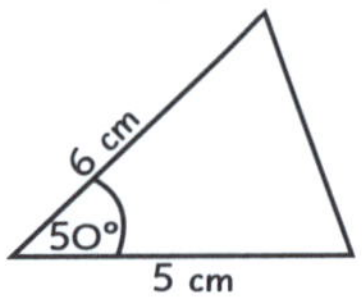

SAS — 2 sides and the angle between them.

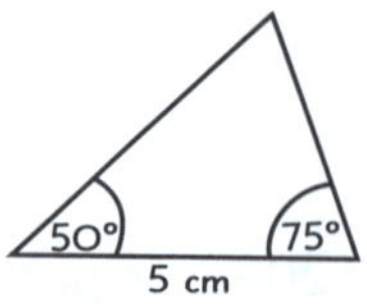

ASA — 2 angles and the side between them.

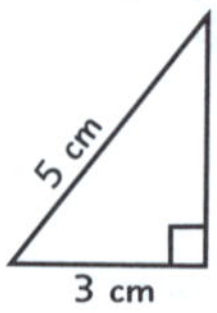

RHS — right angle, the hypotenuse and another side.

However, if you're given 2 sides and an angle which isn't between them, there are TWO possible triangles you could draw.

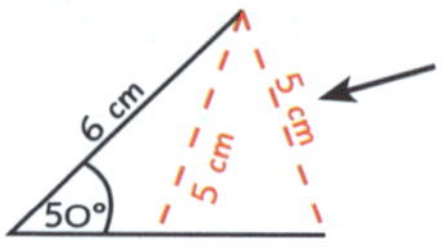

The 5 cm side could be in either of the positions shown.

Don't forget your compasses and protractor for the exam

And make sure to leave your construction marks and label your triangle.

Q1 Construct an equilateral triangle with sides of 5 cm. [2 marks] GRADE 3

Q2 Construct triangle ABC: angle ABC = 45°, angle BCA = 40°, side BC = 7.5 cm. [2 marks] GRADE 3

Q1 Video Solution

Loci and Construction

*A **LOCUS** (another ridiculous maths word) is simply:*

A LINE or REGION that shows all the points which fit a given rule.

*Make sure you learn how to do these **PROPERLY** using a **ruler** and **compasses** as shown on the next few pages.*

The **Four** Different Types of **Loci**

GRADE 4

Loci is just the plural of locus.

1) The locus of points which are 'A FIXED DISTANCE from a given POINT'.

This locus is simply a CIRCLE.

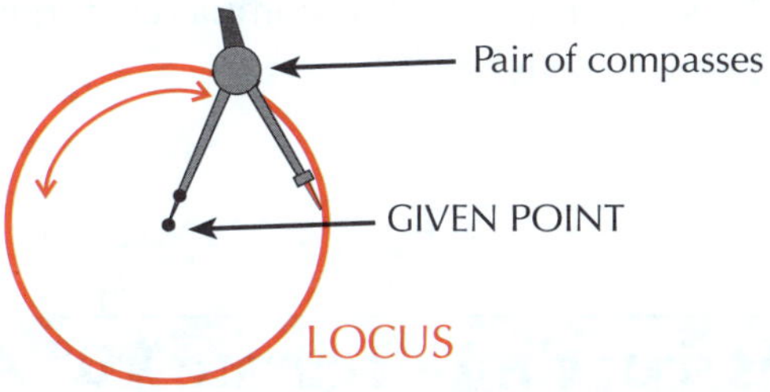

2) The locus of points which are 'A FIXED DISTANCE from a given LINE'.

This locus is a SAUSAGE SHAPE.

It has straight sides (drawn with a ruler) and ends which are perfect semicircles (drawn with compasses).

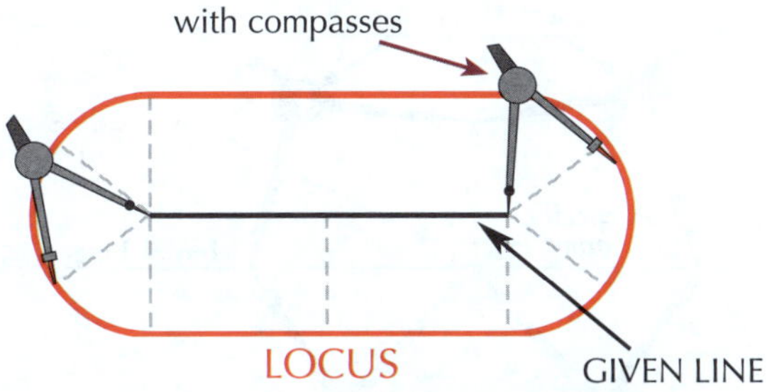

3) The locus of points which are 'EQUIDISTANT from TWO GIVEN LINES'.

1) Keep the compass setting THE SAME while you make all four marks.
2) Make sure you leave your compass marks showing.
3) You get two equal angles — i.e. this LOCUS is actually an ANGLE BISECTOR.

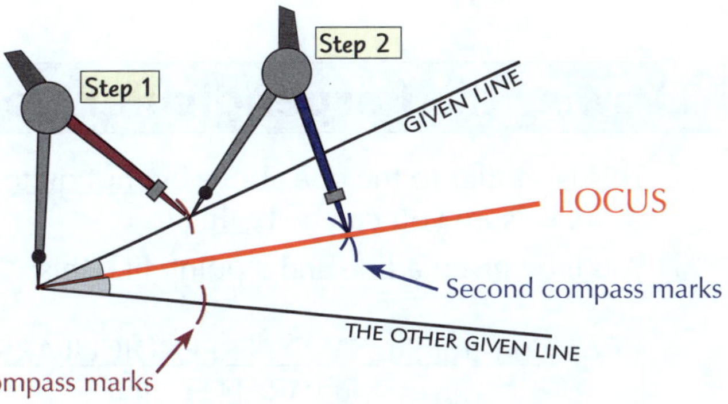

4) The locus of points which are 'EQUIDISTANT from TWO GIVEN POINTS'.

(In the diagram below, A and B are the two given points.)

This LOCUS is all points which are the same distance from A as they are from B.

This time the locus is actually the PERPENDICULAR BISECTOR of the line joining the two points.

The perpendicular bisector of line segment AB is a line at right angles to AB, passing through the midpoint of AB. This is the method to use if you're asked to draw it.

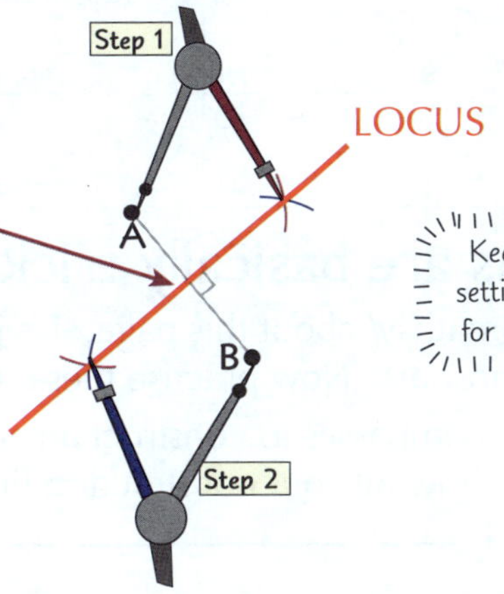

Keep the compass setting THE SAME for all of these arcs.

Loci and Construction

Don't just read the page through once and hope you'll remember it — get your ruler, compasses and pencil out and have a go. It's the only way of testing whether you really know this stuff.

Constructing Accurate 60° Angles

GRADE 4

1) They may well ask you to draw an accurate 60° angle without a protractor.
2) Follow the method shown in this diagram (make sure you leave the compass settings the same for each step).

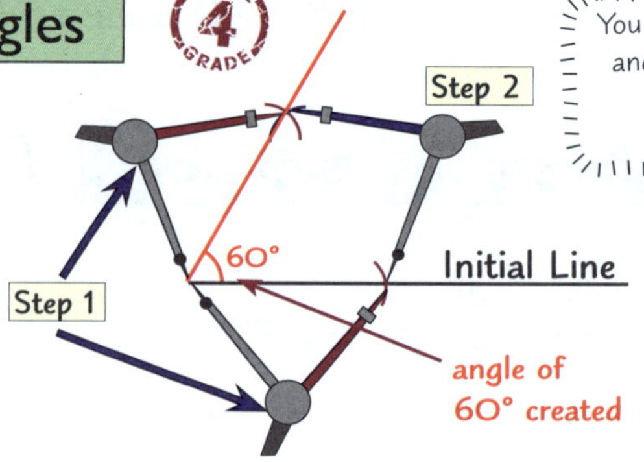

You can construct 30° angles and 45° angles by bisecting 60° and 90° angles (see previous page).

Constructing Accurate 90° Angles

GRADE 4

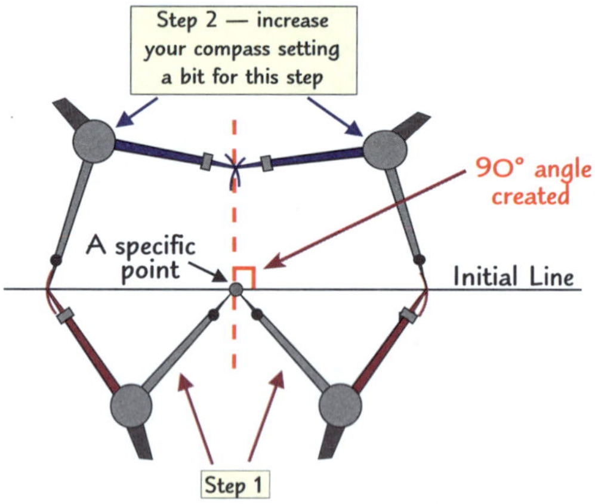

1) They might want you to construct an accurate 90° angle.
2) Make sure you can follow the method shown in this diagram.

The examiners WON'T accept any of these constructions done 'by eye' or with a protractor. You've got to do them the PROPER WAY, with COMPASSES. DON'T rub out your compass marks, or the examiner won't know you used the proper method.

Drawing the Perpendicular from a Point to a Line

GRADE 4

1) This is similar to the one above but not quite the same — make sure you can do both.
2) You'll be given a line and a point, like this:

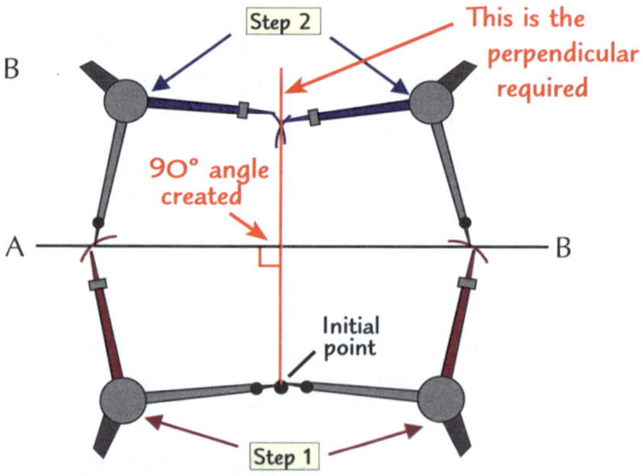

Constructing TWO PERPENDICULARS gives you PARALLEL LINES

If you draw another line perpendicular to the line you've just drawn, it'll be parallel to the initial line:

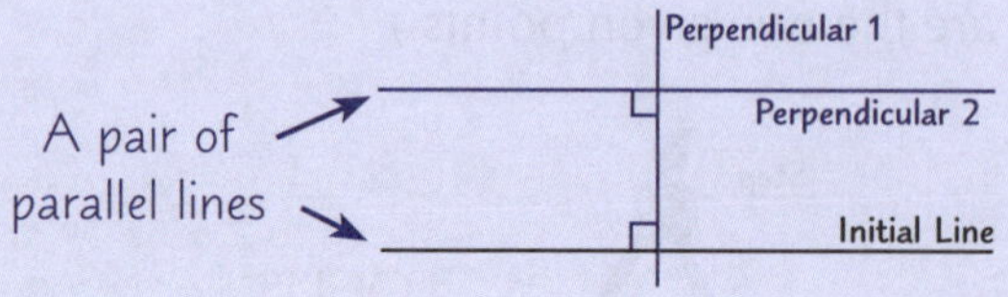

The perpendicular is also the shortest path from the point to the line.

Constructions are basically tricks for maths drawings

There's nothing too 'mathsy' about this page. It's just a few simple tricks to use to draw different angles accurately — it's almost art. Now practise these tricks on this question.

Q1 Use a ruler and compasses to construct an accurate 60° angle at T.
Make sure you show all your construction lines.

T ________________________________

[2 marks] GRADE 4

Loci and Construction — Worked Examples

*Now you know what **loci** are, and how to do all the **constructions** you need, it's time to put them all together.*

Finding a **Locus** that Satisfies **Lots of Rules**

Grade 4

In the exam, you might be given a situation with lots of different conditions, and asked to find the region that satisfies all the conditions. To do this, just draw each locus, then see which bit you want.

EXAMPLE

On the square below, shade the region that is within 3 cm of vertex A and closer to vertex B than vertex D.

The shaded area is the region you want.

Construct a quarter circle 3 cm from A using compasses — you want the region within it.

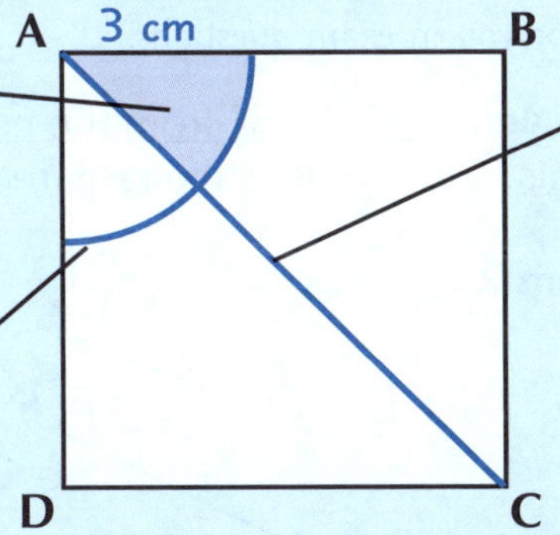

It's a square, so this diagonal is equidistant from B and D. The bit above the line is closer to B than D.

If it wasn't a square you'd have to CONSTRUCT the equidistant line with compasses using the method on p.137.

You might be given the information as a wordy problem — work out what you're being asked for and draw it.

EXAMPLE

Tessa is organising a village fete. The fete will take place on a rectangular field, shown in the diagram below. Tessa is deciding where an ice cream van can go. It has to be <u>at least 1 m away from each edge</u> of the field, and <u>closer to side AB than side CD</u>. There is a maypole at M, and the ice cream van must be <u>at least 2 m away from the maypole</u>. The diagram is drawn to a scale of 1 cm = 1 m. Show on it where the ice cream van can go.

Start by drawing lines 1 cm away from each side (to represent 1 m) — use a ruler to measure along each edge. The ice cream van must go within these lines.

Use compasses to draw a circle 2 cm away from M. The ice cream van has to go outside the circle.

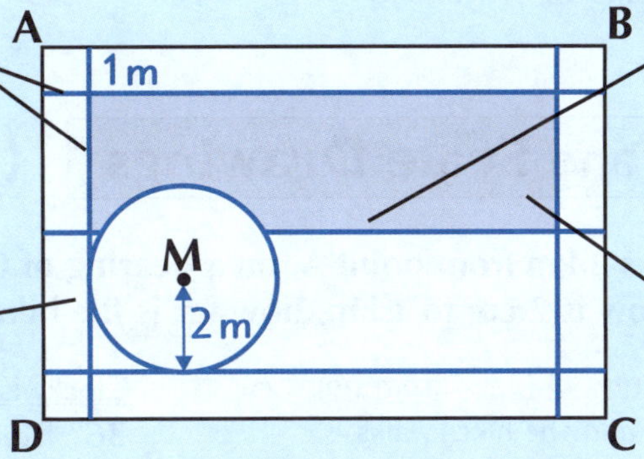

Draw a line equidistant from AB and CD (measure the length of side BC and divide it by two). The ice cream van has to go above this line.

The shaded area shows where the ice cream van can go.

In the examples above, the lines were all at right angles to each other, so you could just measure with a ruler rather than do constructions with compasses. If the question says "Leave your construction lines clearly visible", you'll definitely need to get your compasses out and use some of the methods on p.137-138.

If there are several rules, start by drawing each locus

Don't panic if you get a wordy loci question — deal with one condition at a time and then work out which bit you need to shade. Make sure you draw your loci accurately — it's really important.

Q1 Video Solution

Q1 The gardens of a stately home are shown on the diagram. The public can visit the gardens, but must stay at least 2 m away from the rectangular pond and at least 2 m away from each of the statues (labelled A and B). Make a copy of this diagram using a scale of 1 cm = 2 m and indicate on it the areas where the public can go. [4 marks]

Grade 4

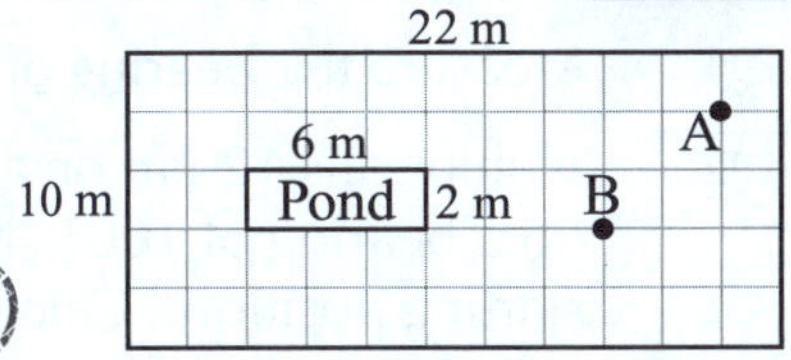

Bearings

Bearings. They'll be useful next time you're off sailing. Or in your Maths exam.

Bearings (Grade 3)

To find or plot a bearing you must remember the three key words:

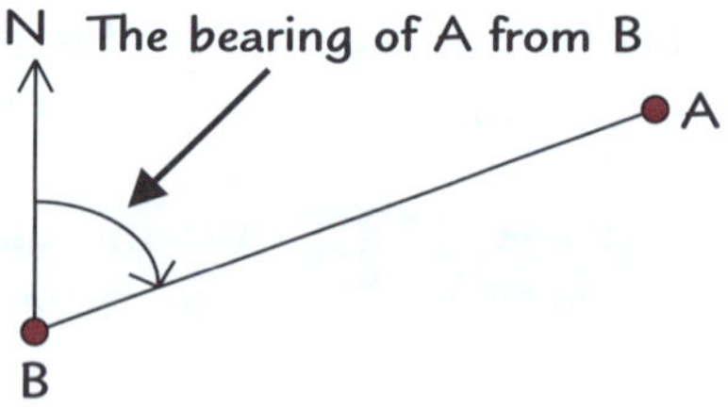

1) 'FROM' Find the word 'FROM' in the question, and put your pencil on the diagram at the point you are going 'from'.

2) NORTH LINE At the point you are going FROM, draw in a NORTH LINE. (There'll often be one drawn for you in exam questions.)

3) CLOCKWISE Now draw in the angle CLOCKWISE from the north line to the line joining the two points. This angle is the required bearing.

EXAMPLE **Find the bearing of Q from P.**

ALL BEARINGS SHOULD BE GIVEN AS 3 FIGURES
e.g. 176°, 034° (not 34°), 005° (not 5°), 018° etc.

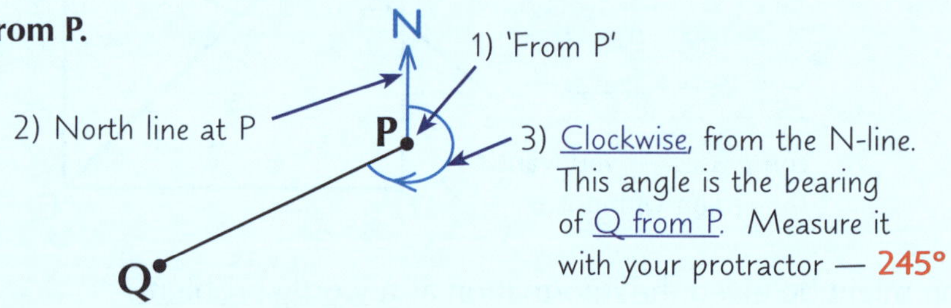

EXAMPLE **The bearing of Z from Y is 110°. Find the bearing of Y from Z.**

First sketch a diagram so you can see what's going on.
Angles a and b are allied, so they add up to 180°.

See page 110 for allied angles.

Angle b = 180° – 110° = 70°
So bearing of Y from Z = 360° – 70° = 290°.

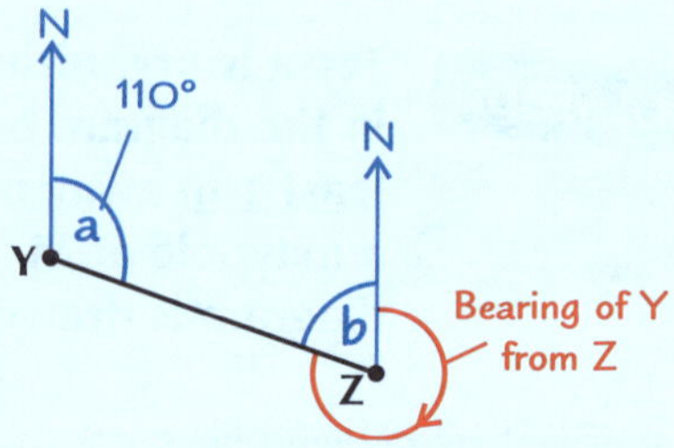

Bearings Questions and Scale Drawings (Grade 3)

EXAMPLE **A hiker walks 2 km from point A, on a bearing of 036°. If the scale of the map below is 2 cm to 1 km, how far is the hiker now from his car?**

First, draw a line at a bearing of 036° from point A. 1 km is 2 cm on the map and the hiker walks 2 km, so make the line from A 4 cm long.

You want the distance of the hiker from the car, so use a ruler to measure it on the map, then use the scale to work out the real distance it represents.

Distance to car on map = 3 cm. 2 cm = 1 km, so 1 cm = 0.5 km, therefore 3 cm = 1.5 km.

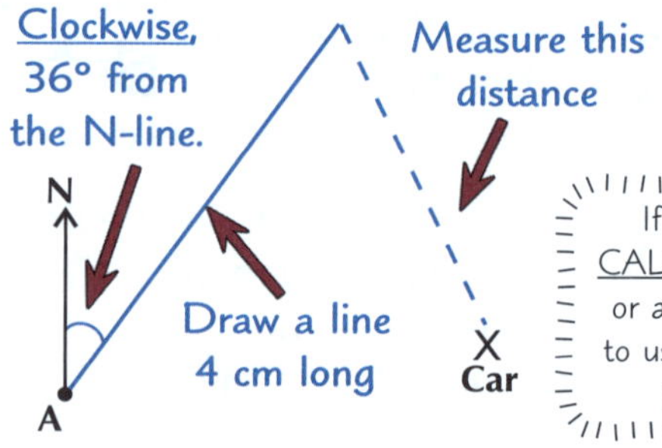

If you are asked to CALCULATE a distance or an angle, you'll need to use the cosine or sine rule (see p.151).

FROM a point draw a NORTH LINE then draw the angle CLOCKWISE

Learn the three key words above and the method for using them. Now try these practice questions.

Q1 Measure the bearing of T from H, shown on the right. [1 mark] (Grade 3)

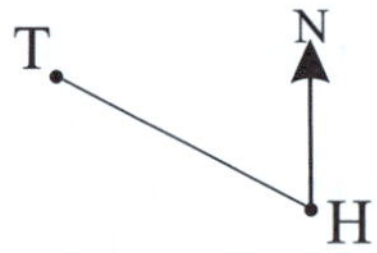

Q2 A ship sails 12 km on a bearing of 050°, then 20 km on a bearing of 100°. It then sails directly back to its starting position. Calculate this distance to 1 d.p. [5 marks] (Grade 8)

Q2 Video Solution

Warm-Up and Worked Exam Questions

You need to work through these one by one and make sure you really know what you're doing with your ruler and compasses. Look back over the last five pages if you get stuck.

Warm-Up Questions

1) Using a ruler and compasses, construct an equilateral triangle with sides of length 4 cm.
2) Construct a triangle with sides 3 cm, 4 cm and 5 cm. Check it by measuring the sides.
3) Jakub owns a piece of land 500 m by 400 m. There is a river on one edge of the land, a road along another edge and a large tree in one corner.

   Jakub wants to build a house on his land. The house must be at least 100 m from the river, at least 100 m from the tree, T, and within 300 m of the road.

   Make a copy of this diagram using a scale of 1 cm = 100 m and indicate on it where the house could be built.

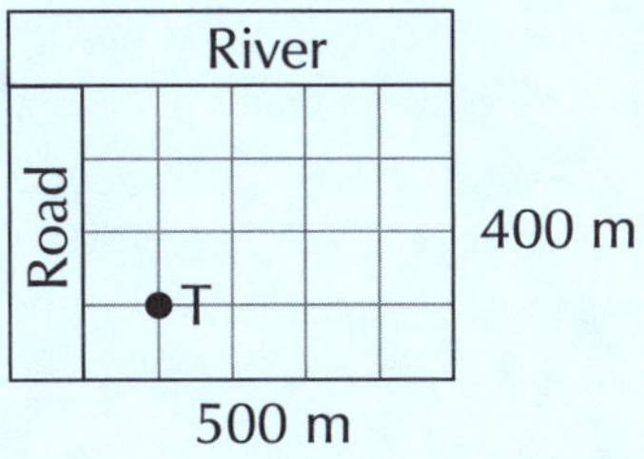

4) Construct four 90° angles to form a square with side length 5.5 cm.
5) Draw a line and a point and construct the perpendicular from the point to the line.
6) A fishing boat leaves the harbour on a bearing of 065° and travels for 3 miles, and then travels 1.5 miles on a bearing of 130°. By drawing a diagram, work out how far the fishing boat is from the harbour now.

Worked Exam Questions

Two worked examples, and then it's over to you.

1 A dog is tied to a beam *AB* by a lead which allows it to run a maximum of 2 m from the beam.

Shade the region on the diagram where the dog may run, using the scale shown.

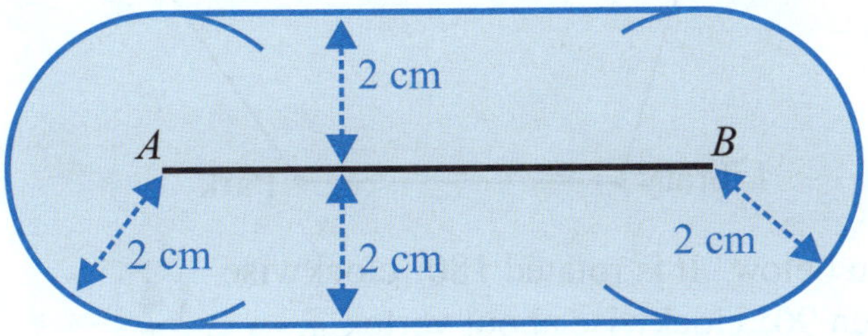

Scale: 1 cm represents 1 m

[2 marks]

2 *ABC* is a triangle.

Find and shade the region inside the triangle which is **both** closer to the line *AB* than the line *BC*, **and** also more than 6.5 cm from the point *C*.

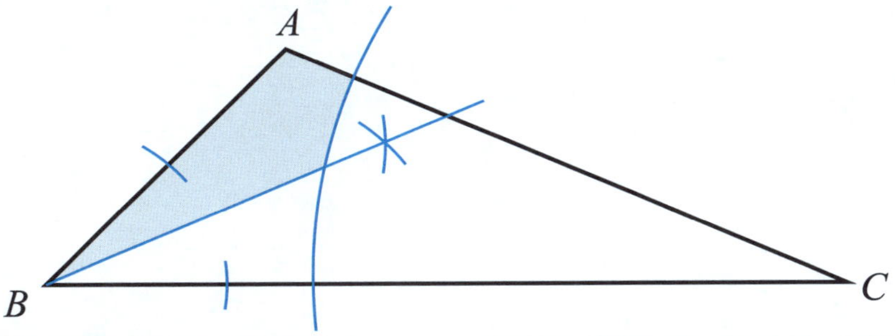

Make sure you remember to leave in your construction lines.

[4 marks]

Exam Questions

3 Two ships leave a port at the same time.
Ship A travels due west for 40 km. Ship B travels 60 km on a bearing of 110°. (Grade 3)

a) Using a scale of 1 cm = 10 km, draw the journeys of the two ships in the space below and clearly mark their final positions.

[4 marks]

b) Measure the final bearing of Ship B from Ship A.

.... °
[1 mark]

c) Calculate the final bearing of Ship A from Ship B.

.... °
[2 marks]

4 A town council are putting up a new visitor information board. They want it to be placed within the area shown, closer to the park than to the library, but also closer to the station than to the park. (Grade 4)

The diagram below shows a scale map of the town centre.
Shade in the region of the town where the board could be placed.

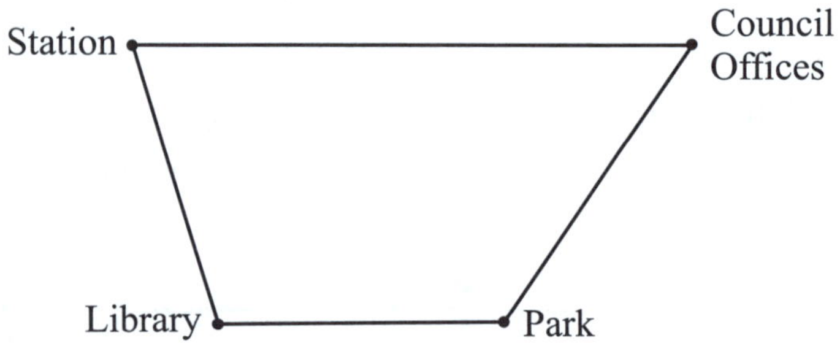

[3 marks]

5 Triangle XYZ is shown below. It is rotated 180° clockwise about vertex X and then 90° clockwise about vertex Z. (Grade 4)
Draw the locus of vertex Y during the two rotations.

Keep an eye on how vertex Y moves during each rotation.

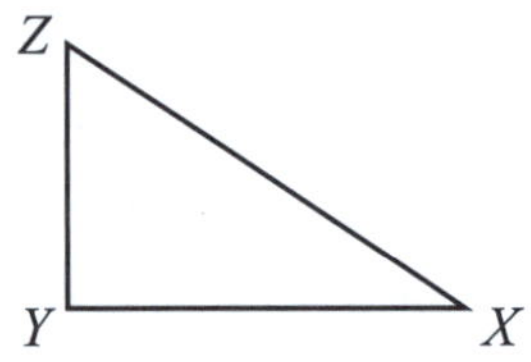

[3 marks]

Revision Questions for Section Five

There are lots of opportunities to show off your artistic skills here (as long as you use them to answer the questions).

- Try these questions and <u>tick off each one</u> when you <u>get it right</u>.
- When you're <u>completely happy</u> with a topic, tick that off too.

For even more practice, try the quiz for Section Five — just scan this QR code!

Section Five Quiz

Angles and Polygons (p109-113)

1) Write down the five simple geometry rules.
2) Find the missing angles in the diagrams below.

a)

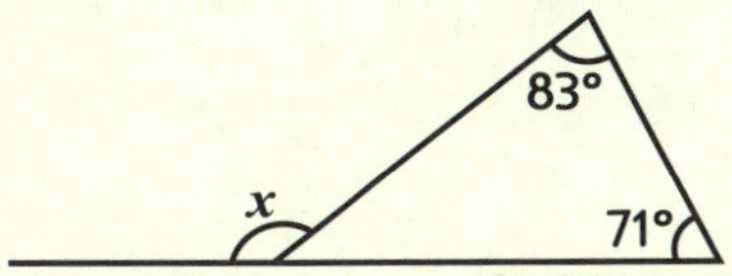

b)

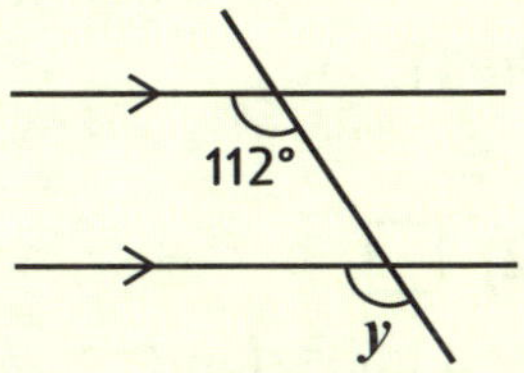

c)

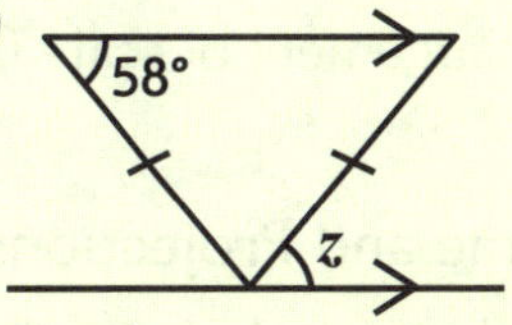

3) Find the exterior angle of a regular hexagon.
What do the interior angles of a regular hexagon add up to?
4) Write down the number of lines of symmetry and the order of rotational symmetry for an equilateral, isosceles and scalene triangle.
5) Name two quadrilaterals that have two pairs of equal angles.

Circle Geometry (p116-118)

6) Write down the nine rules of circle geometry.
7) Find the missing angle in each of the diagrams below.

a)

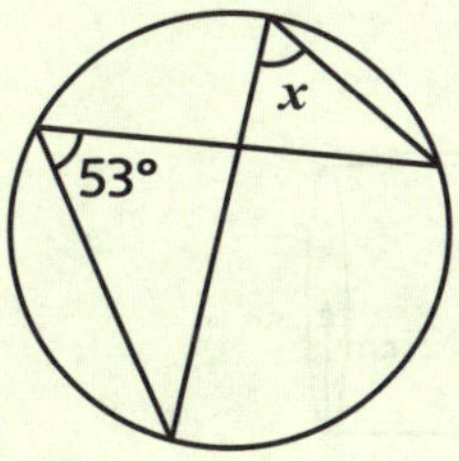

b)

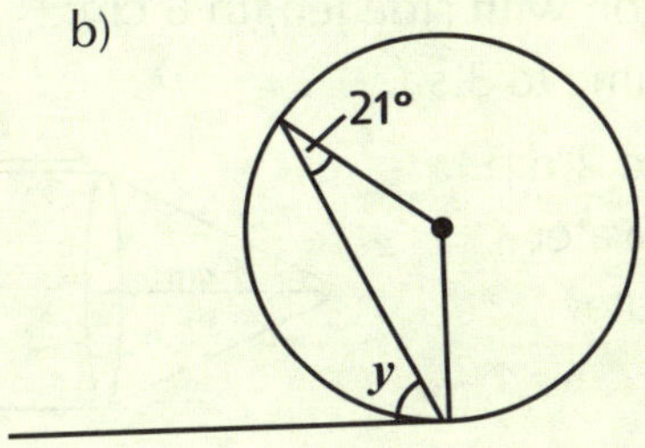

c)

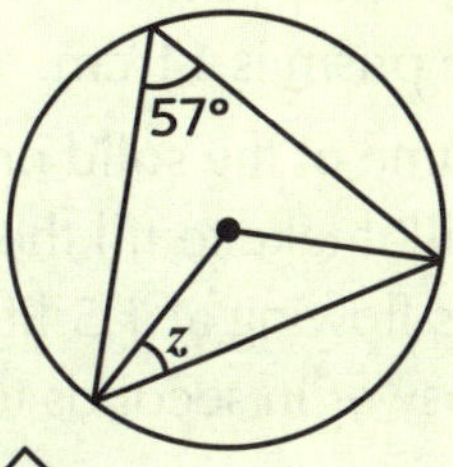

8) Is the quadrilateral on the right cyclic? Explain your answer.

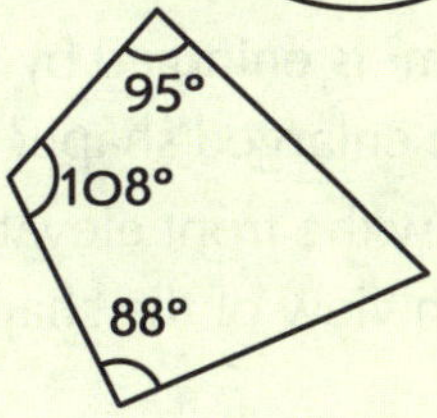

Congruence and Similarity (p121-122)

9) State the four conditions you can use to prove that two triangles are congruent.
10) Prove that triangles ABC and ACD on the right are congruent.

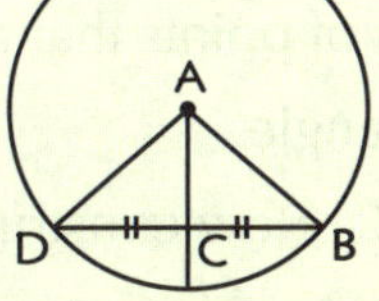

11) The shapes below are similar. What is the length of side x?

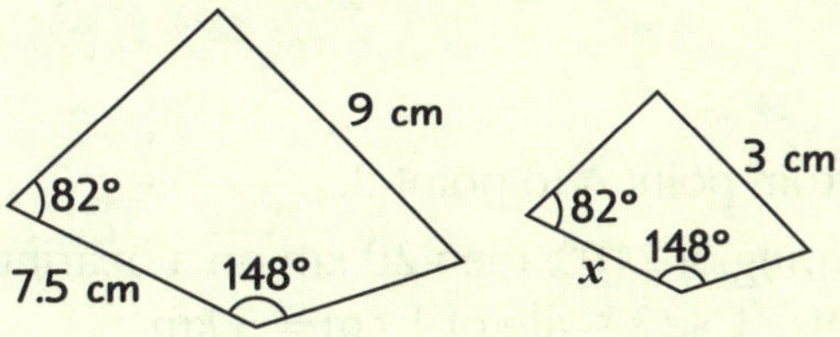

Revision Questions for Section Five

Transformations (p123-124)

12) Describe the transformation that maps:
    a) Shape A onto Shape B
    b) Shape A onto Shape C

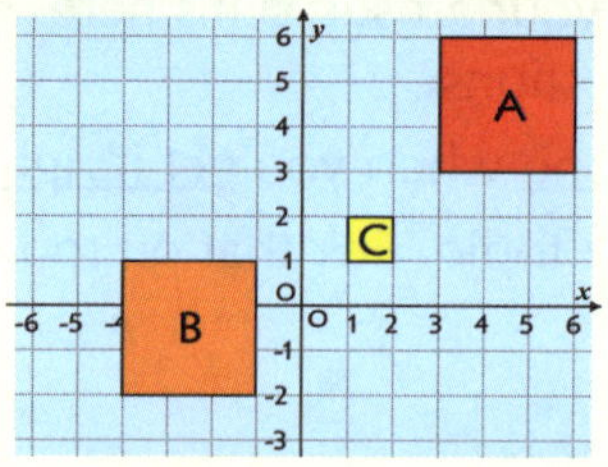

13) Carry out the following transformations on the triangle X, which has vertices (1, 1), (4, 1) and (2, 3):
    a) a rotation of 90° clockwise about (1, 1)
    b) a translation by the vector $\begin{pmatrix}-3\\-4\end{pmatrix}$
    c) an enlargement of scale factor 2, centre (1, 1)

Area, Volume and Projections (p127-132)

14) What is the formula for finding the area of a trapezium?
15) Find the area of the shape on the right.

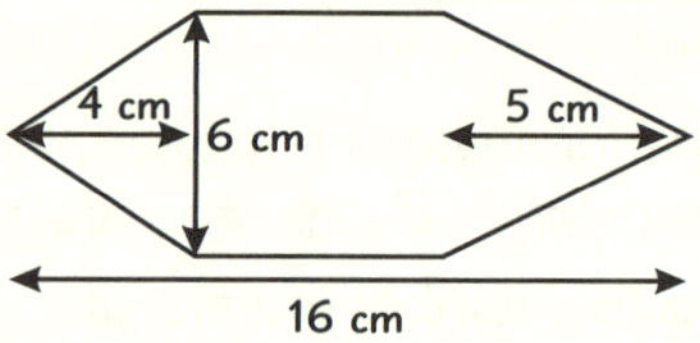

16) A square has an area of 56.25 cm². Find its perimeter.
17) A circle has diameter 16 cm. Find its exact circumference and area.
18) Find the area of the sector with radius 10 cm and angle 45° to 2 d.p.
19) What are the formulas for finding the surface area of a sphere, a cylinder and a cone?
20) The shape on the right is made from a cylinder and a hemisphere. Find its exact surface area.

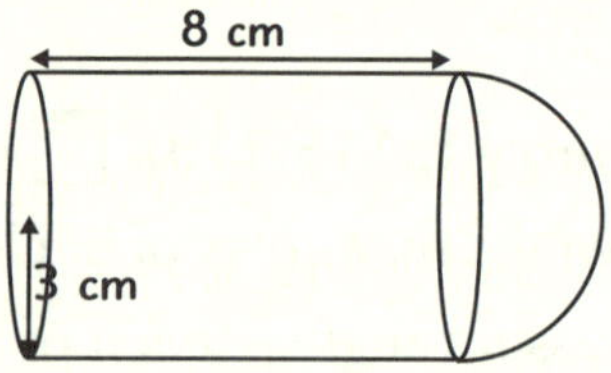

21) The cross-section of a prism is a regular hexagon with side length 6 cm. The length of the prism is 11 cm. Find its volume to 3 s.f.
22) a) Find the volume of the solid on the right (to 2 d.p.).
    b) How long will it take to fill the solid with water if the water is flowing at 1.5 litres per minute? Give your answer in seconds to 1 d.p.

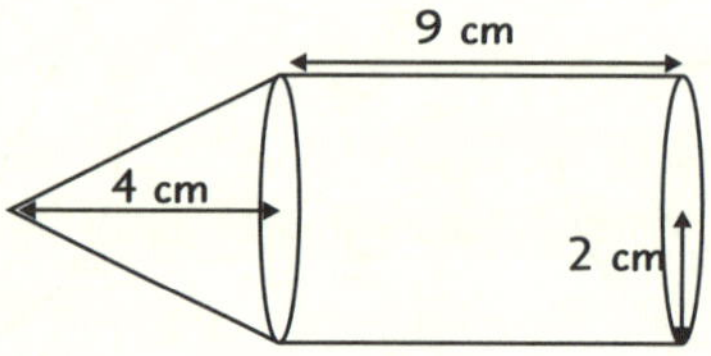

23) A shape with area 5 cm² is enlarged by a scale factor of 4. What is the area of the enlarged shape?
24) On squared paper, draw the front elevation, side elevation and plan view of the shape on the right.

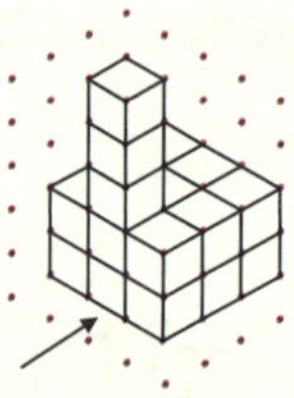

Constructions and Loci (p136-139)

25) Construct triangle XYZ, where XY = 5.6 cm, XZ = 7.2 cm and angle YXZ = 55°.
26) Construct two triangles, ABC, with angle A = 40°, AB = 6 cm, BC = 4.5 cm.
27) What shape does the locus of points that are a fixed distance from a given point make?
28) Construct an accurate 45° angle.
29) Draw a line and label it AB. Now construct the perpendicular bisector of AB.
30) Draw a square with sides of length 6 cm and label it ABCD. Shade the region that is nearer to AB than CD and less than 4 cm from vertex A.

Bearings (p140)

31) Describe how to find a bearing from point A to point B.
32) A helicopter flies 25 km on a bearing of 210°, then 20 km on a bearing of 040°. Draw a scale diagram to show this. Use a scale of 1 cm = 5 km.

Pythagoras' Theorem

*Pythagoras' theorem sounds hard but it's actually **dead simple**.*
It's also dead important, so make sure you really get your teeth into it.

Pythagoras' Theorem — $a^2 + b^2 = c^2$

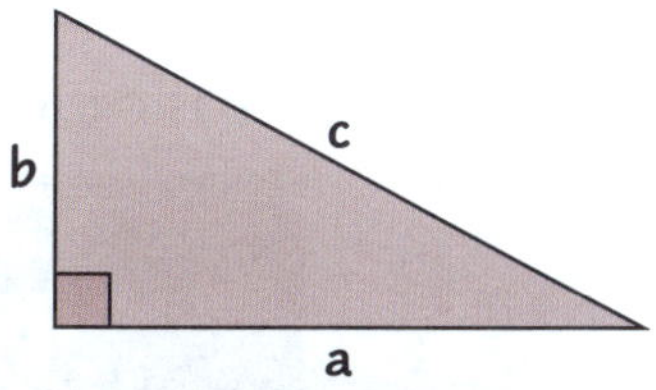

1) PYTHAGORAS' THEOREM only works for RIGHT-ANGLED TRIANGLES.
2) Pythagoras uses two sides to find the third side.
3) The BASIC FORMULA for Pythagoras is $a^2 + b^2 = c^2$
4) Make sure you get the numbers in the RIGHT PLACE. c is the longest side (called the hypotenuse) and it's always opposite the right angle.
5) Always CHECK that your answer is SENSIBLE.

$a^2 + b^2 = c^2$

EXAMPLE **ABC is a right-angled triangle. AB = 6 m and AC = 3 m. Find the exact length of BC.**

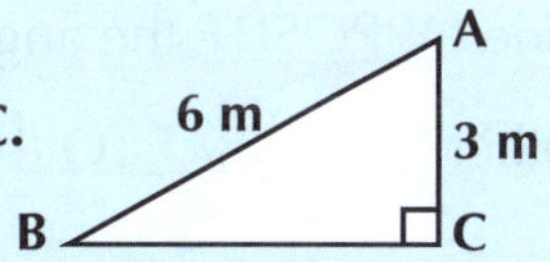

1) Write down the formula. → $a^2 + b^2 = c^2$
2) Put in the numbers. → $BC^2 + 3^2 = 6^2$
3) Rearrange the equation. → $BC^2 = 6^2 - 3^2 = 36 - 9 = 27$
4) Take square roots to find BC. → $BC = \sqrt{27} = 3\sqrt{3}$ m
5) 'Exact length' means you should give your answer as a surd — simplified if possible.

It's not always c you need to find — loads of people go wrong here.

Remember to check the answer's sensible — here it's about 5.2, which is between 3 and 6, so that seems about right...

Use **Pythagoras** to find the **Distance Between Points**

You need to know how to find the straight-line distance between two points on a graph.
If you get a question like this, follow these rules and it'll all become breathtakingly simple:

1) Draw a sketch to show the right-angled triangle.
2) Find the lengths of the shorter sides of the triangle by subtracting the coordinates.
3) Use Pythagoras to find the length of the hypotenuse. (That's your answer.)

EXAMPLE **Point P has coordinates (8, 3) and point Q has coordinates (–4, 8). Find the length of the line PQ.**

1

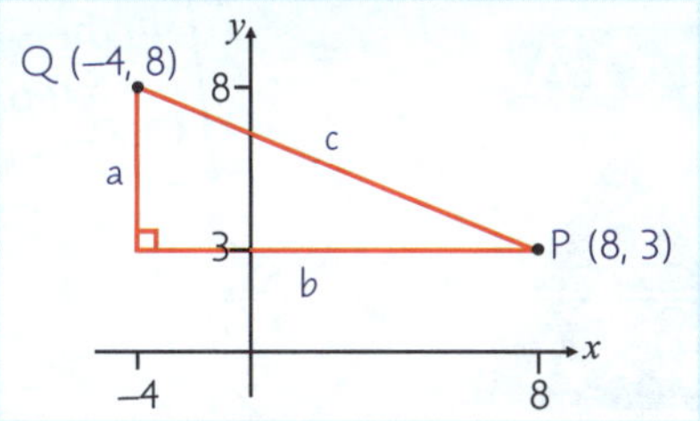

2 Length of side a = 8 – 3 = 5
Length of side b = 8 – –4 = 12

3 Use Pythagoras to find side c:
$c^2 = a^2 + b^2 = 5^2 + 12^2 = 25 + 144 = 169$
So: $c = \sqrt{169} = 13$

Finding lengths in a right-angled triangle? Pythagoras is your man

This is one of the most famous of all maths formulas — learn it, then have a go at these questions.

Q1 Find the length of AC correct to 1 decimal place. [3 marks]

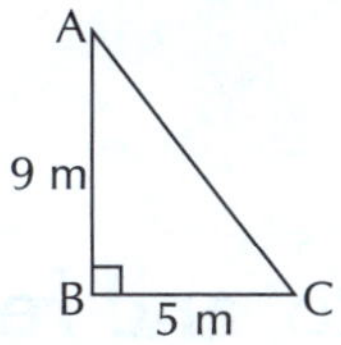

Q2 Point A has coordinates (10, 15) and point B has coordinates (6, 12). Find the length of the line AB. [4 marks]

Q2 Video Solution

Q3 A right-angled triangle has a hypotenuse of length $2\sqrt{10}$ cm. Give possible lengths for the other two sides of the triangle, given that the lengths are integers. [3 marks]

Trigonometry — Sin, Cos, Tan

***Trigonometry** — it's a big scary word. It's **important** and **always cropping up** in exams, but if you just follow the method below it won't be a big scary topic.*

The 3 Trigonometry Formulas

GRADE 6

There are three basic trig formulas — each one links two sides and an angle of a right-angled triangle.

$$\text{Sin } x = \frac{\text{Opposite}}{\text{Hypotenuse}} \qquad \text{Cos } x = \frac{\text{Adjacent}}{\text{Hypotenuse}} \qquad \text{Tan } x = \frac{\text{Opposite}}{\text{Adjacent}}$$

- The Hypotenuse is the LONGEST SIDE.
- The Opposite is the side OPPOSITE the angle being used (x).
- The Adjacent is the (other) side NEXT TO the angle being used.

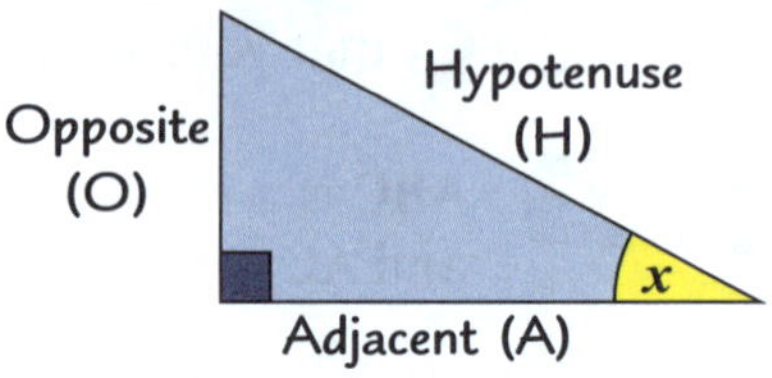

1) Whenever you come across a trig question, work out which two sides of the triangle are involved in that question — then pick the formula that involves those sides.
2) To find the angle — use the inverse, i.e. press SHIFT or 2ndF, followed by sin, cos or tan (and make sure your calculator is in DEG mode) — your calculator will display $\sin^{-1}$, $\cos^{-1}$ or $\tan^{-1}$.
3) Remember, you can only use the sin, cos and tan formulas above on right-angled triangles — you may have to add lines to the diagram to create one.

There's more about formula triangles on p.105 if you need to jog your memory.

Formula Triangles Make Things Simple

GRADE 6

A handy way to tackle trig questions is to convert the formulas into formula triangles. Then you can use the same method every time, no matter which side or angle is being asked for.

1) Label the three sides O, A and H (Opposite, Adjacent and Hypotenuse).
2) Write down from memory 'SOH CAH TOA'.
3) Decide which two sides are involved: O,H A,H or O,A and select SOH, CAH or TOA accordingly.
4) Turn the one you choose into a FORMULA TRIANGLE:

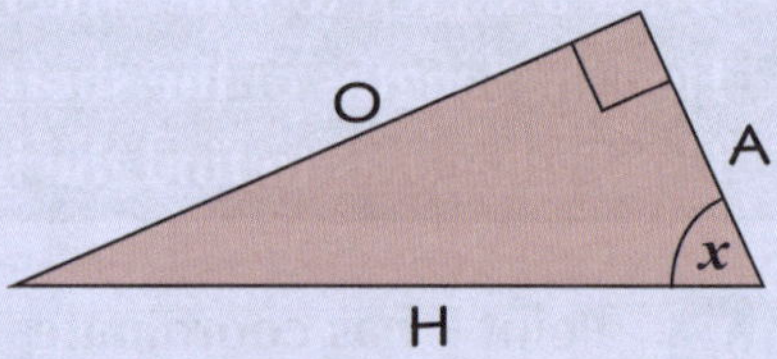

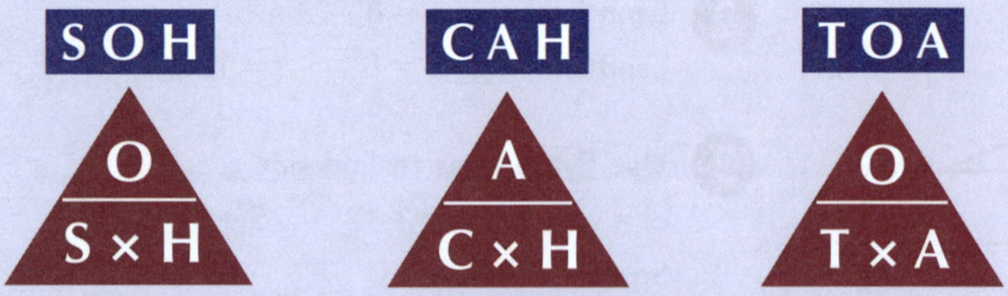

In the formula triangles, S represents sin x, C is cos x, and T is tan x.

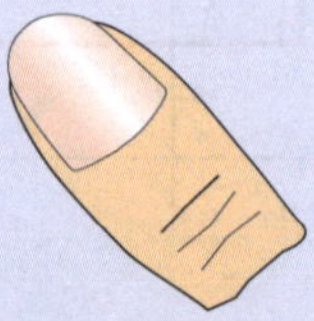

5) Cover up the thing you want to find (with your finger), and write down whatever is left showing.
6) Translate into numbers and work it out.
7) Finally, check that your answer is sensible.

If you can't make SOH CAH TOA stick, try using a mnemonic like 'Strange Orange Hamsters Creep Around Houses Tripping Over Ants'.

H = longest, O = opposite, A = next to, and remember SOH CAH TOA

You'll get the formulas at the top of the page in your exam, but the formula triangles are still super handy to remember — so go over this page a few times until you're familiar with them. All set? Good.

Trigonometry — Examples

Here are some lovely examples using the method from p.146 to help you through the trials of trig.

Examples:

GRADE 6

1 **Find the length of p in the triangle shown to 3 s.f.**

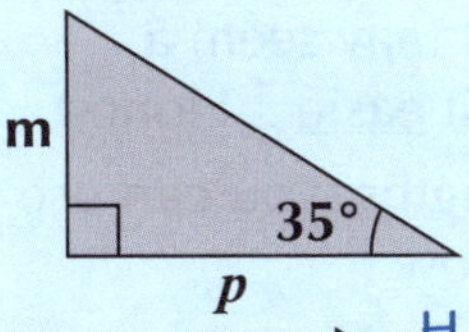

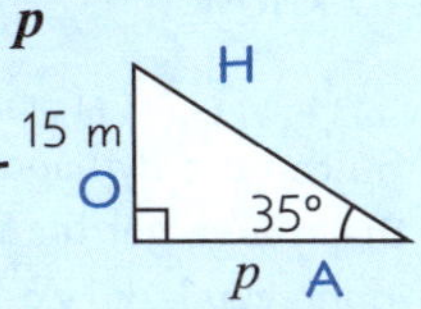

1) Label the sides
2) Write down — SOH CAH TOA (TOA circled)
3) O and A involved
4) Write down the formula triangle

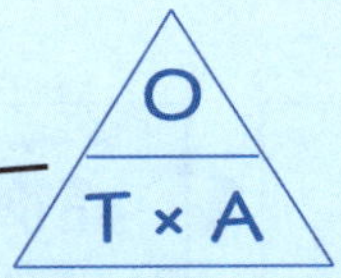

5) You want A so cover it up to give — $A = \frac{O}{T}$
6) Put in the numbers — $p = \frac{15}{\tan 35°} = 21.422...$

$= 21.4 \text{ m (3 s.f.)}$

Is it sensible? Yes, it's a bit bigger than 15, as the diagram suggests.

2 **Find the angle x in this triangle to 1 d.p.**

It's an isosceles triangle so split it down the middle to get a right-angled triangle.

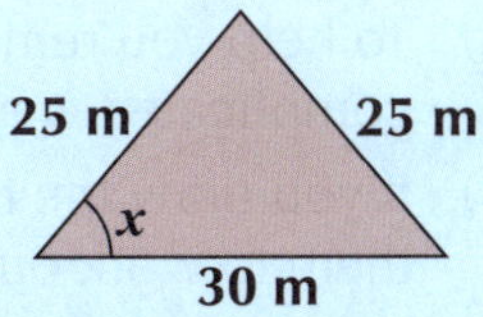

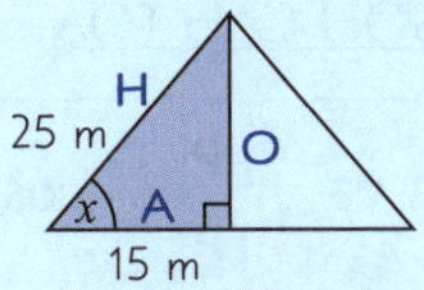

1) Label the sides
2) Write down — SOH CAH TOA (CAH circled)
3) A and H involved
4) Write down the formula triangle

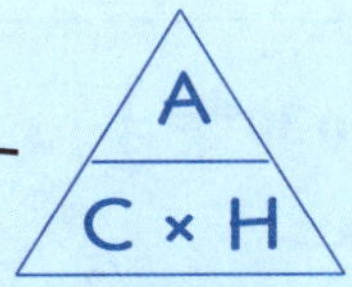

5) You want the angle so cover up C to give — $C = \frac{A}{H}$
6) Put in the numbers — $\cos x = \frac{15}{25} = 0.6$
7) Find the inverse — $\Rightarrow x = \cos^{-1}(0.6) = 53.1301...°$

$= 53.1° \text{ (1 d.p.)}$

Is it sensible? Yes, the angle looks about 50°.

3 **Triangle A and triangle B are similar. Triangle A is such that $\sin x = 0.4$. Find the length of side y.**

GRADE 7

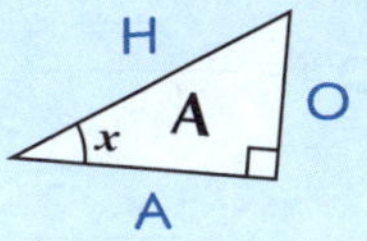

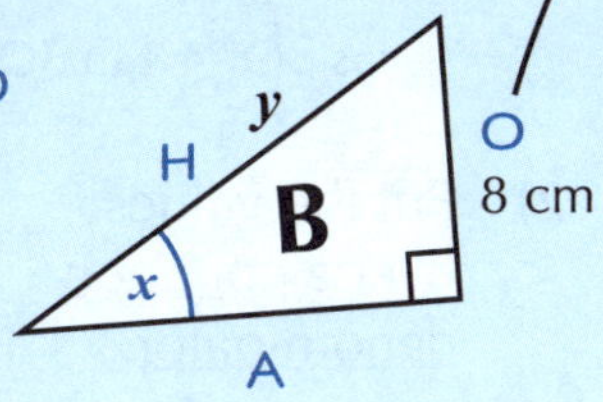

1) Label the sides on both triangles.
2) As the triangles are similar, the small angle in B is also x (see p.122) — label this.
3) Write down — SOH CAH TOA (SOH circled)
4) O and H involved

$\sin x = \frac{O}{H}$ so $0.4 = \frac{O}{H}$

5) Put in the numbers and rearrange to find y. — $0.4 = \frac{8}{y}$

$y = \frac{8}{0.4} = 20 \text{ cm}$

You need to have learnt all seven steps on page 146

Here you can see the seven steps from the last page being put into action. It's easy to apply those steps, but only if you can remember them and practise using them — so practise.

Q1 Find the value of x and give your answer to 1 decimal place. [2 marks] GRADE 6

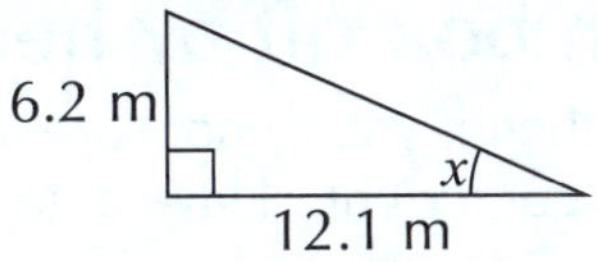

Q2 A 3.2 m ladder leans against a vertical wall at an angle of 68° to the horizontal ground. How far does the ladder reach up the wall? Give your answer to 3 s.f. [2 marks] GRADE 6

Q2 Video Solution

Trigonometry — Common Values

Trig questions quite often use the same angles — so it'll make life ***easier*** *if you know the sin, cos and tan of these* ***commonly used*** *angles. You might need to use them in your non-calculator exam — so* ***learn*** *them.*

Use these **Two Triangles** to **Learn** the **Trig Values**

GRADE 7

1) You need to know the values of sin, cos and tan at 30°, 60° and 45°.
2) To help you remember, you can draw these two triangles. It may seem a complicated way to learn a few numbers, but it does make it easier. Honest.
3) If you draw the triangles, putting in their angles and side lengths, you can use them to work out the special trig values that you need to know.
4) Use SOH CAH TOA...

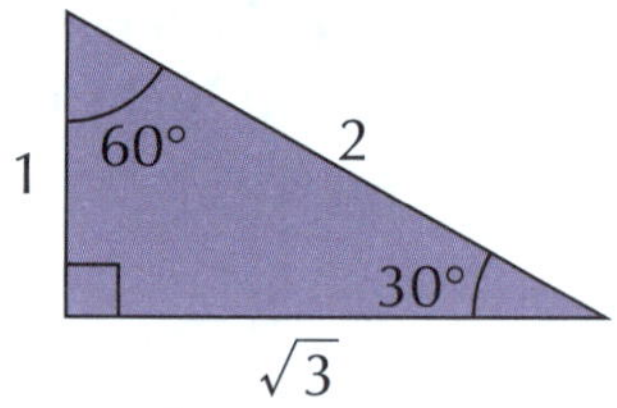

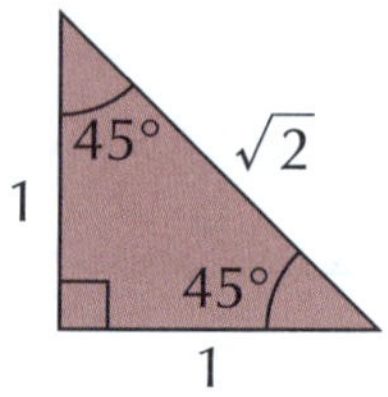

$$\sin x = \frac{\text{opp}}{\text{hyp}} \qquad \cos x = \frac{\text{adj}}{\text{hyp}} \qquad \tan x = \frac{\text{opp}}{\text{adj}}$$

You can use Pythagoras to check that you've got the side lengths right, e.g. $1^2 + (\sqrt{3})^2 = 4 = 2^2$

5) ...to learn these trig values:

$$\sin 30° = \frac{1}{2} \qquad \sin 60° = \frac{\sqrt{3}}{2} \qquad \sin 45° = \frac{1}{\sqrt{2}}$$

$$\cos 30° = \frac{\sqrt{3}}{2} \qquad \cos 60° = \frac{1}{2} \qquad \cos 45° = \frac{1}{\sqrt{2}}$$

$$\tan 30° = \frac{1}{\sqrt{3}} \qquad \tan 60° = \sqrt{3} \qquad \tan 45° = 1$$

$$\sin 0° = 0 \qquad \sin 90° = 1$$

$$\cos 0° = 1 \qquad \cos 90° = 0$$

$$\tan 0° = 0$$

You can't use triangles to work these ones out sadly — you just have to learn them.

EXAMPLES

1. Without using a calculator, find the exact length of side b in the right-angled triangle shown.

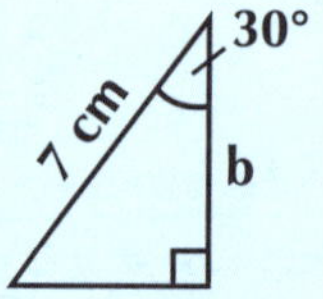

1) It's a right-angled triangle so use SOH CAH TOA to pick the correct trig formula to use.

$C = \frac{A}{H}$

2) Put in the numbers from the diagram in the question.

$\cos 30° = \frac{b}{7}$

3) You know the value of cos 30°, so substitute this in.

$\frac{\sqrt{3}}{2} = \frac{b}{7}$

$b = \frac{7\sqrt{3}}{2}$ cm

2. Without using a calculator, show that $\cos 30° + \tan 30° = \frac{5\sqrt{3}}{6}$

GRADE 8

1) Put the right values into the question.

$\cos 30° + \tan 30° = \frac{\sqrt{3}}{2} + \frac{1}{\sqrt{3}}$

2) Put the values over a common denominator.

$= \frac{\sqrt{3} \times \sqrt{3}}{2\sqrt{3}} + \frac{2}{2\sqrt{3}}$

$= \frac{3+2}{2\sqrt{3}}$

3) Rationalise the denominator — see p.29

$= \frac{5}{2\sqrt{3}}$

$= \frac{5\sqrt{3}}{2\sqrt{3}\sqrt{3}} = \frac{5\sqrt{3}}{6}$

Learn the angles in the green box off by heart

Use the triangles to learn the trig values — then if you're not sure about a trig value in the exam, you can sketch the triangle to check you've got it right. Have a go at this Exam Practice Question.

Q1 Find the exact area of this triangle.

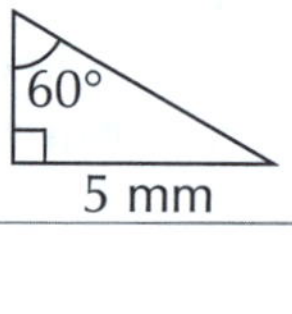

[4 marks] GRADE 7

Q1 Video Solution

Warm-Up and Worked Exam Questions

Learning facts and practising exam questions is the only recipe for success.
That's what the questions on these pages are all about. All you have to do... is do them.

Warm-Up Questions

1) In a right-angled triangle, the two shorter sides are 10 cm and 8.4 cm. Find
   a) the length of the longest side, correct to 3 significant figures.
   b) the smallest angle, correct to the nearest degree.
2) Find the length of x on the triangle to the right without using a calculator.

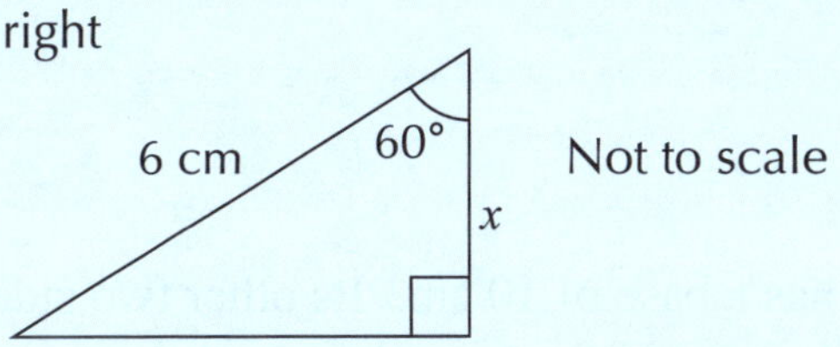

Worked Exam Questions

There's a knack to using the facts you've stored away in your brain box to get marks in the exam.
These worked examples will really help you see how...

1 The diagram shows a right-angled triangle *ABC*.
AC is 4 cm long. *BC* is 8 cm long.

Calculate the length of *AB*.
Give your answer to 2 decimal places.

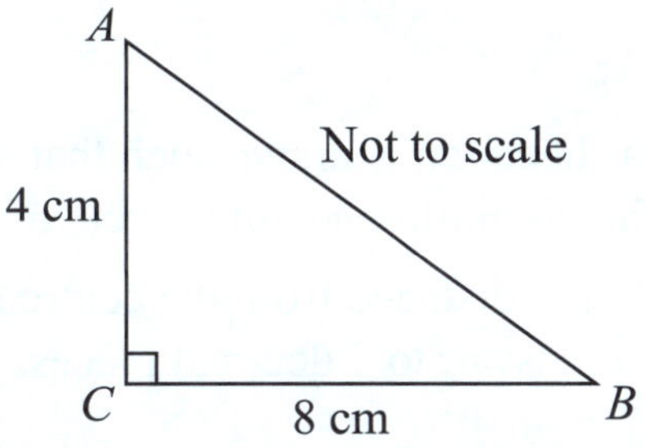

$4^2 + 8^2 = AB^2$

Use Pythagoras' theorem: $a^2 + b^2 = c^2$

$16 + 64 = 80 = AB^2$

$\sqrt{80} = AB$, so $AB = 8.94427... = 8.94$ cm (to 2 d.p.)

..........8.94.......... cm
[3 marks]

2 The diagram shows a right-angled triangle.
Find the size of the angle marked x.
Give your answer to 1 decimal place.

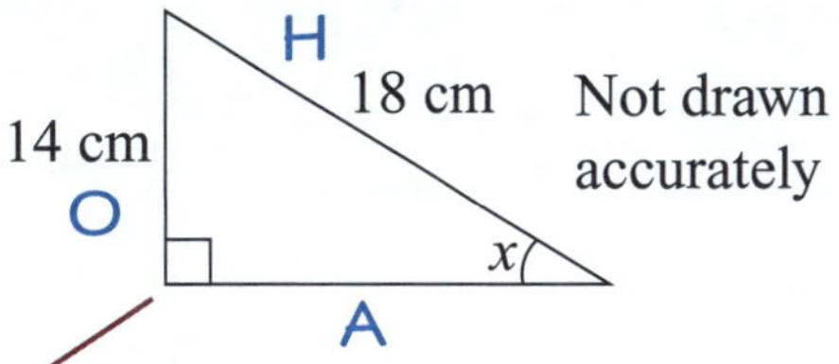

Remember to label your triangle

(SOH) CAH TOA

$S = \frac{O}{H}$

$\sin x = \frac{14}{18}$, so $x = \sin^{-1}(14 \div 18) = 51.05755... = 51.1°$ (to 1 d.p.)

..........51.1.......... °
[2 marks]

Exam Questions

3 Point A has coordinates (2, –1). Point B has coordinates (8, 8).
Find the exact length of the line segment AB. Simplify your answer as much as possible.

....
[3 marks]

4 A triangle has a base of 10 cm. Its other two sides are both 13 cm long.
Calculate the area of the triangle.

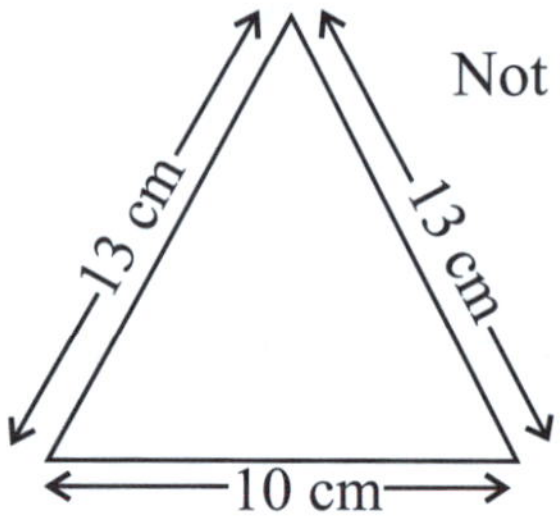

Not to scale

.... cm^2
[4 marks]

5 A regular hexagon is drawn such that all of its vertices are on the circumference of a circle of radius 8.5 cm.
Calculate the distance from the centre of the circle to the centre of one edge of the hexagon.
Give your answer to 2 decimal places.

The sum of interior angles in a polygon = (number of sides – 2) × 180°

.... cm
[5 marks]

6 Show that $\sin 60° + \tan 60° = \frac{3\sqrt{3}}{2}$

[3 marks]

The Sine and Cosine Rules

Normal trigonometry using SOH CAH TOA etc. can only be applied to ***right-angled*** *triangles. Which leaves us with the question of what to do with other-angled triangles. Step forward the* ***Sine and Cosine Rules****...*

Labelling the Triangle

This is very important. You must label the sides and angles properly so that the letters for the sides and angles correspond with each other. Use lower case letters for the sides and capitals for the angles.

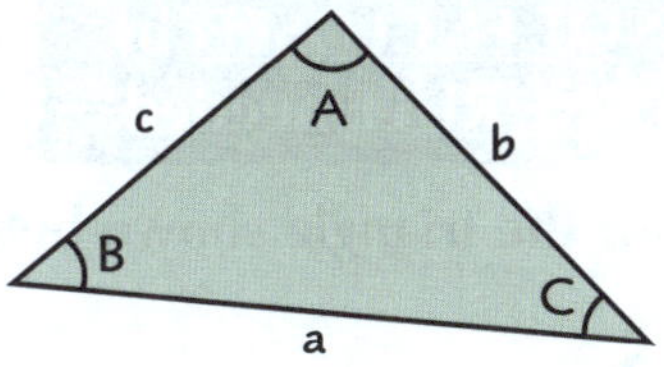

Remember, side 'a' is opposite angle A etc.

It doesn't matter which sides you decide to call a, b and c, just as long as the angles are then labelled properly.

Three Formulas:

The Sine Rule

$$\frac{a}{\sin A} = \frac{b}{\sin B} = \frac{c}{\sin C}$$

You don't use the whole thing with both '=' signs of course, so it's not half as bad as it looks — you just choose the two bits that you want:

e.g. $\frac{b}{\sin B} = \frac{c}{\sin C}$ or $\frac{a}{\sin A} = \frac{b}{\sin B}$

The Cosine Rule

The 'normal' form is...

$$a^2 = b^2 + c^2 - 2bc\cos A$$

...or this form is good for finding an angle (you get it by rearranging the 'normal' version):

$$\text{or} \quad \cos A = \frac{b^2 + c^2 - a^2}{2bc}$$

Area of the Triangle

This formula comes in handy when you know two sides and the angle between them:

$$\text{Area of triangle} = \tfrac{1}{2}ab\sin C$$

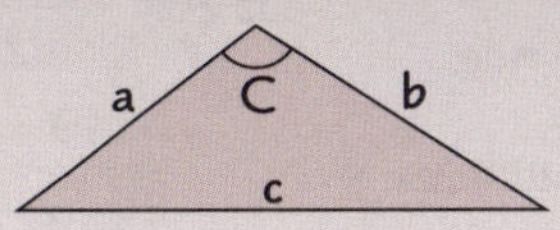

Of course, you already know a simple formula for calculating the area using the base length and height (see p.127). The formula here is for when you don't know those values.

EXAMPLE

Triangle XYZ has XZ = 18 cm, YZ = 13 cm and angle XZY = 58°. Find the area of the triangle, giving your answer correct to 3 significant figures.

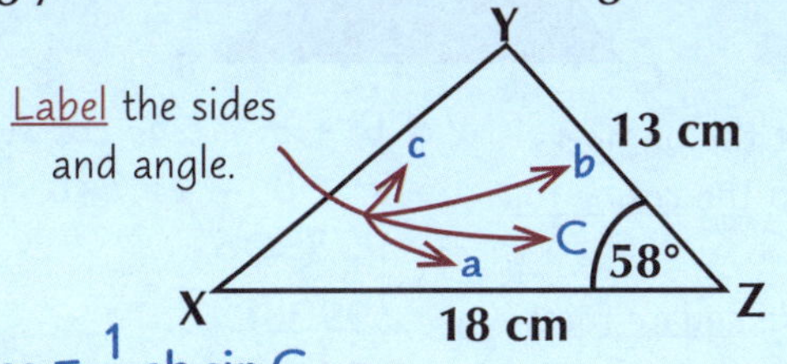

Area $= \frac{1}{2}ab\sin C$

$= \frac{1}{2} \times 18 \times 13 \times \sin 58°$

$= 99.2\ \text{cm}^2$ (3 s.f.)

Don't forget the units.

Make sure you label each side and angle of the triangle correctly

The formulas won't work if your labels don't match up. You'll get these formulas in the exam... but you'll still need to get plenty of practice using them.

Q1 Triangle FGH has FG = 9 cm, FH = 12 cm and angle GFH = 37°. Find its area, giving your answer correct to 3 significant figures. [2 marks]

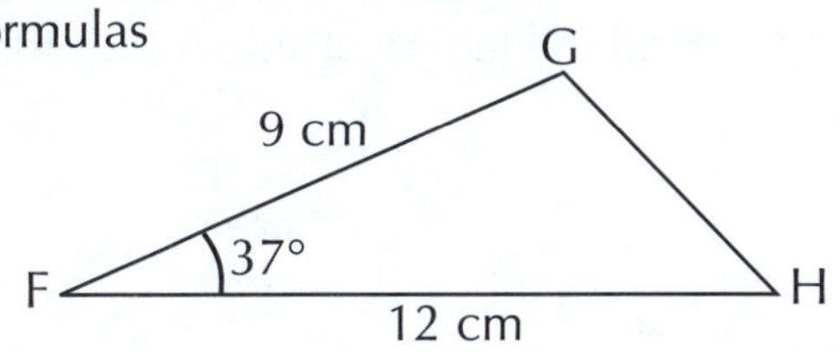

The Sine and Cosine Rules

*There are four main question types where the **sine** and **cosine** rules would be applied. So learn the exact details of these four examples and you'll be laughing.*

The Four **Examples**

Grade 7

1 TWO ANGLES given plus ANY SIDE — SINE RULE needed.

Find the length of AB for the triangle below.

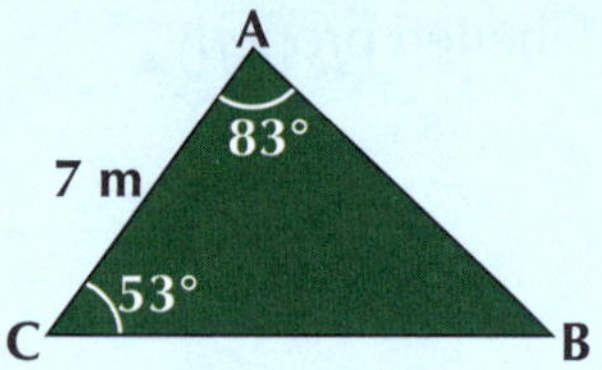

1) Don't forget the obvious... $B = 180° - 83° - 53° = 44°$
2) Put the numbers into the sine rule. $\frac{b}{\sin B} = \frac{c}{\sin C} \Rightarrow \frac{7}{\sin 44°} = \frac{c}{\sin 53°}$
3) Rearrange to find c. $\Rightarrow c = \frac{7 \times \sin 53°}{\sin 44°} = 8.05\text{ m (3 s.f.)}$

2 TWO SIDES given plus an ANGLE NOT ENCLOSED by them — SINE RULE needed.

Find angle ABC for the triangle shown below.

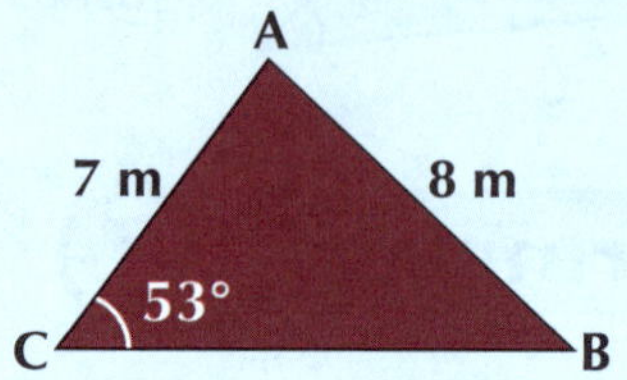

1) Put the numbers into the sine rule. $\frac{b}{\sin B} = \frac{c}{\sin C} \Rightarrow \frac{7}{\sin B} = \frac{8}{\sin 53°}$
2) Rearrange to find sin B. $\Rightarrow \sin B = \frac{7 \times \sin 53°}{8} = 0.6988...$
3) Find the inverse. $\Rightarrow B = \sin^{-1}(0.6988...) = 44.3°\text{ (1 d.p.)}$

3 TWO SIDES given plus the ANGLE ENCLOSED by them — COSINE RULE needed.

Find the length CB for the triangle shown below.

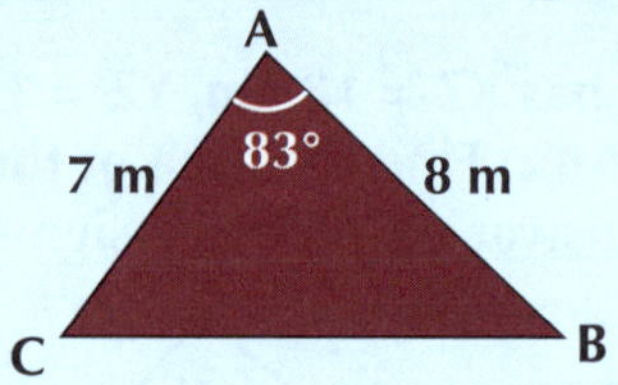

1) Put the numbers into the cosine rule. $a^2 = b^2 + c^2 - 2bc \cos A$
$= 7^2 + 8^2 - 2 \times 7 \times 8 \times \cos 83°$
$= 99.3506...$
2) Take square roots to find a. $a = \sqrt{99.3506...}$
$= 9.97\text{ m (3 s.f.)}$

You might come across a triangle that isn't labelled ABC — just relabel it yourself to match the sine and cosine rules.

4 ALL THREE SIDES given but NO ANGLES — COSINE RULE needed.

Find angle CAB for the triangle shown.

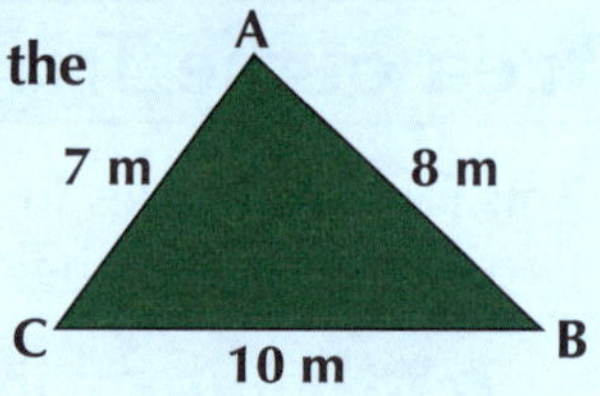

1) Use this version of the cosine rule. $\cos A = \frac{b^2 + c^2 - a^2}{2bc}$
2) Put in the numbers. $= \frac{49 + 64 - 100}{2 \times 7 \times 8}$
3) Take the inverse to find A. $= \frac{13}{112} = 0.11607...$
$\Rightarrow A = \cos^{-1}(0.11607...)$
$= 83.3°\text{ (1 d.p.)}$

Learn which rule you need for each question type

You need to get really good at spotting which of the four methods to use, so try these practice questions.

Q1 Find the length of side AB for triangle ABC.

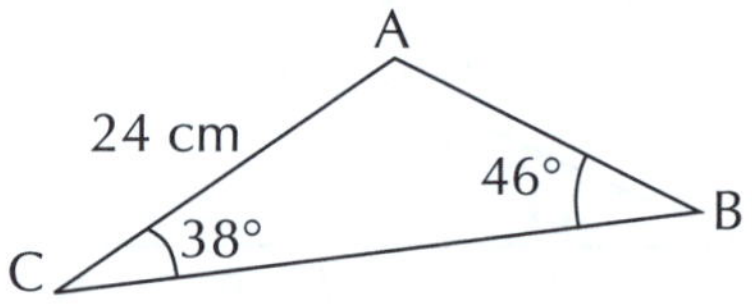

[3 marks] Grade 7

Q2 Find the size of angle RPQ for triangle PQR.

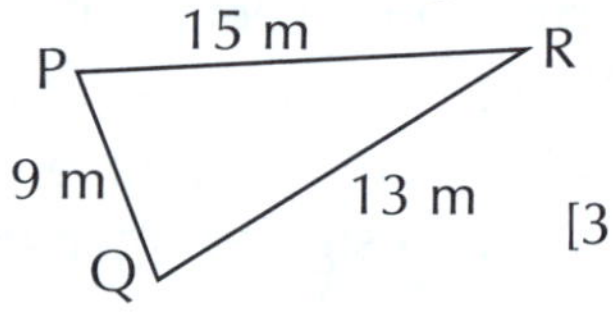

[3 marks] Grade 7

Q2 Video Solution

3D Pythagoras

*This is a 3D version of the 2D Pythagoras' theorem you saw on page 145. There's just **one simple formula** — learn it and the world's your oyster...*

3D Pythagoras for Cuboids — $a^2 + b^2 + c^2 = d^2$

Grade 7

Cuboids have their own formula for calculating the length of their longest diagonal:

$$a^2 + b^2 + c^2 = d^2$$

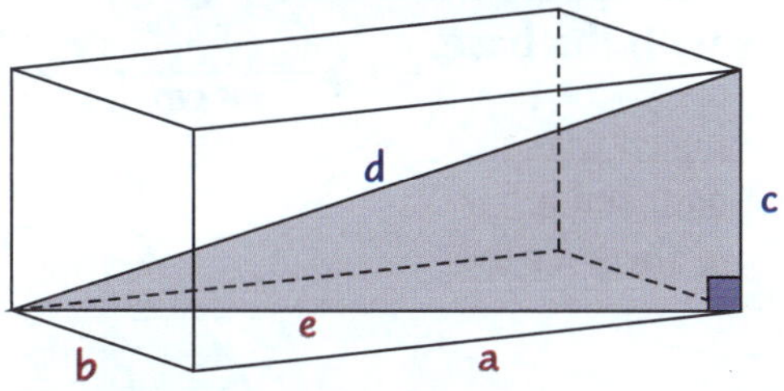

In reality it's nothing you haven't seen before — it's just 2D Pythagoras' theorem being used twice:

1) a, b and e make a right-angled triangle so

$$e^2 = a^2 + b^2$$

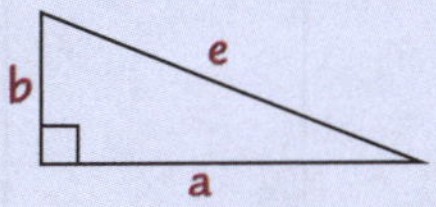

2) Now look at the right-angled triangle formed by e, c and d:

$$d^2 = e^2 + c^2 = a^2 + b^2 + c^2$$

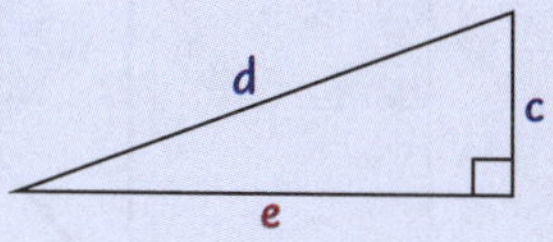

EXAMPLE **Find the exact length of the diagonal BH for the cube in the diagram.**

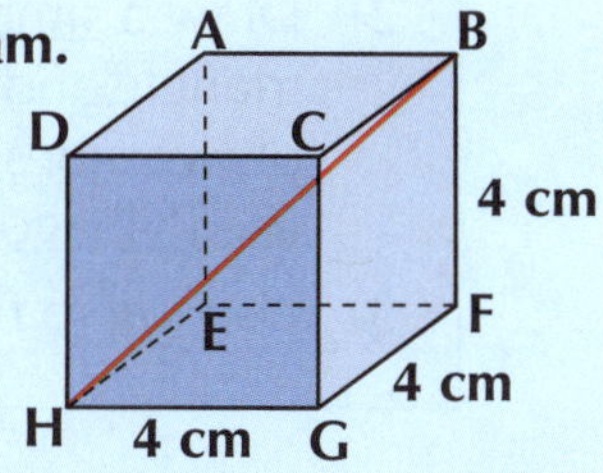

1) Write down the formula. $a^2 + b^2 + c^2 = d^2$

2) Put in the numbers. $4^2 + 4^2 + 4^2 = BH^2$

3) Take the square root to find BH. $\Rightarrow BH = \sqrt{48} = 4\sqrt{3}$ cm

The Cuboid Formula can be used in Other 3D Shapes

Grade 8

EXAMPLE **In the square-based pyramid shown, M is the midpoint of the base. Find the vertical height AM.**

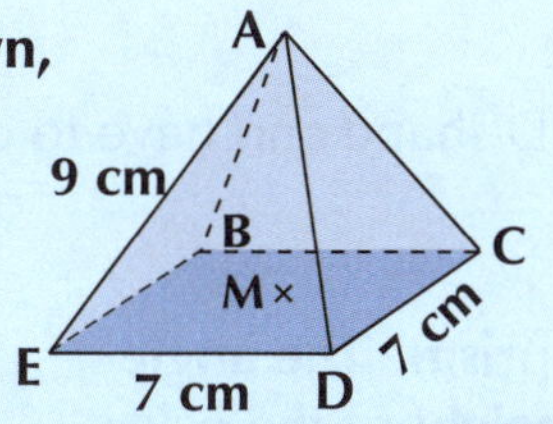

1) Label N as the midpoint of ED.
Then think of EN, NM and AM as three edges of a cuboid, and AE as the longest diagonal in the cuboid (like d in the section above).

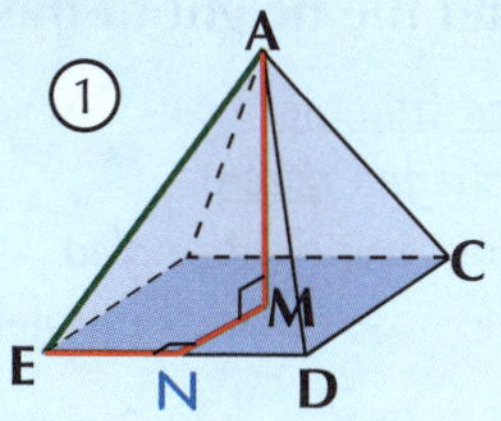

2) Sketch the full cuboid.

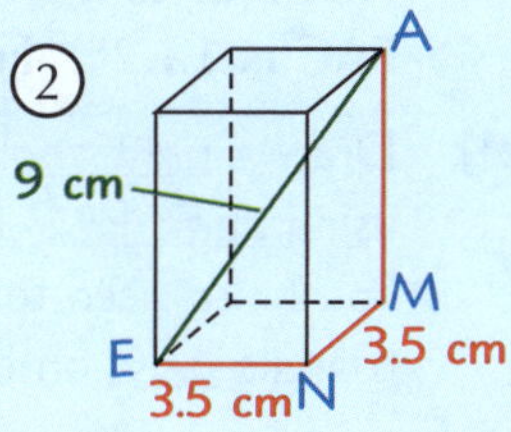

3) Write down the 3D Pythagoras formula. $a^2 + b^2 + c^2 = d^2$

4) Rewrite it using edge labels. $EN^2 + NM^2 + AM^2 = AE^2$

5) Put in the numbers and solve for AM. $\Rightarrow 3.5^2 + 3.5^2 + AM^2 = 9^2$

$\Rightarrow AM = \sqrt{81 - 2 \times 12.25} = 7.52$ cm (3 s.f.)

$a^2 + b^2 + c^2 = d^2$ gives you the longest diagonal from the 3 edge lengths

Finding the length of the longest diagonal of a cuboid is pretty easy as long as you learn the formula. Other 3D shapes can be a little more tricky, but really you just need to work out where the formula fits.

Q1 Find the length AH in the cuboid shown to 3 s.f.

[3 marks] Grade 7

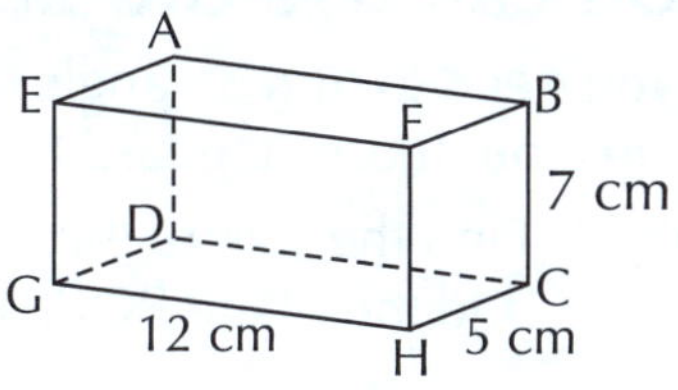

3D Trigonometry

3D trig may sound tricky, and I suppose it is a bit... but it's actually just using the ***same old rules****.*

Angle Between Line and Plane — Use a Diagram

GRADE 8

Learn the 3-Step Method

1) Make a right-angled triangle between the line and the plane.

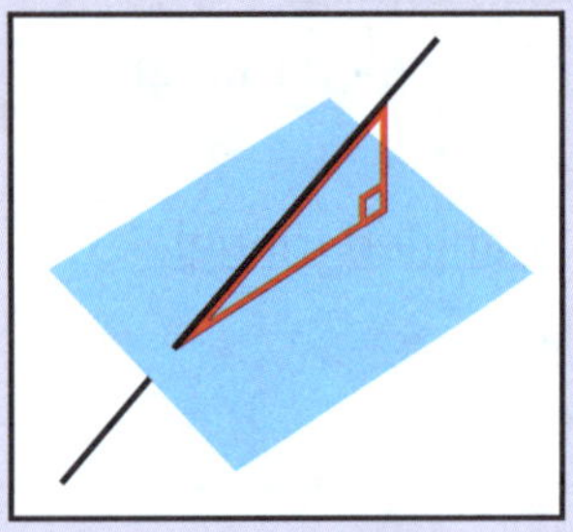

2) Draw a simple 2D sketch of this triangle and mark on the lengths of two sides (you might have to use Pythagoras to find one).
3) Use trig to find the angle.

Have a look at p.145-148 to jog your memory about Pythagoras and trig.

EXAMPLE

ABCDE is a square-based pyramid with M as the midpoint of its base. Find the angle the edge AE makes with the base.

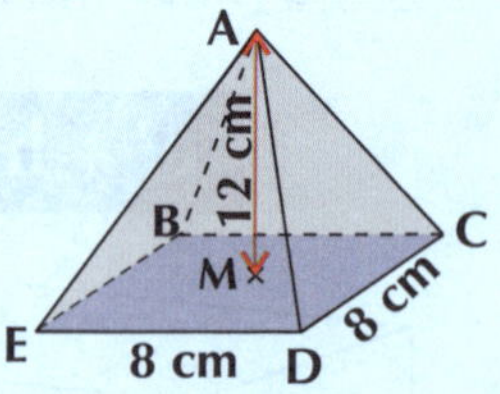

1) Draw a right-angled triangle using AE, the base and a line between the two (here it's the vertical height).

   Label the angle you need to find.

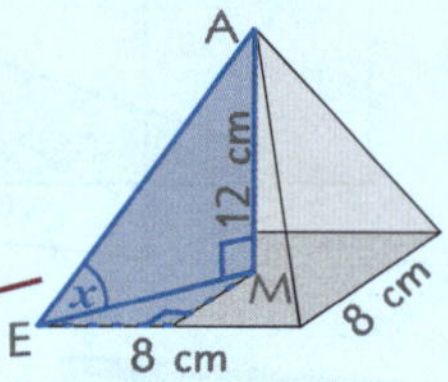

2) Now sketch this triangle in 2D and label it.

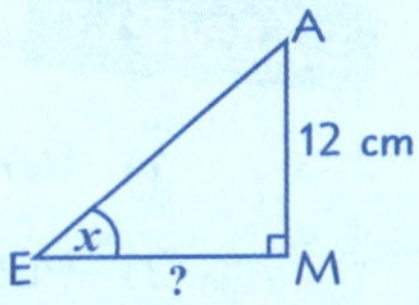

   Use Pythagoras (on the base triangle) to find EM.

   $EM^2 = 4^2 + 4^2 = 32$

   $\Rightarrow EM = \sqrt{32}$ cm

3) Finally, use trigonometry to find x — you know the opposite and adjacent sides so use tan.

   $\tan x = \dfrac{12}{\sqrt{32}} = 2.1213...$

   $x = \tan^{-1}(2.1213...)$

   $= 64.8°$ (1 d.p.)

You may be given the angle in a 3D shape and have to use it to find a length or other quantity.

EXAMPLE

ABCDEF is a triangular prism. The angle FAC is 13.2°. Find the height of the prism.

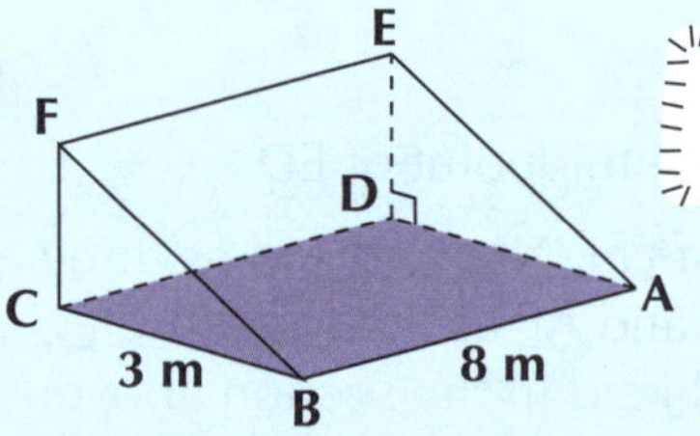

The length of CF is the quantity you need to find.

1) Draw a right-angled triangle using angle FAC (shaded green). You'll also need to use the right-angled triangle ABC on the base (shaded purple).

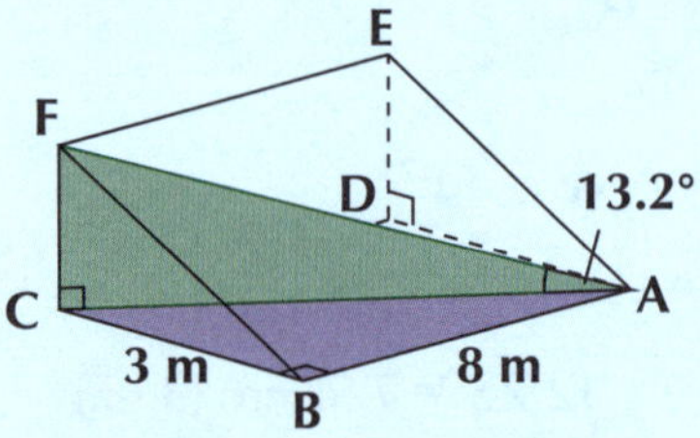

2) Use Pythagoras on the purple triangle to find AC.

   $AC^2 = 3^2 + 8^2 = 73$

   $\Rightarrow AC = \sqrt{73}$ m

3) Use SOH CAH TOA on the green triangle to find the height.

   $T = \dfrac{O}{A}$ $\tan 13.2° = \dfrac{CF}{\sqrt{73}}$

   $CF = \sqrt{73}\tan 13.2$

   $= 2.0$ m (1 d.p.)

You can use the 2D trigonometry rules in 3D shapes too

If you need to find an angle in a 3D question, don't panic — just put those standard trig formulas to work.

Q1 Find the size of the angle between the line PV and the plane PQRS in the cuboid shown.

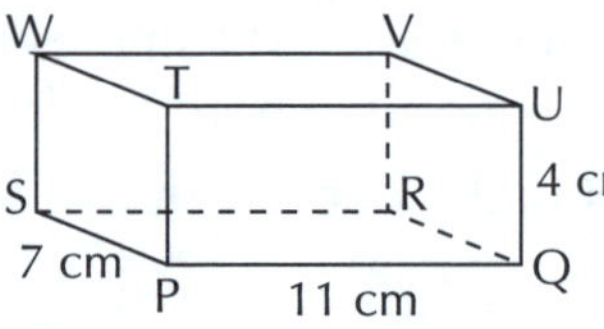

[4 marks] GRADE 8

Q1 Video Solution

Vectors

*Vectors represent a movement of a certain **size** in a certain **direction**.*
They might seem a bit weird at first, but there are really just a few facts to get to grips with...

The Vector **Notations**

GRADE 6

There are several ways to write vectors...

1) Column vectors: $\begin{pmatrix} 2 \\ -5 \end{pmatrix}$ — 2 units right, 5 units down; $\begin{pmatrix} -7 \\ 4 \end{pmatrix}$ — 7 units left, 4 units up
2) **a** —— exam questions use bold like this
3) $\underset{\sim}{a}$ or $\underline{a}$ — you should always underline them
4) $\overrightarrow{AB}$ —— this means the vector from point A to point B

They're represented on a diagram by an arrow.

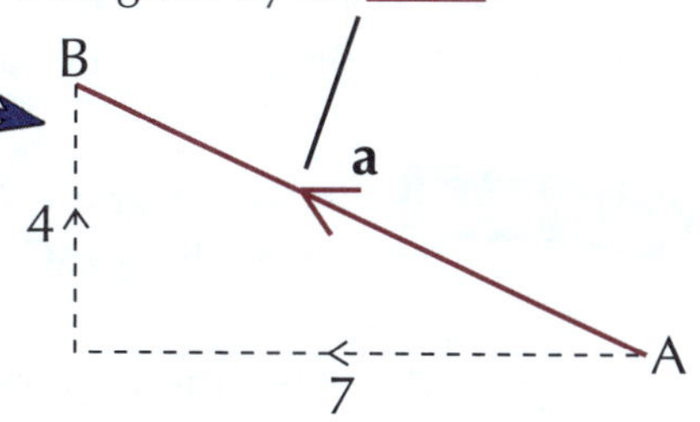

Multiplying a Vector **by a Scalar**

GRADE 6

Scalars are just normal numbers (i.e. not vectors).

Multiplying a vector by a positive number changes the vector's size but not its direction — it scales the vector. If the number's negative then the direction gets switched.

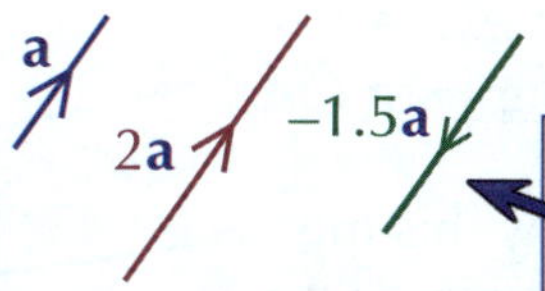

Vectors that are scalar multiples of each other are parallel.

Adding and **Subtracting** Vectors

GRADE 6

You can describe movements between points by adding and subtracting known vectors. Loads of vector exam questions are based around this.

"$\underset{\sim}{a} + \underset{\sim}{b}$" means 'go along $\underset{\sim}{a}$ then $\underset{\sim}{b}$'.

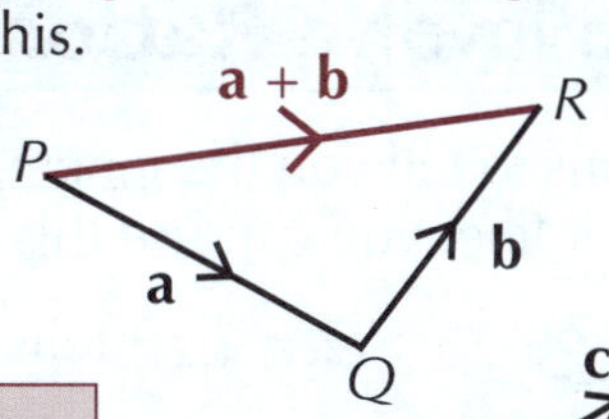

In the diagrams, $\overrightarrow{PR} = \underset{\sim}{a} + \underset{\sim}{b}$ and $\overrightarrow{XZ} = \underset{\sim}{c} - \underset{\sim}{d}$.

"$\underset{\sim}{c} - \underset{\sim}{d}$" means 'go along $\underset{\sim}{c}$ then backwards along $\underset{\sim}{d}$' (the minus sign means go the opposite way).

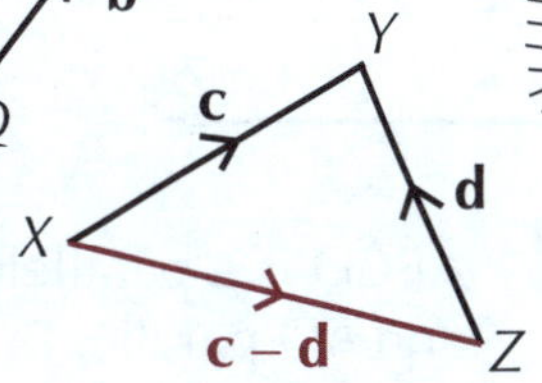

When adding column vectors, add the top to the top and the bottom to the bottom. The same goes when subtracting. $\begin{pmatrix} 3 \\ -1 \end{pmatrix} + \begin{pmatrix} 5 \\ 3 \end{pmatrix} = \begin{pmatrix} 8 \\ 2 \end{pmatrix}$

EXAMPLE

In the diagram below, M is the midpoint of BC.
Find vectors $\overrightarrow{AM}$, $\overrightarrow{OC}$ and $\overrightarrow{AC}$ in terms of **a**, **b** and **m**.

To obtain the unknown vector just 'get there' by any route made up of known vectors.

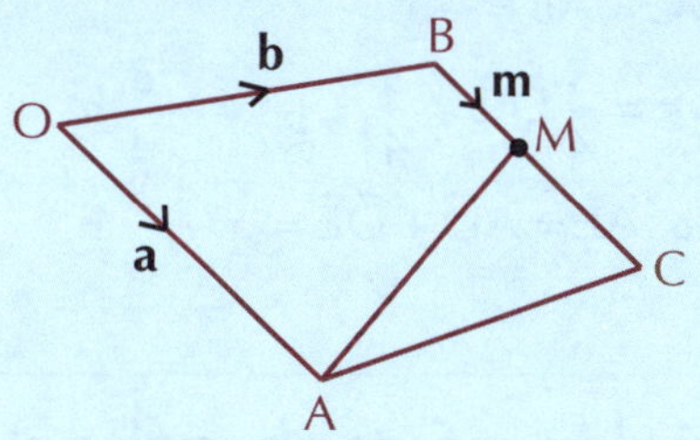

$\overrightarrow{AM} = -\underset{\sim}{a} + \underset{\sim}{b} + \underset{\sim}{m}$ —— A to M via O and B

$\overrightarrow{OC} = \underset{\sim}{b} + 2\underset{\sim}{m}$ —— O to C via B and M — M is halfway between B and C so $\overrightarrow{BC} = 2\underset{\sim}{m}$

$\overrightarrow{AC} = -\underset{\sim}{a} + \underset{\sim}{b} + 2\underset{\sim}{m}$ —— A to C via O, B and M

That's three vital vector facts done

You need to get to grips with questions like the one above, so here's one to have a go at...

Q1 In triangle ABC, M is the midpoint of BC and N is the midpoint of AB. $\overrightarrow{AC} = \mathbf{p}$ and $\overrightarrow{BM} = \mathbf{q}$.
Find $\overrightarrow{AB}$ and $\overrightarrow{NA}$ in terms of **p** and **q**. [3 marks] GRADE 7

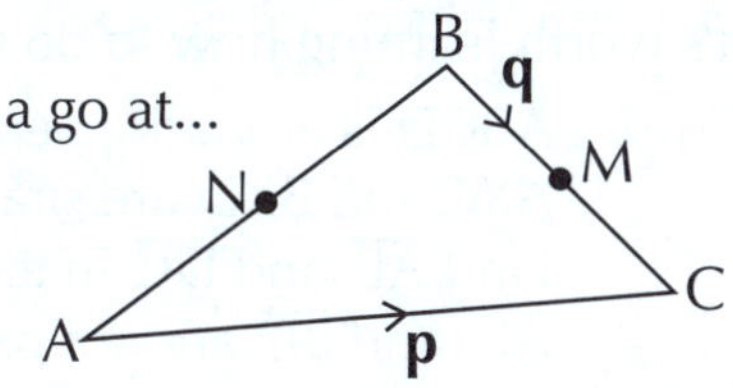

Vectors

Extra bits and pieces can crop up in vector questions — these examples will show you how to tackle them...

Vectors Along a Straight Line

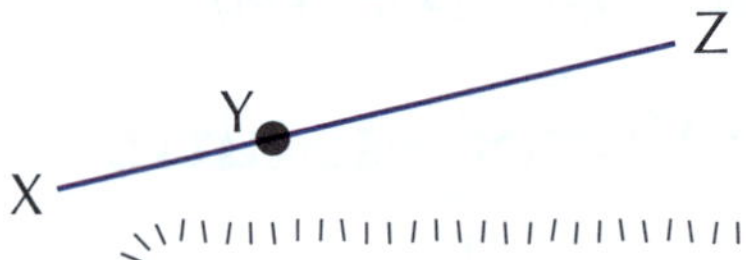

If XYZ is a straight line then $\overrightarrow{XY}$ must be a scalar multiple of $\overrightarrow{YZ}$.

1) You can use vectors to show that points lie on a straight line.
2) You need to show that the vectors along each part of the line point in the same direction — i.e. they're scalar multiples of each other.

EXAMPLE

In the diagram, $\overrightarrow{OB} = \mathbf{a}$, $\overrightarrow{AB} = 2\mathbf{b}$, $\overrightarrow{BD} = \mathbf{a} - \mathbf{b}$ and $\overrightarrow{DC} = \frac{1}{2}\mathbf{a} - 4\mathbf{b}$.
Show that OAC is a straight line.

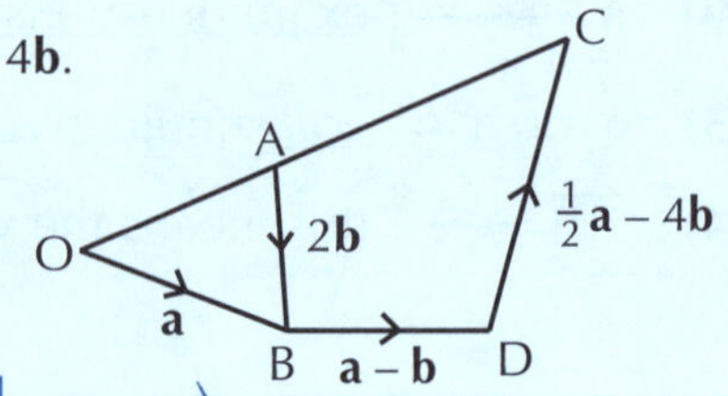

1) Work out the vectors along the two parts of OAC (OA and AC) using the vectors you know.

$\overrightarrow{OA} = \underset{\sim}{a} - 2\underset{\sim}{b}$

$\overrightarrow{AC} = 2\underset{\sim}{b} + (\underset{\sim}{a} - \underset{\sim}{b}) + \left(\frac{1}{2}\underset{\sim}{a} - 4\underset{\sim}{b}\right)$

2) Check that $\overrightarrow{AC}$ is a scalar multiple of $\overrightarrow{OA}$.

$= \frac{3}{2}\underset{\sim}{a} - 3\underset{\sim}{b} = \frac{3}{2}(\underset{\sim}{a} - 2\underset{\sim}{b})$

So, $\overrightarrow{AC} = \frac{3}{2}\overrightarrow{OA}$.

3) Explain why this means OAC is a straight line.

$\overrightarrow{AC}$ is a scalar multiple of $\overrightarrow{OA}$, so OAC must be a straight line.

Vector Questions Can Involve Ratios

Ratios are used in vector questions to tell you the lengths of different sections of a straight line. If you know the vector along part of that line, you can use this information to find other vectors along the line.

E.g.

XY : YZ = 2 : 3 tells you that $\overrightarrow{XY} = \frac{2}{5}\overrightarrow{XZ}$ and $\overrightarrow{YZ} = \frac{3}{5}\overrightarrow{XZ}$.

EXAMPLE

ABCD is a parallelogram, with AB parallel to DC and AD parallel to BC.
Point E lies on DC, such that DE : EC = 3 : 1.
$\overrightarrow{BC} = \mathbf{a}$ and $\overrightarrow{BA} = \mathbf{b}$.
Find $\overrightarrow{AE}$ in terms of **a** and **b**.

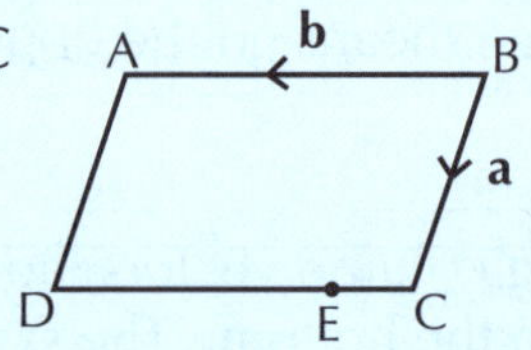

1) Write $\overrightarrow{AE}$ as a route along the parallelogram.

$\overrightarrow{AE} = \overrightarrow{AD} + \overrightarrow{DE}$

2) Use the parallel sides to find $\overrightarrow{AD}$ and $\overrightarrow{DC}$.

$\overrightarrow{AD} = \overrightarrow{BC} = \underset{\sim}{a}$

$\overrightarrow{DC} = \overrightarrow{AB} = -\underset{\sim}{b}$

3) Use the ratio to find $\overrightarrow{DE}$.

$\overrightarrow{DE} = \frac{3}{4}\overrightarrow{DC} = \frac{3}{4}(-\underset{\sim}{b}) = -\frac{3}{4}\underset{\sim}{b}$

4) Now use $\overrightarrow{AD}$ and $\overrightarrow{DE}$ to find $\overrightarrow{AE}$.

So $\overrightarrow{AE} = \overrightarrow{AD} + \overrightarrow{DE} = \underset{\sim}{a} - \frac{3}{4}\underset{\sim}{b}$

Remember — parallel vectors are scalar multiples of each other

For these types of questions, just use the vectors you're given to find what you're asked for.
It's worth learning how to do these questions, because you could get very similar ones in the exam.

Q1 Video Solution

Q1 ABCD is a quadrilateral. $\overrightarrow{AX} = \mathbf{a}$ and $\overrightarrow{BX} = \mathbf{b}$.
AXC and BXD are straight lines, with AX : XC = BX : XD = 2 : 3.
Find $\overrightarrow{AB}$ and $\overrightarrow{DC}$ in terms of **a** and **b**.
$\overrightarrow{AD}$ and $\overrightarrow{BC}$ are not parallel. What sort of quadrilateral is ABCD?

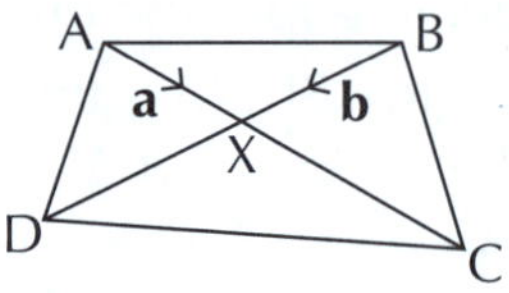

[6 marks]

Warm-Up and Worked Exam Questions

Trigonometry and vector questions can be pretty tricky until you get your head around the basics. That's what these warm-up questions are all about — work through them carefully and check any bits you don't know.

Warm-Up Questions

1) In the triangle on the right, find the length of AC, correct to 1 decimal place.

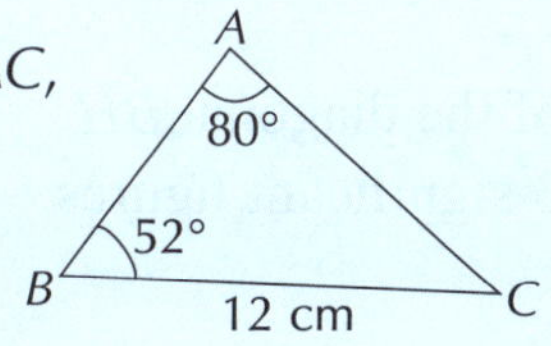

2) A triangle has sides of 4 cm, 6 cm and 8 cm. Calculate the largest angle, correct to 1 d.p.

3) The diagram to the right shows a cuboid $ABCDEFGH$. $FG = 5$ cm, $CD = 2$ cm and $CG = 8$ cm. Calculate the size of the angle FDG.

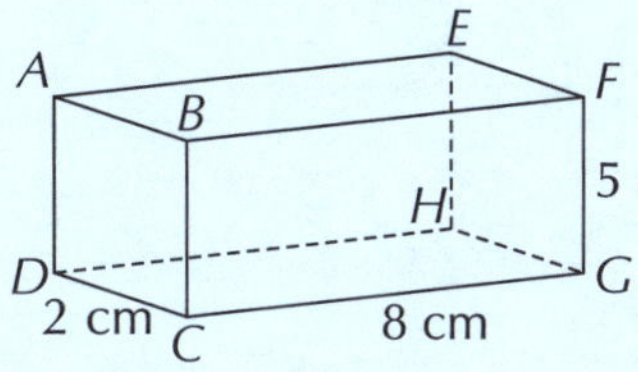

4) $ABCD$ is a parallelogram. $\overrightarrow{AB} = 2\mathbf{a}$ and $\overrightarrow{AD} = 2\mathbf{d}$. L is the midpoint of AC, and M is the midpoint of BC. Write each of the following in terms of $\mathbf{a}$ and $\mathbf{d}$.

a) $\overrightarrow{CD}$ b) $\overrightarrow{AC}$ c) $\overrightarrow{BL}$

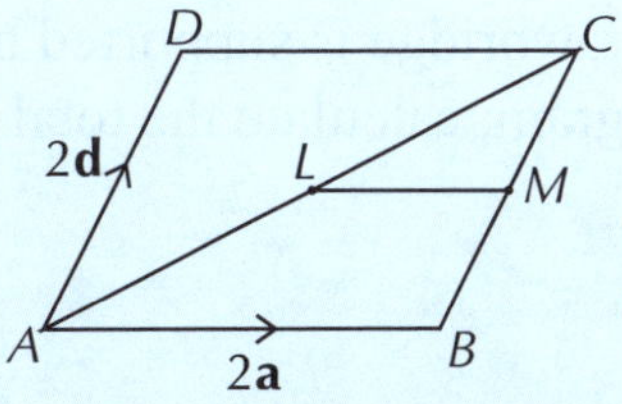

Worked Exam Question

Take the time to go through this example and make sure you understand it all.
If any of the facts are baffling you, it's not too late to take another peek over the section.

1 In the triangle below, $AB = 10$ cm, $BC = 7$ cm and angle $ABC = 85°$. (GRADE 7)

Label the sides of the triangle.

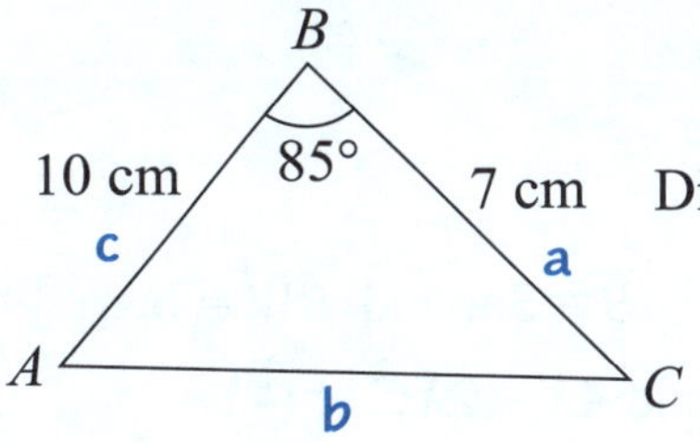

Diagram not accurately drawn

a) Calculate the length of AC.
Give your answer to 3 significant figures.

Use the cosine rule to find b:

$b^2 = a^2 + c^2 - 2ac \cos B$,

$AC^2 = 7^2 + 10^2 - (2 \times 7 \times 10 \times \cos 85°)$

$AC = \sqrt{149 - 140 \times \cos 85°} = 11.69607... = 11.7$ cm (to 3 s.f.)

11.7 cm
[2 marks]

b) Calculate the area of triangle ABC.
Give your answer to 3 significant figures.

You know the length of two sides and the angle between them, so use the area formula from p.151.

$\text{Area} = \frac{1}{2}ac \sin B$

$= \frac{1}{2} \times 7 \times 10 \times \sin 85°$

$= 34.86681... = 34.9$ cm^2 (to 3 s.f)

34.9 cm^2
[2 marks]

Exam Questions

2 The diagram on the right is a cuboid *ABCDEFGH*. (Grade 7)

The cuboid has edges of length 6 cm, 4 cm and 3 cm.

Calculate the length of the diagonal *BH*.
Give your answer to 3 significant figures.

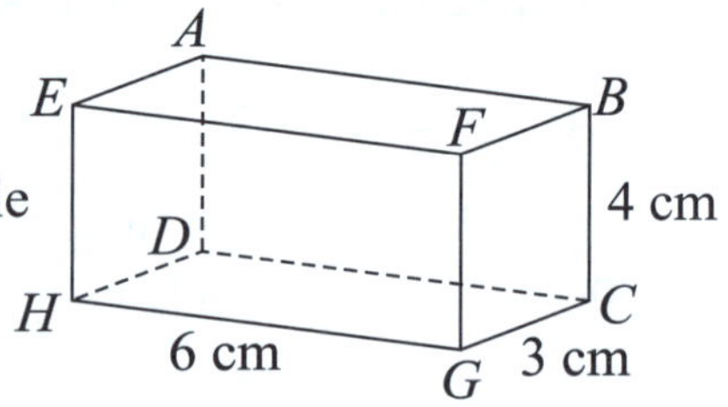

.... cm
[3 marks]

3 A castle drawbridge is supported by two chains, *AB* and *AC*. Using the information on the diagram, calculate the total length of the drawbridge, *BD*, correct to 3 s.f. (Grade 8)

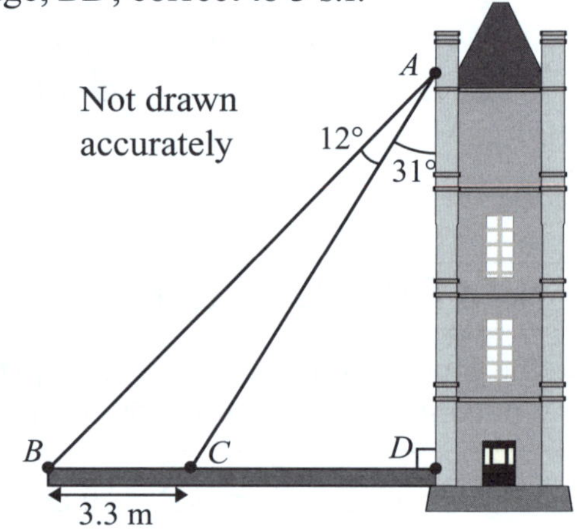

.... m
[5 marks]

4 *ABCD* is a parallelogram. $\overrightarrow{AB} = 3\mathbf{a}$, and $\overrightarrow{BW} = \mathbf{b}$.
M is the midpoint of *CD* and $AX = 2XC$. (Grade 9)
$BW : WC = 1 : 5$

a) Find $\overrightarrow{BX}$ in terms of **a** and **b**.

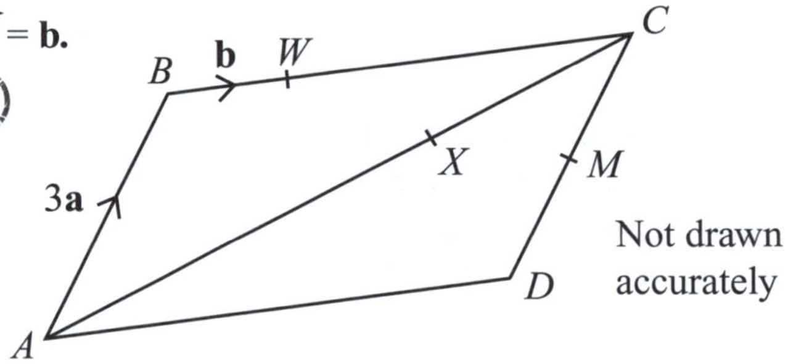

....
[4 marks]

b) Hence show that *B*, *X* and *M* are three points on a straight line.

[4 marks]

Revision Questions for Section Six

It's time to check you've got all these facts and formulas sorted...

- Try these questions and tick off each one when you get it right.
- When you're completely happy with a topic, tick that off too.

For even more practice, try the quiz for Section Six — just scan this QR code!

Section Six Quiz

Pythagoras' Theorem (p145)

1) What is the formula for Pythagoras' theorem? What do you use it for?
2) A museum has a flight of stairs up to its front door (see diagram). A ramp is to be put over the top of the steps for wheelchair users. Calculate the length that the ramp would need to be to 3 s.f.

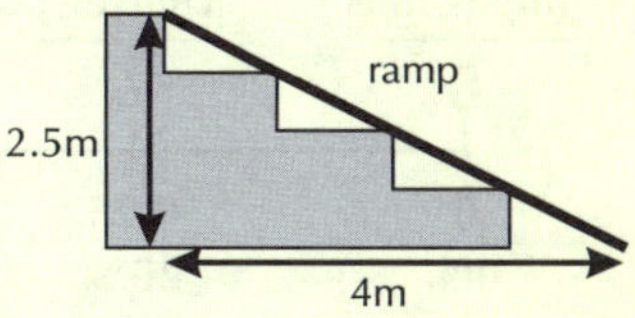

3) Point P has coordinates (–3, –2) and point Q has coordinates (2, 4). Calculate the length of the line PQ to 1 d.p.

Trigonometry — Sin, Cos, Tan (p146-148)

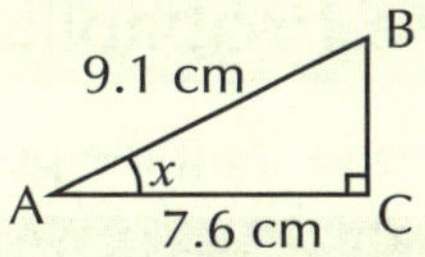

4) Write down the three trigonometry formula triangles.
5) Find the size of angle x in triangle ABC to 1 d.p.

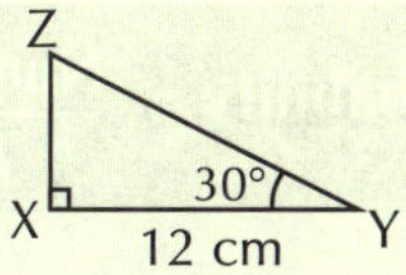

6) Draw two triangles and use them to write down the values of sin, cos and tan for 30°, 60° and 45°.
7) Find the exact length of side XZ in triangle XYZ.

The Sine and Cosine Rules (p151-152)

8) Write down the sine and cosine rules and the formula (involving sin) for the area of any triangle.
9) List the 4 different types of sine/cosine rule questions and which rule you need for each.
10) Triangle JKL has side JK = 7 cm, side JL = 11 cm and angle JLK = 32°. Find angle JKL.
11) In triangle FGH side FH = 8 cm, side GH = 9 cm and angle FHG = 47°. Find the length of side FG.
12) Triangle PQR has side PQ = 12 cm, side QR = 9 cm and angle PQR = 63°. Find its area.
13) WXYZ is a quadrilateral.
    a) Find the length of side XY to 3 s.f.
    b) Find the area of the quadrilateral to 3 s.f.

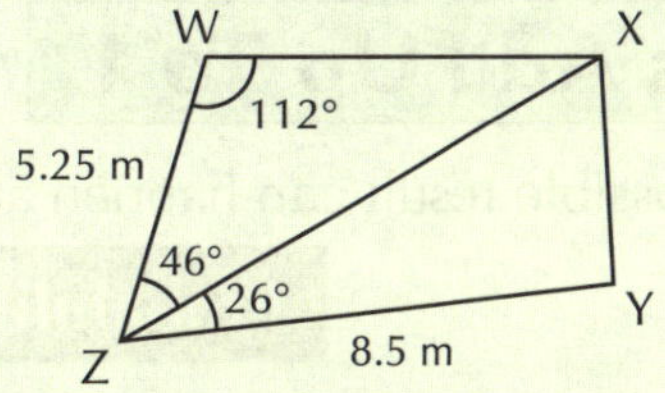

3D Pythagoras (p153)

14) What is the formula for finding the length of the longest diagonal in a cuboid?
15) Find the length of the longest diagonal in the cuboid measuring 5 m × 6 m × 9 m.

3D Trigonometry (p154)

16) Find the angle between the line BH and the plane ABCD in this cuboid.

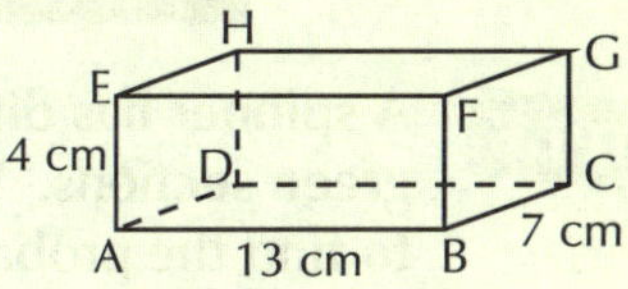

17) Find the size of angle WQU in the cuboid on the right to 1 d.p.

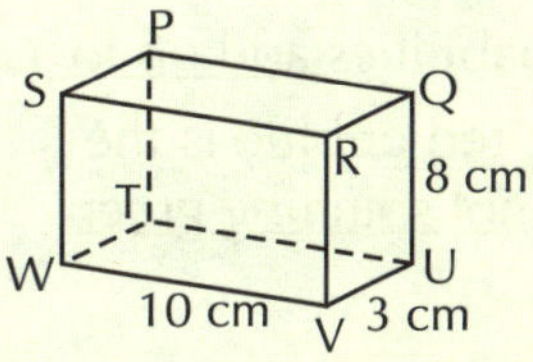

Vectors (p155-156)

18) What is the effect of multiplying a vector by a scalar?
19) $\underset{\sim}{a}$ and $\underset{\sim}{b}$ are column vectors, where $\underset{\sim}{a} = \begin{pmatrix}4\\-2\end{pmatrix}$ and $\underset{\sim}{b} = \begin{pmatrix}7\\6\end{pmatrix}$.
    a) Find $\underset{\sim}{a} - \underset{\sim}{b}$
    b) Find $5\underset{\sim}{a}$
    c) Find $3\underset{\sim}{a} + \underset{\sim}{b}$
    d) Find $-4\underset{\sim}{a} - 2\underset{\sim}{b}$
20) ABCD is a quadrilateral. AXC is a straight line with AX : XC = 1 : 3.
    a) Find $\overrightarrow{AX}$.
    b) Find $\overrightarrow{DX}$ and $\overrightarrow{XB}$.
    c) Is DXB a straight line? Explain your answer.

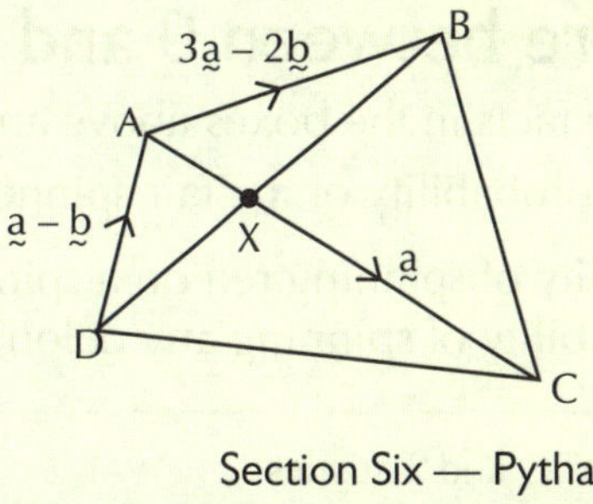

Probability Basics

A lot of people reckon ***probability*** *is pretty tough. But learn the* ***basics*** *well, and it'll all make sense.*

All **Probabilities** are **Between 0 and 1**

1) Probabilities are always between 0 and 1. The higher the probability of something, the more likely it is.
2) A probability of ZERO means it will NEVER HAPPEN and a probability of ONE means it DEFINITELY WILL.

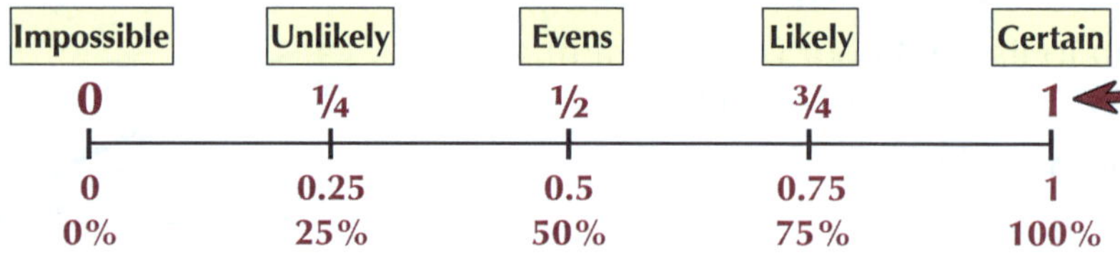

Probabilities can be given as fractions, decimals or percentages.

You Can Find **Some** Probabilities Using a **Formula**

Careful... this formula only works if all the possible outcomes (things that could happen) are equally likely.

$$\text{Probability} = \frac{\text{Number of ways for something to happen}}{\text{Total number of possible outcomes}}$$

Words like 'fair' and 'at random' show possible outcomes are all equally likely. 'Biased' and 'unfair' mean the opposite.

EXAMPLE **Work out the probability of randomly picking a letter 'P' from the tiles below.**

A P P L E P I E

1) There are 3 P's — so there are 3 different ways to 'pick a letter P'.
2) And there are 8 tiles altogether — each of these is a possible outcome.

$$\text{Probability} = \frac{\text{number of ways to pick a P}}{\text{total number of possible outcomes}} = \frac{3}{8} \text{ (or 0.375)}$$

Probabilities **Add Up To 1**

1) If only one possible result can happen at a time, then the probabilities of all the results add up to 1.

Probabilities always ADD UP to 1

2) So since something must either happen or not happen (i.e. only one of these can happen at a time):

P(event happens) + P(event doesn't happen) = 1

EXAMPLE **A spinner has different numbers of red, blue and green sections. Work out the value of x and use it to find the probability of spinning red or blue.**

Colour	red	blue	green
Probability	$3x$	$2x$	$5x$

1) The probabilities add up to 1. $3x + 2x + 5x = 1$ so $10x = 1$ and so $x = 0.1$
2) Spinning red or blue is the same as not spinning green. P(red or blue) = 1 – P(green) $= 1 - (5 \times 0.1) = 0.5$

'P(result)' just means the probability of that result.

Probabilities are between 0 and 1

You need to know the facts in the boxes above and how to use them. Now try these practice questions.

Q1 Calculate the probability of the fair spinner on the right landing on 4. [2 marks]

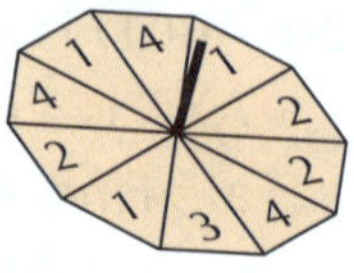

Q2 If the probability of spinning red on a spinner is $1 - 3x$, find the probability of spinning any colour except red. [1 mark]

Counting Outcomes

With a lot of probability questions, a good place to start is with a list of all the ***possible outcomes****.*
Once you've got a ***list of outcomes****, the rest of the question should be straightforward.*

Listing **All Outcomes**

GRADE 4

A sample space diagram shows all the possible outcomes. It can be a simple list, but a two-way table works well if there are two activities going on (e.g. two coins being tossed, or a dice being thrown and a spinner being spun).

EXAMPLE **The spinners on the right are spun, and the scores added together.**

a) Make a sample space diagram showing all the possible outcomes.

1) All the scores from one spinner go along the top. All the scores from the other spinner go down the side.
2) Add the two scores together to get the different possible totals (the outcomes).

+	3	4	5
1	4	5	6
2	5	6	7
3	6	7	8

There are 9 outcomes here — even though some of the actual totals are repeated.

b) Find the probability of spinning a total of 6.

There are 9 possible outcomes altogether, and 3 ways to score 6.

$$P(\text{total} = 6) = \frac{\text{number of ways to score 6}}{\text{total number of possible outcomes}} = \frac{3}{9} = \frac{1}{3}$$

Use the **Product Rule** to **Count Outcomes**

GRADE 6

1) Sometimes it'll be difficult to list all the outcomes (e.g. if the number of outcomes is large or if there are more than two activities going on).
2) Luckily, you can count outcomes using the product rule.

The number of ways to carry out a combination of activities equals the number of ways to carry out each activity multiplied together.

EXAMPLE **Jason rolls four fair six-sided dice.**

a) How many different ways are there to roll the four dice?

Each dice has 6 different ways that it can land (on 1, 2, 3, 4, 5 or 6).

Total number of ways of rolling four dice = $6 \times 6 \times 6 \times 6 = 1296$

b) How many different ways are there to only get even numbers when rolling the four dice?

Each dice has 3 different ways that it can land on an even number (on 2, 4, or 6).

Number of ways of only rolling even numbers = $3 \times 3 \times 3 \times 3 = 81$

c) What is the probability of only getting even numbers when rolling four dice?

$$P(\text{only even numbers}) = \frac{\text{number of ways to only get even numbers}}{\text{total number of ways to roll the dice}} = \frac{81}{1296} = \frac{1}{16}$$

Learn the product rule for counting the number of outcomes

Just find the number of ways of doing each activity and multiply them all together. Simple.
Now, have a go at these questions to see if you've got it...

Q1 Video Solution

Q1 Three fair coins are tossed: a) List all the possible outcomes. [1 mark]
b) Find the probability of getting exactly 2 heads. [1 mark] GRADE 4

Q2 Ten fair coins are tossed. Find the probability of not getting any heads. [2 marks] GRADE 6

Probability Experiments

You need to know how to use probability experiments to find relative frequencies.

Fair or Biased?

The probability of rolling a three on a normal dice is $\frac{1}{6}$ — you know that each of the 6 numbers on the dice is equally likely to be rolled, and there's only 1 three.

BUT this only works if it's a fair dice. If the dice is a bit wonky (the technical term is 'biased') then each number won't have an equal chance of being rolled. This is where relative frequency comes in — you can use it to estimate probabilities when things might be wonky.

Do the Experiment Again and Again and Again...

You need to do an experiment over and over again and count how many times each outcome happens (its frequency). Then you can calculate the relative frequency using this formula:

$$\text{Relative frequency} = \frac{\text{Frequency}}{\text{Number of times you tried the experiment}}$$

An experiment could just mean rolling a dice.

You can use the relative frequency of a result as an estimate of its probability.

EXAMPLE **The spinner on the right was spun 100 times. Use the results in the table below to estimate the probability of getting each of the scores.**

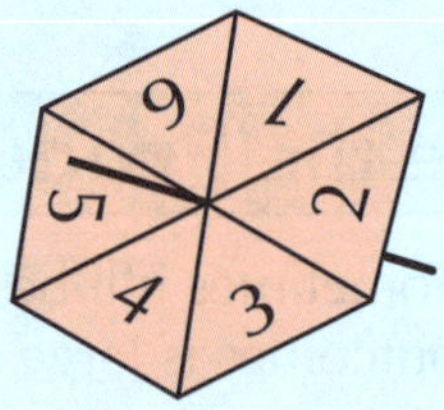

Score	1	2	3	4	5	6
Frequency	3	14	41	20	18	4

Divide each of the frequencies by 100 to find the relative frequencies.

Score	1	2	3	4	5	6
Relative Frequency	$\frac{3}{100} = 0.03$	$\frac{14}{100} = 0.14$	$\frac{41}{100} = 0.41$	$\frac{20}{100} = 0.2$	$\frac{18}{100} = 0.18$	$\frac{4}{100} = 0.04$

The MORE TIMES you do the experiment, the MORE ACCURATE your estimate of the probability should be.

E.g. if you spun the above spinner 1000 times, you'd get a better estimate of the probability for each score.

If the relative frequency of a result is far away from what you'd expect, then you can say that the dice/spinner/coin/etc. is probably biased. If not, you can say it's probably not biased or it seems fair.

EXAMPLE **Do the above results suggest that the spinner is biased?**

Yes, because the relative frequency of 3 is much higher than you'd expect, while the relative frequencies of 1 and 6 are much lower.

For a fair 6-sided spinner, you'd expect all the relative frequencies to be about 1 ÷ 6 = 0.17(ish).

More experiments mean a more accurate probability estimate

Learn the relative frequency formula. If something is biased, this just means it isn't fair. Even with a fair dice you're unlikely to get exactly the expected result, but the more experiments you do, the closer you'll get.

Q1 Sandro rolled a dice 1000 times and got the results shown in the table below.

Score	1	2	3	4	5	6
Frequency	140	137	138	259	161	165

a) Find the relative frequencies for each of the scores 1-6. [2 marks]

b) Do these results suggest that the dice is biased? Give a reason for your answer. [1 mark]

Q1 Video Solution

Probability Experiments

One more page on probability experiments — frequency trees can be used to help find relative frequencies when an experiment has two steps or more.

Record Results in **Frequency Trees**

When an experiment has two or more steps, you can record the results using frequency trees.

EXAMPLE

120 GCSE maths students were asked if they would go on to do A-level maths.

- **45 of them said they would go on to do A-level maths.**
- **30 of the students who said they would do A-level maths actually did.**
- **9 of the students who said they wouldn't do A-level maths actually did.**

a) Complete the frequency tree below.

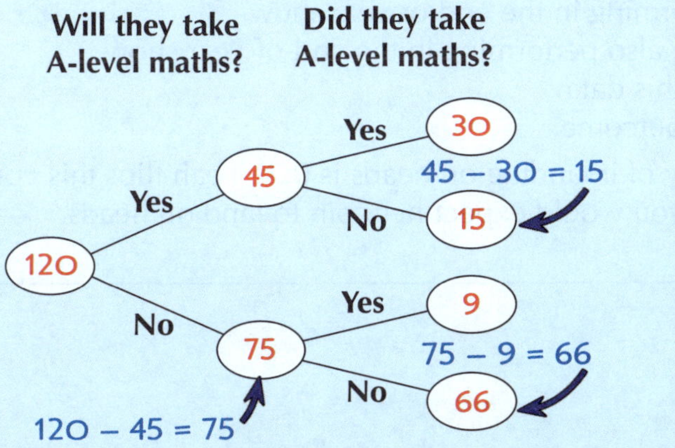

b) Use the data to find the relative frequency of each outcome.

Relative Frequency

Yes, Yes = $\frac{30}{120}$ = 0.25

Yes, No = $\frac{15}{120}$ = 0.125

No, Yes = $\frac{9}{120}$ = 0.075

No, No = $\frac{66}{120}$ = 0.55

Use Probability to Find an "**Expected Frequency**"

1) You can estimate how many times you'd expect something to happen if you do an experiment n times.
2) This expected frequency is based on the probability of the result happening.

Expected frequency of a result = probability × number of trials

EXAMPLE

A game involves throwing a fair six-sided dice. The player wins if they score either a 5 or a 6. If one person plays the game 180 times, estimate the number of times they will win.

1) First calculate the probability that they win each game.

Probability of winning = $\frac{\text{number of ways to win}}{\text{total number of possible outcomes}} = \frac{2}{6} = \frac{1}{3}$

2) Then estimate the number of times they'll win in 180 separate attempts.

Expected number of wins = probability of winning × number of trials

$= \frac{1}{3} \times 180 = 60$

If you don't know the probability of a result, fear not...
... you can estimate the probability using the relative frequency of the result in past experiments.

Expected frequency is how many times you'd expect something to happen

Make sure you can remember the formula for expected frequency in the box above. Try to get your head around where you can use frequency trees — they're really useful for experiments with two or more steps.

Q1 Using the frequency tree above, estimate how many out of 600 GCSE maths students:

a) You'd expect to say they're going to do A-level maths but then don't. [2 marks]

b) You'd expect to take A-level maths. [3 marks]

Warm-Up and Worked Exam Questions

Keep it up — this is the last section, then you've got some fun practice papers to do. Warm up your grey matter with the questions below.

Warm-Up Questions

1) Calculate the probability of a fair spinner numbered 1 to 8 landing on a number greater than 6.
2) The probability of winning a competition is $2x$, the probability of losing is x and the probability of drawing is 0.1. What is the value of x?
3) 2 fair dice are rolled and the numbers shown are added together. Make a sample space diagram showing all the possible outcomes and calculate the probability of scoring a total of 3.
4) 200 dancers are in a dance school. 120 of the dancers are learning jazz. 52 of the dancers who do jazz are performing in the end-of-year show. 63 of the dancers who don't do jazz are also performing in the end-of-year show.
   a) Draw a frequency tree to represent this data.
   b) Find the relative frequency of each outcome.
5) Leah flips a biased coin. The probability of it landing on heads is 0.2. Leah flips this coin 500 times. Calculate how many times you would expect her coin to land on heads.

Worked Exam Question

Look through this worked exam question and make sure you understand it — don't leave it to chance.

1 Suda has a six-sided dice. The sides are numbered 1 to 6. (GRADE 4)

Suda rolls the dice 50 times. Her results are shown in the table below.

Number	1	2	3	4	5	6
Relative Frequency	0.32	0.12	0.24	0.14	0.06	0.12

a) How many times did she roll a 6?

$50 \times 0.12 = 6$

6

[2 marks]

b) Is Suda's dice fair? Explain your answer.

On a fair dice, the theoretical probability of throwing each number is 0.166..., so as 1 has a much higher relative frequency and 5 has a much lower relative frequency, the dice is probably not fair.

[2 marks]

c) She rolls the dice another 50 times. Should she expect the same results? Explain your answer.

No, each dice roll is random, so in a small number of trials like 50 she is likely to get different results.

[1 mark]

Exam Questions

2 Arthur has stripy, spotty and plain socks in his drawer.
He picks a sock from the drawer at random.

The probability of picking a plain sock is 0.4, and of picking a spotty sock is y.
He is twice as likely to pick a stripy sock as a spotty sock.
Find the value of y. Give your answer as a decimal.

y = ...
[3 marks]

3 Alvar has a fair six-sided dice and a set of five cards numbered 2, 4, 6, 8 and 10.
He rolls the dice and chooses a card at random.
Alvar adds the number on the dice to the number on the card to calculate his total score.

Find the probability that Alvar will score more than 4.

....
[3 marks]

4 Eimear has a bag containing a large number of counters.
Each counter is numbered either 1, 2, 3, 4 or 5.

She selects one counter from the bag at random, makes a note of its number, and then puts it back in the bag. Eimear does this 100 times. Her results are shown in the table below.

Number on counter	1	2	3	4	5
Frequency	23	25	22	21	9
Relative Frequency					

a) Complete the table, giving the relative frequencies of each counter being selected.
[2 marks]

b) Elvin says that he thinks that the bag contains the same number of counters with each number. Do you agree? Give a reason for your answer.

....
[1 mark]

c) Using Eimear's results, estimate the probability of selecting an odd number when one counter is picked from the bag at random.

....
[2 marks]

5 A shop sells three different meal deals. The possible meal deal options are:
- sandwich and drink
- sandwich and snack
- sandwich, snack and drink

There are 5 different sandwiches, 8 different drinks and 4 different snacks.
How many possible meal deal combinations are there?

....
[3 marks]

The AND/OR Rules

*This page will show you how to find probabilities when **more than one** thing is happening at a time.*

Independent and **Dependent** Events

1) You need to know the difference between independent and dependent events, if you're going to use the AND / OR rules properly.
2) Two events are independent if one event happening doesn't affect the probability of the other happening. E.g. rolling a 6 both times on two dice rolls or picking a blue ball, replacing it, then picking a red ball.
3) If one event happening does affect the probability of the other happening, the events are dependent. E.g. picking a blue ball then picking a red ball without replacing the blue ball first.

The **AND Rule** gives **P(Both Events Happen)**

If two events, call them A and B, are independent then...

P(A and B) = P(A) × P(B)

If they're dependent, use the conditional probability rule (p.168).

The probability of events A AND B BOTH happening is equal to the two separate probabilities MULTIPLIED together.

EXAMPLE **Dave picks one ball at random from each of bags X and Y. Find the probability that he picks a yellow ball from both bags.**

1) Write down the probabilities of the different events.
   P(Dave picks a yellow ball from bag X) = $\frac{4}{10}$ = 0.4
   P(Dave picks a yellow ball from bag Y) = $\frac{2}{8}$ = 0.25
2) Use the formula.
   So P(Dave picks a yellow ball from both bags) = 0.4 × 0.25 = 0.1

The **OR Rule** gives **P(At Least One Event Happens)**

For two events, A and B...

P(A or B) = P(A) + P(B) – P(A and B)

The probability of EITHER event A OR event B happening is equal to the two separate probabilities ADDED together MINUS the probability of events A AND B BOTH happening.

If the events A and B can't happen together then P(A and B) = 0 and the OR rule becomes:

When events can't happen together they're called mutually exclusive.

P(A or B) = P(A) + P(B)

EXAMPLE **A spinner with red, blue, green and yellow sections is spun — the probability of it landing on each colour is shown in the table below. Find the probability of spinning either red or green.**

Colour	red	blue	yellow	green
Probability	0.25	0.3	0.35	0.1

The spinner can't land on both red and green so use the simpler OR rule. Just put in the probabilities.

P(red or green) = P(red) + P(green)
= 0.25 + 0.1 = 0.35

Two rules to learn here

You won't go far if you don't learn the AND/OR rules. Remember, '× with AND' and '+ with OR'.

Q1 Lee rolls 2 fair six-sided dice. Find the probability that he rolls two odd numbers. [2 marks]

Q2 On any day, the probability that Azi wears these clothes is P(hoodie) = 0.25, P(jeans) = 0.6, and P(hoodie and jeans) = 0.2. What is the probability she's wearing a hoodie or jeans today? [2 marks]

Q2 Video Solution

Tree Diagrams

Tree diagrams can really help you work out probabilities when you have a ***combination of events****.*

Remember These **Four** Key **Tree Diagram Facts**

GRADE 7

1) On any set of branches which meet at a point, the probabilities must add up to 1.

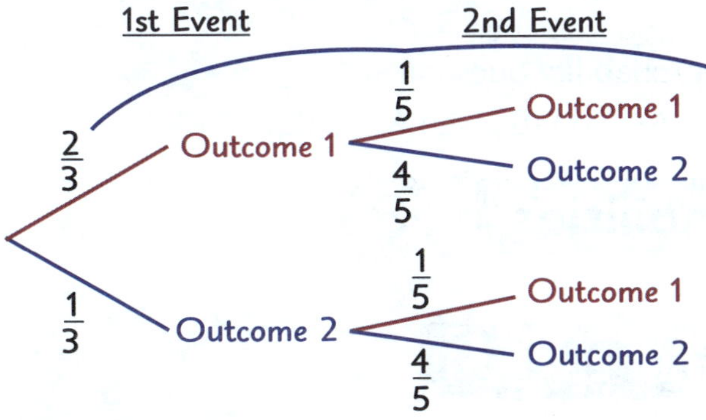

$\frac{2}{3} \times \frac{1}{5} = \frac{2}{15}$

$\frac{2}{3} \times \frac{4}{5} = \frac{8}{15}$

$\frac{1}{3} \times \frac{1}{5} = \frac{1}{15}$

$\frac{1}{3} \times \frac{4}{5} = \frac{4}{15}$

Total = 1

2) Multiply along the branches to get the end probabilities.

3) Check your diagram — the end probabilities must add up to 1.

4) To answer any question, add up the relevant end probabilities (see below).

EXAMPLE

A box contains 5 red discs and 3 green discs. One disc is taken at random and its colour noted before being replaced. A second disc is then taken. Find the probability that both discs are the same colour.

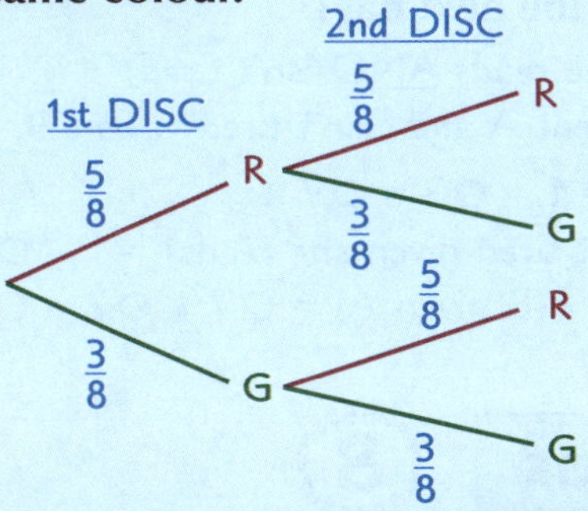

The probabilities for the 1st and 2nd discs are the same. This is because the 1st disc is replaced — so the events are independent.

P(both discs are red) = P(R and R) = $\frac{5}{8} \times \frac{5}{8} = \frac{25}{64}$

P(both discs are green) = P(G and G) = $\frac{3}{8} \times \frac{3}{8} = \frac{9}{64}$

P(both discs are same colour) = P(R and R or G and G)

$= \frac{25}{64} + \frac{9}{64} = \frac{34}{64} = \frac{17}{32}$

Look Out for 'At Least' Questions

GRADE 8

When a question asks for 'at least' a certain number of things happening, it's usually easier to work out (1 – probability of 'less than that number of things happening').

EXAMPLE **I roll 3 fair six-sided dice. Find the probability that I roll at least 1 six.**

1) Rewrite this as 1 minus a probability. P(at least 1 six) = 1 – P(less than 1 six)
= 1 – P(no sixes)

2) Work out P(no sixes). You can use a tree diagram — don't draw the whole thing, just the part you need.

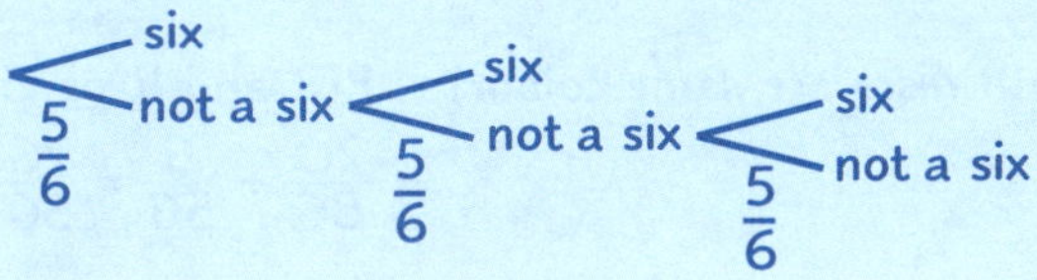

P(no sixes) = $\frac{5}{6} \times \frac{5}{6} \times \frac{5}{6} = \frac{125}{216}$

So P(at least 1 six) = $1 - \frac{125}{216} = \frac{91}{216}$

See how useful tree diagrams are

With probability questions that seem hard, drawing a tree diagram can be a good place to start. It takes some thinking to decide how to draw it and which bits you need, but after that it's plain sailing.

Q1 A bag contains 6 red balls and 4 black ones. If two balls are picked at random (with replacement), find the probability that they're different colours. [3 marks] GRADE 7

Conditional Probability

Conditional probabilities *crop up when you have* ***dependent events*** *— where one event affects another.*

Using **Conditional Probabilities**

You might see 'A given B' written as A|B.

1) The conditional probability of A given B is the probability of event A happening given that event B happens.
2) Keep an eye out in questions for items being picked 'without replacement' — it's a tell-tale sign that it's going to be a conditional probability question.
3) If events A and B are independent then P(A given B) = P(A) and P(B given A) = P(B).

The **AND** rule for **Conditional Probabilities**

If events A and B are dependent (see p.166) then...

P(A and B) = P(A) × P(B given A)

The probability of events A AND B BOTH happening is equal to the probability of event A happening MULTIPLIED by the probability of event B happening GIVEN that event A happens.

EXAMPLE

Alia either watches TV or reads before bed. The probability she watches TV is 0.3. If she reads, the probability she is tired the next day is 0.8. What is the probability that Alia reads and isn't tired the next day?

1) Label the events A and B.

We want to find P(she reads AND isn't tired)
So call "she reads" event A and "isn't tired" event B.

2) Use the information given in the question to work out the probabilities that you'll need to use the formula.

P(A) = P(she reads) = 1 – 0.3 = 0.7
P(B given A) = P(isn't tired given she reads) = 1 – 0.8 = 0.2
P(A and B) = P(A) × P(B given A) = 0.7 × 0.2 = 0.14

Conditional Probabilities on **Tree Diagrams**

A good way to deal with conditional probability questions is to draw a tree diagram. The probabilities on a set of branches will change depending on the previous event.

EXAMPLE

A box contains 5 red discs and 3 green discs.
Two discs are taken at random without replacement.
Find the probability that both discs are the same colour.

This example was done 'with replacement' on p.167.

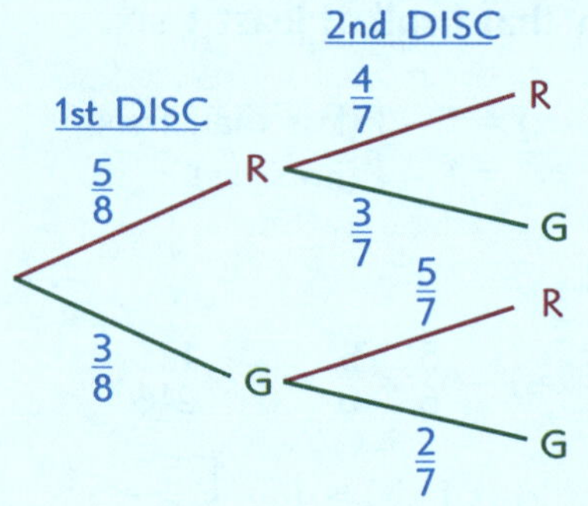

The probabilities for the 2nd pick depend on the colour of the 1st disc picked. This is because the 1st disc is not replaced.

P(both discs are red) = P(R and R) = $\frac{5}{8} \times \frac{4}{7} = \frac{20}{56}$

P(both discs are green) = P(G and G) = $\frac{3}{8} \times \frac{2}{7} = \frac{6}{56}$

P(both discs are same colour) = P(R and R or G and G)

$= \frac{20}{56} + \frac{6}{56} = \frac{26}{56} = \frac{13}{28}$

Learn the AND rule for conditional probability

With probability questions that seem quite hard, drawing a tree diagram is usually a good place to start. Try it with the (quite hard) Exam Practice Question below...

Q1 There are 21 numbers, 1-21, in a lottery draw. A machine selects the numbers randomly. Find the probability that out of the first two numbers selected:
a) at least one is even. [3 marks]
b) one is odd and one is even [3 marks]

Q1 Video Solution

Sets and Venn Diagrams

Venn diagrams *are a way of displaying sets in* ***intersecting circles****.*

Showing **Sets** on **Venn Diagrams**

1) Sets are just collections of things (e.g. numbers) — we call these 'things' elements.
2) Sets can be written in different ways but they'll always be in a pair of curly brackets {}.
   E.g. {2, 3, 5, 7}, {prime numbers less than 10}, or {x : x is a prime number less than 10}.
3) n(A) just means 'the number of elements in set A'. E.g. if A = {1, 5, 9, 11}, n(A) = 4.
4) On a Venn diagram, each set is represented by a circle containing the elements of the set or the number of elements in the set.

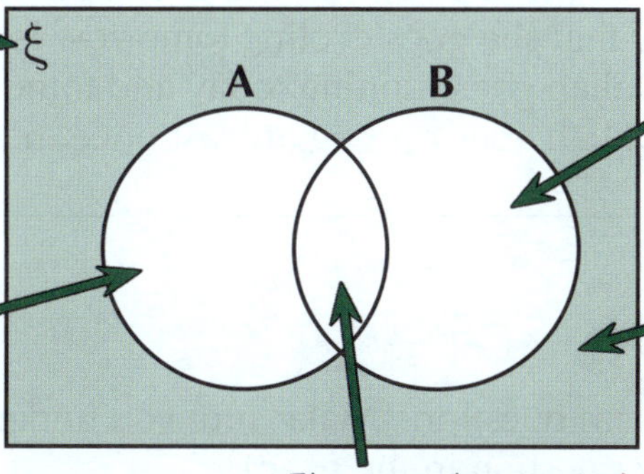

Finding probabilities from **Venn Diagrams**

EXAMPLE **The Venn diagram on the right shows the number of Year 10 pupils going on the History (H) and Geography (G) school trips.**

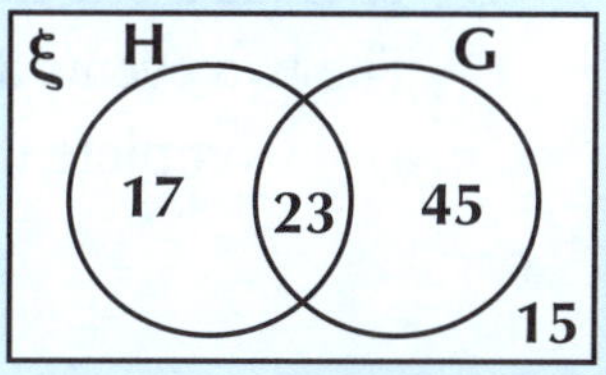

Find the probability that a randomly selected Year 10 pupil is:

a) not going on the History trip.

n(Year 10 pupils) = 17 + 23 + 45 + 15 = 100

n(Not going on History trip) = 45 + 15 = 60

P(Not going on History trip) = $\frac{60}{100} = \frac{3}{5} = 0.6$

Use the formula from p.160 to find the probabilities.

b) not going on the History trip but going on the Geography trip.

n(Not going on History trip but going on Geography trip) = 45

P(Not going on History trip but going on Geography trip) = $\frac{45}{100} = \frac{9}{20} = 0.45$

c) going on the Geography trip given that they're not going on the History trip.

Careful here — think of this as selecting a pupil going on the Geography trip from those not going on the History trip.

P(Going on Geography trip given not going on History trip) = $\frac{45}{45 + 15}$

$= \frac{45}{60} = \frac{3}{4} = 0.75$

You could also use the conditional probability formula and your answers to parts a) and b).

Learn how to write things in set notation

If you're asked to write something in set notation, you need to use curly brackets. n(A) is the number of elements in set A. The weird ξ symbol is the universal set (where all the elements come from).

Q1 Video Solution

Q1 Out of 80 customers at an ice cream van, 48 had syrup, 28 had sprinkles and 16 had both toppings on their ice cream. Use a Venn diagram to find the probability that a randomly selected customer doesn't have either topping given that they don't have sprinkles. [3 marks]

Warm-Up and Worked Exam Questions

Probability is really not that difficult once you get the hang of it, but it's important to get loads of practice, so try these warm-up questions. Take a look back at anything you're unsure about.

Warm-Up Questions

1) What is the probability of rolling a six three times in a row with a fair six-sided dice?
2) Nicola has 16 individual socks in her sock drawer. 6 of them are red and 10 are green. Nicola picks two socks from her drawer at random (without replacement). Find the probability that they're the same colour.
3) Each day Daliah either goes running or cycling. The probability that Daliah goes running today is 0.7. If she goes running today the probability that she goes cycling tomorrow is 0.1. If she goes cycling today the probability that she goes cycling tomorrow is 0.6.
   a) What is the probability that Daliah goes running today and tomorrow?
   b) What is the probability that Daliah goes cycling at least once in the next two days?

Worked Exam Question

Have a good hard look at this worked exam question. Make sure you understand each bit — you'll usually get at least one probability question in the exam.

1 Jo and Heather are meeting for coffee.

The probability that Jo will wear burgundy trousers is $\frac{2}{5}$. (GRADE 5)

There is a one in four chance that Heather will wear burgundy trousers.

The two events are independent.

a) Complete the tree diagram below.

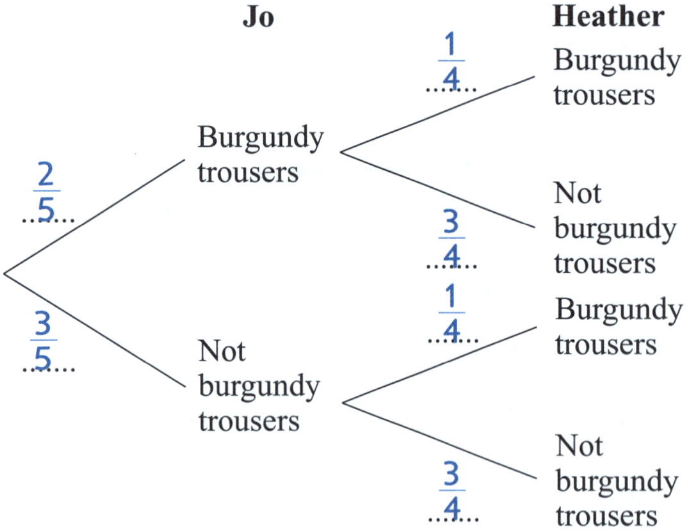

Check that the probabilities on each set of branches add up to 1.

[2 marks]

b) What is the probability that neither of them wear burgundy trousers?

$$\text{P(neither wear burgundy trousers)} = \frac{3}{5} \times \frac{3}{4} = \frac{3 \times 3}{5 \times 4} = \frac{9}{20}$$

$\frac{9}{20}$

[2 marks]

Exam Questions

2 The Venn diagram below shows the number of elements in sets A and B.

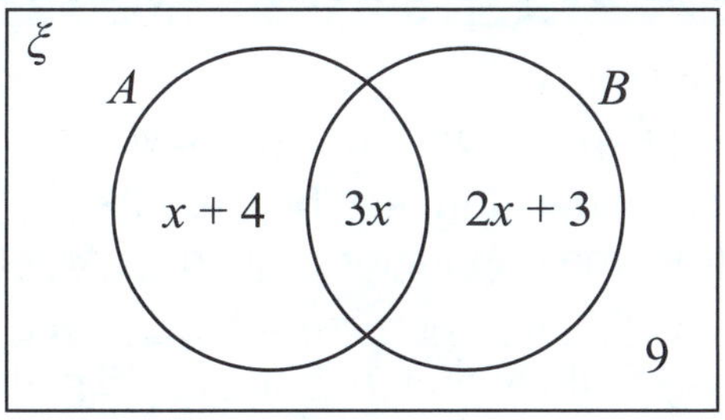

a) Given that $n(\xi) = 40$, find the value of x.

$x =$ ....

[2 marks]

b) An element is randomly chosen from the universal set, ξ.
Find the probability that the chosen element is in both set A and set B.

....

[2 marks]

3 Shaun is playing the game ‘hook-a-duck’. The probability that he wins a prize is 0.3, independent of what has happened in previous games.

a) If he plays three games, what is the probability that he doesn’t win a prize?

....

[1 mark]

b) If he plays two games, what is the probability that he wins at least one prize?

....

[2 marks]

4 A box of chocolates contains 12 chocolates. 7 are milk chocolate and 5 are white chocolate. Two chocolates are chosen at random without replacement.

a) Given that the first chocolate is a milk chocolate, what is the probability that the second chocolate is a milk chocolate?

....

[1 mark]

b) Calculate the probability of at least one milk chocolate being chosen.

....

[2 marks]

c) Calculate the probability that one milk chocolate and one white chocolate are chosen in any order.

....

[2 marks]

Sampling and Bias

***Sampling** is about using what you know about **smaller** groups to tell you about **bigger** groups.*

Use a **Sample** to **Find Out About** a **Population**

1) The whole group you want to find out about is called the POPULATION. It can be a group of anything — people, plants, penguins, you name it.
2) Often you can't survey the whole population, e.g. because it's too big. So you select a smaller group from the population, called a SAMPLE, instead.
3) It's really important that your sample fairly represents the WHOLE population. This allows you to apply any conclusions from your survey to the whole population. E.g. if you find that ¾ of the people in your sample like cheese, you can estimate that ¾ of the people in the whole population like cheese.

For a sample to be representative, it needs to be:

1. A RANDOM SAMPLE — which means every member of the population has an equal chance of being in it.
2. BIG ENOUGH for the size of the population. The bigger the sample, the more reliable it should be.

Simple Random Sampling — choosing a **Random** Sample

To SELECT a SIMPLE RANDOM SAMPLE...

1. Assign a number to every member of the population.
2. Create a list of random numbers.
3. Match the random numbers to members of the population.

E.g. by using a computer, calculator or picking numbers out of a bag.

You Need to **Spot Problems** with **Sampling Methods**

A BIASED sample (or survey) is one that doesn't properly represent the whole population.

To SPOT BIAS, you need to think about:

1) WHEN, WHERE and HOW the sample is taken.
2) HOW MANY members are in it.

If certain groups are excluded, the SAMPLE ISN'T RANDOM. And that can lead to BIAS from things like age, gender, different interests, etc. If the sample is too small, it's also likely to be biased.

EXAMPLE

Samir's school has 800 pupils. Samir is interested in whether these pupils would like to have more music lessons. For his sample he selects 5 members of the school orchestra to ask. Explain why the opinions Samir collects from his sample might not represent the whole school.

The sample isn't random — only members of the orchestra are included, so it's likely to be biased in favour of more music lessons. Also, a sample of 5 is too small to represent the whole school.

Always watch out for sampling bias

Make sure you understand why samples should be representative and how to spot when they're not — always think about the two key points above. Then you'll be ready to take on this Exam Practice Question.

Q1 Tina wants to find out how often people in the UK travel by train. She decides to ask 20 people waiting for trains at her local train station one morning. Comment on whether Tina can use the results of her survey to draw conclusions about the whole population. [2 marks]

Collecting Data

*Data you **collect yourself** is called **primary** data. If you use data that **someone else has collected**, e.g. you get it from a website, it's called **secondary** data.*

There are Different Types of Data

GRADE 3

QUALITATIVE DATA is descriptive. It uses words, not numbers.

E.g. pets' names — Smudge, Snowy, Dave, etc. Favourite flavours of ice cream — 'vanilla', 'chocolate', 'raspberry ripple', etc.

QUANTITATIVE DATA measures quantities using numbers.

E.g. heights of people, times taken to finish a race, numbers of goals scored in football matches, and so on.

There are two types of quantitative data.

Discrete Data

1) It's discrete if the numbers can only take certain exact values.
2) E.g. the number of customers in a shop each day has to be a whole number — you can't have half a person.

Continuous Data

1) If the numbers can take any value in a range, it's called continuous data.
2) E.g. heights and weights are continuous measurements.

You can Organise your Data into Classes

GRADE 3

1) To record data in a table, you often need to group it into classes to make it more manageable. Discrete data classes should have 'gaps' between them, e.g. '0-1 goals', '2-3 goals' (jump from 1 to 2 because there are no values in between). Continuous data classes should have no 'gaps', so are often written using inequalities (see p.178).
2) Make sure none of the classes overlap and that they cover all the possible values.

EXAMPLE **Jonty wants to find out about the ages (in whole years) of people who use his local library. Design a table he could use to collect his data.**

Include columns for: the data values, 'Tally' to count the data and 'Frequency' to show the totals.

The data's discrete so leave gaps between classes.

Include classes like '...or over', '...or less' or 'other' to cover all options in a sensible number of classes.

Age (whole years)	Tally	Frequency
0-19		
20-39		
40-59		
60-79		
80 or over		

Grouping data loses some accuracy because you don't know the exact values any more.

Questionnaires should be Designed Carefully

GRADE 3

Another way to record data is to ask people to fill in a questionnaire. Your questions should be:

1) Clear and easy to understand
2) Easy to answer
3) Fair — not leading or biased

Leading questions suggest an answer.

Watch out for response boxes that could be interpreted in different ways, that overlap, or that don't allow for all possible answers.

Data can be qualitative or quantitative

You need to know what type of data you've got so you can record and display it in a suitable way.

Q1 James asks some students how many times they went to the cinema in the last year. Say whether this data is qualitative, discrete or continuous and design a table to record it in. [2 marks] GRADE 3

Mean, Median, Mode and Range

***Mean**, **median**, **mode** and **range** pop up all the time in statistics questions — make sure you know what they are.*

The **Four Definitions** (Grade 3)

MODE = MOST common

MEDIAN = MIDDLE value (when values are in order of size)

MEAN = TOTAL of items ÷ NUMBER of items

RANGE = Difference between highest and lowest

REMEMBER:
Mode = most (emphasise the 'mo' in each when you say them)
Median = mid (emphasise the m*d in each when you say them)
Mean is just the average, but it's mean 'cos you have to work it out.

There's one Golden Rule for finding the median that lots of people forget:

Always REARRANGE the data in ASCENDING ORDER (and check you have the same number of entries!)

You absolutely must do this when finding the median, but it's also really useful for working out the mode too.

EXAMPLE **Find the median, mode, mean, and range of these numbers:**

2, 5, 3, 2, 6, -4, 0, 9, -3, 1, 6, 3, -2, 3

The MEDIAN is the middle value, so rearrange the numbers in order of size.

Check that you still have the same number of entries after you've rearranged them.

When there are two middle numbers, the median is halfway between the two.

-4, -3, -2, 0, 1, 2, 2, 3, 3, 3, 5, 6, 6, 9

← seven numbers this side | seven numbers this side →

Median = 2.5

To find the position of the median of n values, you can use the formula (n + 1) ÷ 2. Here, (14 + 1) ÷ 2 = position 7.5 — that's halfway between the 7th and 8th values.

MODE (or modal value) is the most common value. ⟶ Mode = 3

Data sets can have more than one mode.

$$\text{MEAN} = \frac{\text{total of items}}{\text{number of items}} \longrightarrow \frac{-4-3-2+0+1+2+2+3+3+3+5+6+6+9}{14}$$

$= 31 \div 14 = 2.214... = 2.21$ (3 s.f.)

RANGE = distance from lowest to highest value, i.e. from –4 up to 9. ⟶ $9 - (-4) = 13$

A **Trickier Example** (Grade 4)

EXAMPLE **The heights (to the nearest cm) of 8 penguins at a zoo are 41, 43, 44, 44, 47, 48, 50 and 51. Two of the penguins are moved to a different zoo. If the mean height of the remaining penguins is 44.5 cm, find the heights of the two penguins that moved.**

$$\text{Mean} = \frac{\text{total height}}{\text{no. of penguins}}$$

So total height = no. of penguins × mean

Total height of 8 penguins = 368 cm.

Total height of remaining 6 penguins = 6 × 44.5 = 267 cm.

Combined height of penguins that moved = 368 – 267 = 101 cm.

So the heights must be 50 cm and 51 cm.

Mean, median, mode & range — easy marks for learning four words

The maths involved in working these out is so simple that you'd be mad not to learn the definitions.

Q1 Find the mean, median, mode and range for the set of data below:
1, 3, 14, –5, 6, –12, 18, 7, 23, 10, –5, –14, 0, 25, 8. [4 marks] (Grade 3)

Q2 Another value is added to the data in Q1.
If the mean is now 5.5, find the new value. [3 marks] (Grade 4)

Q1 Video Solution

Frequency Tables — Finding Averages

*The word **FREQUENCY** means **HOW MANY**, so a frequency table is just a **'How many in each category' table**. You saw how to find **averages and range** on p.174 — it's similar here, but with the data in a table.*

Find **Averages** from **Frequency Tables**

1) The MODE is just the CATEGORY with the MOST ENTRIES.
2) The RANGE is found from the extremes of the first column.
3) The MEDIAN is the CATEGORY containing the middle value.
4) To find the MEAN, you have to WORK OUT A THIRD COLUMN yourself.

The MEAN is then: **3rd Column Total ÷ 2nd Column Total**

Categories — How many

Number of cats	Frequency	
0	17	
1	22	
2	15	
3	7	

Third column for you to fill in.

EXAMPLE **Some people were asked how many sisters they have. The table opposite shows the results.**

Find the mode, the range, the mean and the median of the data.

Number of sisters	Frequency
0	7
1	15
2	12
3	8
4	4
5	0

1. The MODE is the category with the most entries — i.e. the one with the highest frequency:

   The highest frequency is 15 for '1 sister', so MODE = 1

2. The RANGE is the difference between the highest and lowest numbers of sisters — that's 4 sisters (no one has 5 sisters) and no sisters, so:

   RANGE = 4 – 0 = 4

3. To find the MEAN, add a 3rd column to the table showing 'number of sisters × frequency'. Add up these values to find the total number of sisters of all the people asked.

   You can label the first column x and the frequency column f, then the third column is $f \times x$.

Number of sisters (x)	Frequency (f)	No. of sisters × Frequency ($f \times x$)
0	7	0
1	15	15
2	12	24
3	8	24
4	4	16
5	0	0
Total	46	79

$$\text{MEAN} = \frac{\text{total number of sisters}}{\text{total number of people asked}} = \frac{79}{46} = 1.72 \text{ (3 s.f.)}$$

(79: 3rd column total; 46: 2nd column total)

4. The MEDIAN is the category of the middle value. Work out its position, then count through the 2nd column to find it.

   It helps to imagine the data set out in an ordered list:
   0000000111111111111111222222222222333333334444
   (median)

   The median is in position (n + 1) ÷ 2 = (46 + 1) ÷ 2 = 23.5 — halfway between the 23rd and 24th values. There are a total of (7 + 15) = 22 values in the first two categories, and another 12 in the third category takes you to 34. So the 23rd and 24th values must both be in the category '2 sisters', which means the MEDIAN is 2.

Remember — mode is most, median is middle...

... and mean is average. Now have a go at this exciting practice question.

Q1 50 people were asked how many times a week they play sport. The table opposite shows the results.

a) Find the median. [2 marks]

b) Calculate the mean. [3 marks]

No. of times sport played	Frequency
0	8
1	15
2	17
3	6
4	4
5 or more	0

Q1 Video Solution

Warm-Up and Worked Exam Questions

A few lovely warm-up questions for you here. If you have any problems with these, flick back and have another look at the last few pages before looking at the exam questions.

Warm-Up Questions

1) The following situation involves a population and a sample. Identify both and also identify the source of probable bias:

   A flour company wants to know what proportion of Birmingham households bake some or all of their own bread. A sample of 600 residential addresses in Birmingham is taken and interviewers are sent to these addresses. The interviewers are employed during regular working hours on weekdays and interview only during these hours.

2) Write down the 4 basic definitions of the following: Mode, Median, Mean and Range.

3) The data shows the number of cars owned by 124 households in a survey. Find the:
   a) Mean; b) Median; c) Mode; d) Range

Number of cars	0	1	2	3	4	5	6
Frequency	1	24	36	31	22	9	1

Worked Exam Question

Take a good look at this worked exam question. There are usually loads of different correct answers for questions like part b), so have a think about other answers you could give.

1 Maria asked 50 people at a football match how they travelled there. She found that 22 of them travelled by car. There were 5000 people at the match altogether. (GRADE 4)

a) Use the information above to estimate the number of people who travelled to the match by car.

Proportion of people in sample who travelled by car
$= 22 \div 50 = 0.44$

Estimate of number of people at match who travelled by car
$= 0.44 \times 5000 = 2200$

2200

[3 marks]

b) Daisy was at a different football match on the same day. She uses Maria's sample data to estimate that 374 of the 850 people at her match travelled there by car.

Explain the assumption Daisy has made and comment on the reliability of her estimate.

Daisy has made the assumption that Maria's sample is a fair representation of the people at her match. Her estimate is unreliable because she hasn't sampled people from the correct population.

[2 marks]

Remember, to get reliable estimates, a sample needs to fairly represent the population.

Exam Questions

2 Faye is investigating how many chocolate bars teenagers buy each week.
She is going to collect data by asking her teenage friends how many they buy.

a) Design a table that Faye could use to record her data.

[2 marks]

b) Comment on whether she can use her results to draw conclusions about teenagers in the UK.

....

....

....

[2 marks]

3 Lottie has 6 pygmy goats. Their weights, in kg, are listed below.

32 23 31 28 36 26

a) Which three weights, from the list above, would have a range which is half the value of the median of the three weights? Write down the range and median with your answer.

.................., ..., ...

range = ..., median = ...

[2 marks]

b) Two of the goats wander off and don't return. The mean weight of the herd is now 27.25 kg. Find the weights of the two goats who wandered off.

.... kg and .... kg

[3 marks]

4 Show that the difference between the mean and the median of five consecutive integers is always zero.

[3 marks]

Grouped Frequency Tables

Grouped frequency tables *group together the data into* ***classes****. They look like ordinary frequency tables, but they're a* ***slightly trickier*** *kettle of fish...*

See p.173 for grouped discrete data.

Grade 4

NO GAPS BETWEEN CLASSES

- Use inequality symbols to cover all possible values.
- Here, 10 would go in the 1st class, but 10.1 would go in the 2nd class.

Height (h millimetres)	Frequency
$5 < h \le 10$	12
$10 < h \le 15$	15

To find MID-INTERVAL VALUES:

- Add together the end values of the class and divide by 2.
- E.g. $\frac{5+10}{2} = 7.5$

Find Averages from Grouped Frequency Tables

Grade 5

Unlike with ordinary frequency tables, you don't know the actual data values, only the classes they're in. So you have to ESTIMATE THE MEAN, rather than calculate it exactly. Again, you do this by adding columns:

1) **Add a 3RD COLUMN and enter the MID-INTERVAL VALUE for each class.**
2) **Add a 4TH COLUMN to show 'FREQUENCY × MID-INTERVAL VALUE' for each class.**

You'll be asked to find the MODAL CLASS and the CLASS CONTAINING THE MEDIAN, not exact values. And the RANGE can only be estimated too — using the class boundaries.

EXAMPLE

This table shows information about the weights, in kilograms, of 60 school children.

a) Write down the modal class.
b) Write down the class containing the median.
c) Calculate an estimate for the mean weight.
d) Estimate the range of weights.

Weight (w kg)	Frequency
$30 < w \le 40$	8
$40 < w \le 50$	16
$50 < w \le 60$	18
$60 < w \le 70$	12
$70 < w \le 80$	6

a) The modal class is the one with the highest frequency.

Modal class is $50 < w \le 60$

b) Work out the position of the median, then count through the 2nd column.

The median is in position $(n + 1) \div 2 = (60 + 1) \div 2 = 30.5$, halfway between the 30th and 31st values. Both these values are in the third class, so the class containing the median is $50 < w \le 60$.

c) Add extra columns for 'mid-interval value' and 'frequency × mid-interval value'. Add up the values in the 4th column to estimate the total weight of the 60 children.

Weight (w kg)	Frequency (f)	Mid-interval value (x)	$f \times x$
$30 < w \le 40$	8	35	280
$40 < w \le 50$	16	45	720
$50 < w \le 60$	18	55	990
$60 < w \le 70$	12	65	780
$70 < w \le 80$	6	75	450
Total	60	—	3220

Don't add up the mid-interval values.

$$\text{Mean} \approx \frac{\text{total weight}}{\text{number of children}}$$

total weight ← 4th column total
number of children ← 2nd column total

$$= \frac{3220}{60} = 53.7 \text{ kg (3 s.f.)}$$

d) Find the difference between the highest and lowest class boundaries.

Estimated range = 80 – 30
= 50 kg

This is the largest possible range — it assumes there are data values on the class boundaries. The actual range is likely to be smaller, but you can't tell without knowing the individual values.

Now there are two columns to add

Learn what the columns are and how to find them.

Length (l cm)	$15.5 \le l < 16.5$	$16.5 \le l < 17.5$	$17.5 \le l < 18.5$	$18.5 \le l < 19.5$
Frequency	12	18	23	8

Q1 a) Estimate the mean of the data in the table. Give your answer to 3 significant figures. [4 marks]
b) Ana says that 20% of the lengths are below 16.5 cm. Comment on her statement. [2 marks]

Grade 5

Box Plots

*The humble **box plot** might not look very fancy, but it gives you a **useful summary** of a data set.*

Box Plots show the Spread of a Data Set (Grade 6)

1) The lower quartile Q_1, the median Q_2 and the upper quartile Q_3 are the values 25% (¼), 50% (½) and 75% (¾) of the way through an ordered set of data.
So if a set of data has n values, you can work out the positions of the quartiles using these formulas:

Q_1: $(n + 1)/4$ Q_2: $(n + 1)/2$ Q_3: $3(n + 1)/4$

2) The INTERQUARTILE RANGE (IQR) is the difference between the upper quartile and the lower quartile and contains the middle 50% of values.
3) A box plot shows the minimum and maximum values in a data set and the values of the quartiles. But it doesn't tell you the individual data values.

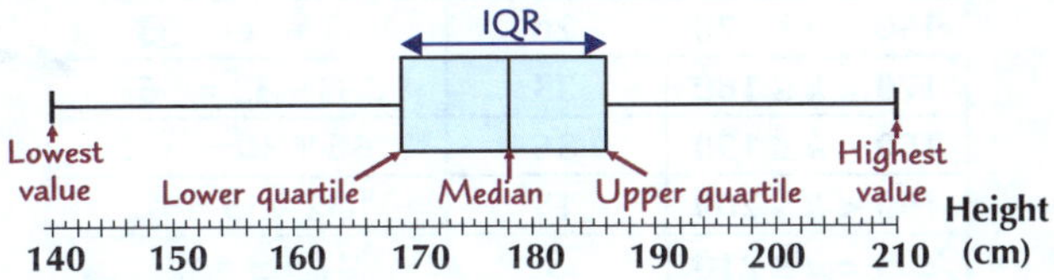

EXAMPLE **This table gives information about the numbers of rainy days last year in some cities. On the grid below, draw a box plot to show the information.**

Minimum number	90
Maximum number	195
Lower quartile	130
Median	150
Upper quartile	175

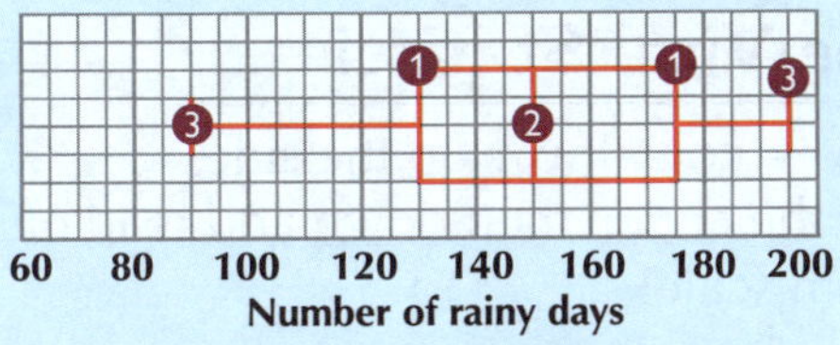

1. Mark on the quartiles and draw the box.
2. Draw a line at the median.
3. Mark on the minimum and maximum points and join them to the box with horizontal lines.

- Box plots show two measures of spread — range (highest – lowest) and interquartile range ($Q_3 - Q_1$).
- The range is based on all of the data values, so it can be affected by outliers — data values that don't fit the general pattern (i.e. that are a long way from the rest of the data).
- The IQR is based on only the middle 50% of the data values, so isn't affected by outliers. This means it can be a more reliable measure of spread than the range.

EXAMPLE **This box plot shows a summary of the heights of a group of gymnasts.**

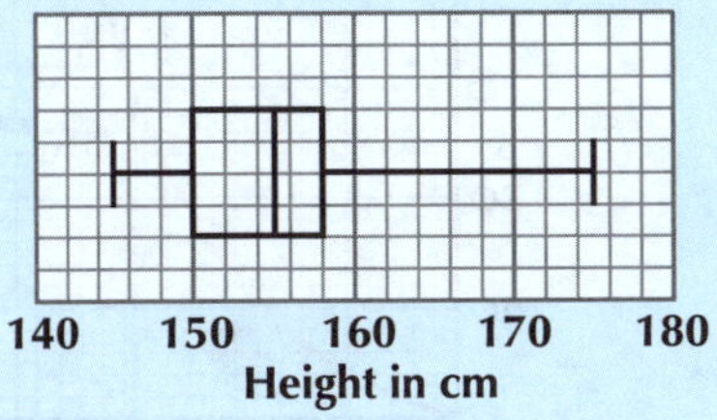

a) Work out the range of the heights.
Range = highest – lowest = 175 – 145 = 30 cm

b) Work out the interquartile range for the heights.
Q_1 = 150 cm and Q_3 = 158 cm, so IQR = 158 – 150 = 8 cm

c) Do you think the range or the interquartile range is a more reliable measure of spread for this data? Give a reason for your answer.
The IQR is small and 75% of the values are less than 158 cm, so it's likely that the tallest height of 175 cm is an outlier. The IQR doesn't include the tallest height, so the IQR should be more reliable.

d) Explain whether or not it is possible to work out the number of gymnasts represented by the box plot.
The box plot gives no information about the number of values it represents, so it isn't possible to work out the number of gymnasts.

The edges of the box show the lower and upper quartiles

Make sure you think carefully about what the values of the average and spread actually mean.

Q1 A large amount of data is analysed and the following conclusions are made: the minimum and maximum values are 5 and 22, 50% of the values are less than 12, 75% of the values are less than 17 and the IQR is 8. Draw a box plot to represent this information. [3 marks] (Grade 6)

Q1 Video Solution

Cumulative Frequency

Cumulative frequency *just means* ***adding it up as you go along*** *— i.e. the* ***total frequency so far****.*
You need to be able to ***draw*** *a* ***cumulative frequency graph*** *and* ***make estimates*** *from it.*

GRADE 6

EXAMPLE

The table below shows information about the heights of a group of people.

a) Draw a cumulative frequency graph for the data.

b) Use your graph to estimate the median and interquartile range of the heights.

Height (h cm)	Frequency	Cumulative Frequency
$140 < h \le 150$	4	4
$150 < h \le 160$	9	4 + 9 = 13
$160 < h \le 170$	20	13 + 20 = 33
$170 < h \le 180$	33	33 + 33 = 66
$180 < h \le 190$	36	66 + 36 = 102
$190 < h \le 200$	15	102 + 15 = 117
$200 < h \le 210$	3	117 + 3 = 120

Total number of people surveyed

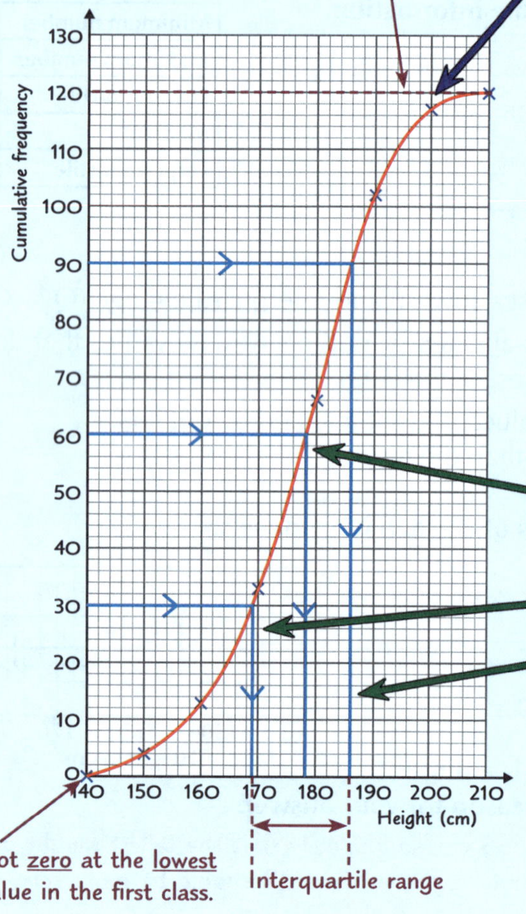

To Draw the Graph...

1) Add a 'CUMULATIVE FREQUENCY' COLUMN to the table — and fill it in with the RUNNING TOTAL of the frequency column.
2) PLOT points using the HIGHEST VALUE in each class and the CUMULATIVE FREQUENCY. (150, 4), (160, 13), etc.
3) Join the points with a smooth curve or straight lines.

To Find the Vital Statistics...

1) MEDIAN — go halfway up the side, across to the curve, then down and read off the bottom scale.
2) LOWER AND UPPER QUARTILES — go ¼ and ¾ up the side, across to the curve, then down and read off the bottom scale.
3) INTERQUARTILE RANGE — the distance between the lower and upper quartiles.

1) The halfway point is at ½ × 120 = 60. Reading across and down gives a median of 178 cm.
2) ¼ of the way up is at ¼ × 120 = 30. Reading across and down gives a lower quartile of 169 cm.
¾ of the way up is at ¾ × 120 = 90. Reading across and down gives an upper quartile of 186 cm.
3) The interquartile range = 186 – 169 = 17 cm.

More Estimating...

To use the graph to estimate the number of values that are less than or greater than a given value:

Go along the bottom scale to the given value, up to the curve, then across to the cumulative frequency.

The values you read from the graph are estimates because they're based on grouped data — you don't know how the actual data values are spread within each class.

Plot the points using the highest value in each class

In the example above, to estimate how many people are 176 cm or shorter, find 176 on the bottom, go up to the curve and across to the answer — 51.

Q1 a) Draw a cumulative frequency diagram for this data. [3 marks]

b) Use your diagram to estimate the percentage of fish that are longer than 50 mm. [2 marks]

GRADE 6

Length of fish (l mm)	Frequency
$0 < l \le 20$	4
$20 < l \le 40$	11
$40 < l \le 60$	20
$60 < l \le 80$	15
$80 < l \le 100$	6

Warm-Up and Worked Exam Questions

By the time the big day comes you need to know all the facts in this mini-section like the back of your hand. It's not easy, but the only way to get good marks is to practise with these questions.

Warm-Up Questions

1) The grouped frequency table below represents data from 79 random people.

Height (cm)	$145 \le x < 155$	$155 \le x < 165$	$165 \le x < 175$	$175 \le x < 185$
Frequency	18	22	24	15

a) Estimate the mean.
b) Which group contains the median?
c) State the modal group.

2) A vet records the ages of 100 parrots. The minimum age is 16 and the maximum age is 67. 75% of them are younger than 45, 50% of them are younger than 38 and 25% are younger than 24. Calculate the range and the interquartile range, then draw a box plot to represent this data.

Worked Exam Question

There's no better preparation for exam questions than doing, err... practice exam questions. Hang on, what's this I see...

1 During a science experiment 10 seeds were planted and their growth measured to the nearest cm after 12 days. The results were recorded in the table below. GRADE 4

Growth in cm	Number of plants
$0 \le x \le 2$	2
$3 \le x \le 5$	4
$6 \le x \le 8$	3
$9 \le x \le 11$	1

Use the table to find:

a) the modal class,

The modal class is the one with the highest frequency, so that's $3 \le x \le 5$

$3 \le x \le 5$
[1 mark]

b) the class which contains the median,

$(10 + 1) \div 2 = 5.5$, so the median is halfway between the 5th and 6th values, so it lies in the group containing the 5th and 6th values, which is $3 \le x \le 5$

$3 \le x \le 5$
[1 mark]

c) an estimate of the mean growth.

This is using the mid-interval values. You could add a couple of extra columns to the table — for mid-interval values and frequency × mid-interval value.

$[(1 \times 2) + (4 \times 4) + (7 \times 3) + (10 \times 1)] \div 10$
$= 49 \div 10 = 4.9$ cm

4.9 cm
[4 marks]

Exam Questions

2 Rachel and Harry record the distance they cycle each week for 26 weeks. The box plots show information about the distances they cycle.

Rachel:

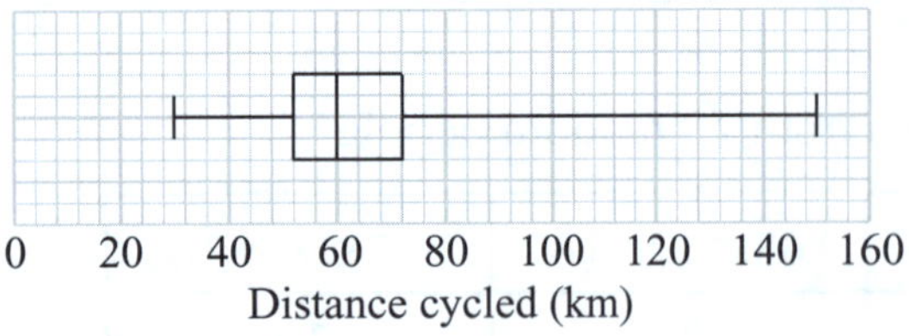

Harry:

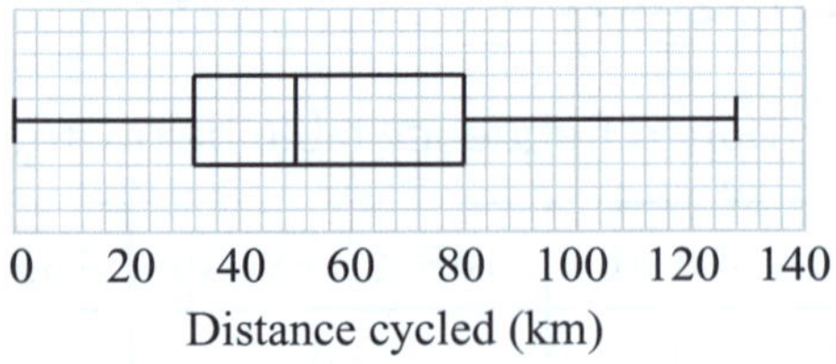

Rachel says that the distances she cycled were more consistent than the distances Harry cycled. Do you agree with her? Explain your answer.

....

....

[2 marks]

3 120 pupils in a year group sit an examination at the end of the year. Their results are given in the table below.

Exam mark (%)	$0 < x \leq 20$	$20 < x \leq 30$	$30 < x \leq 40$	$40 < x \leq 50$	$50 < x \leq 60$	$60 < x \leq 70$	$70 < x \leq 80$	$80 < x \leq 100$
Frequency	3	10	12	24	42	16	9	4

a) Complete the cumulative frequency table below.

Exam mark (%)	≤ 20	≤ 30	≤ 40	≤ 50	≤ 60	≤ 70	≤ 80	≤ 100
Cumulative Frequency								

[1 mark]

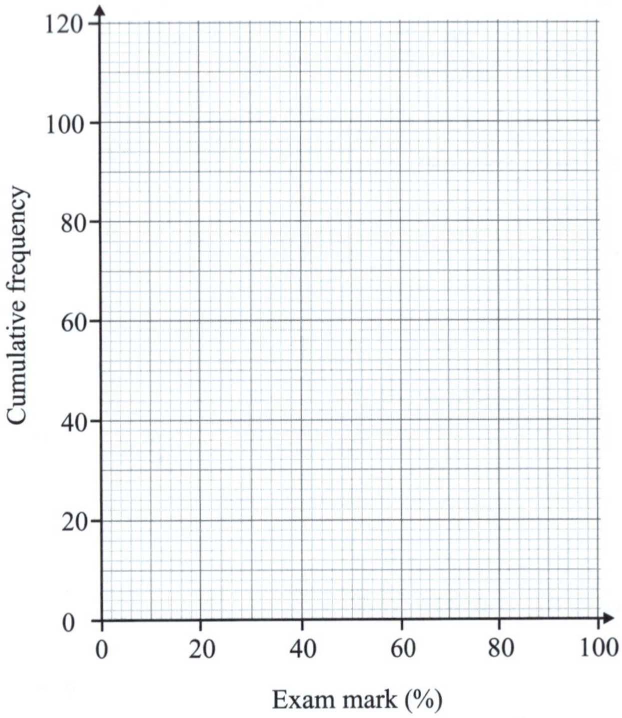

b) Use your table to draw a cumulative frequency graph on the graph paper to the left.

[3 marks]

c) Use your graph to find an estimate for the median.

.... %

[1 mark]

d) Use your graph to find an estimate for the interquartile range.

.... %

[2 marks]

e) Each pupil was awarded a grade based on their mark. 4 times as many pupils achieved grade 5 or higher as those who got a lower grade. Estimate the lowest mark needed to get grade 5.

....

....

[3 marks]

Histograms and Frequency Density

*A **histogram** is just a bar chart where the bars can be of **different widths**. This changes them from nice, easy-to-understand diagrams into seemingly incomprehensible monsters.*

Histograms Show **Frequency Density**

GRADE 7

1) The vertical axis on a histogram is always called frequency density. You work it out using this formula:

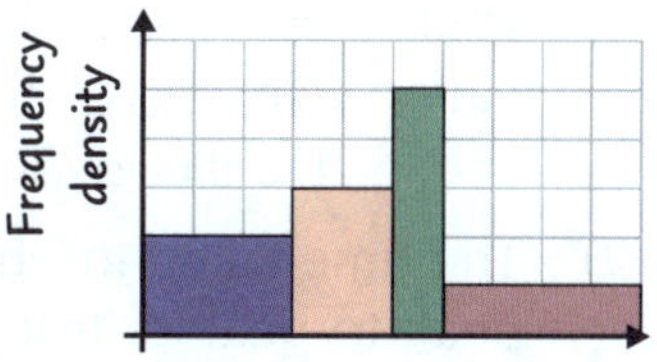

Frequency Density = Frequency ÷ Class Width

Remember... 'frequency' is just another way of saying 'how much' or 'how many'.

2) You can rearrange it to work out how much a bar represents.

Frequency = Frequency Density × Class Width = AREA of bar

EXAMPLE

This table and histogram show the lengths of beetles found in a garden.

GRADE 8

Length (mm)	Frequency
$0 < x \le 10$	32
$10 < x \le 15$	36
$15 < x \le 18$	
$18 < x \le 22$	28
$22 < x \le 30$	16

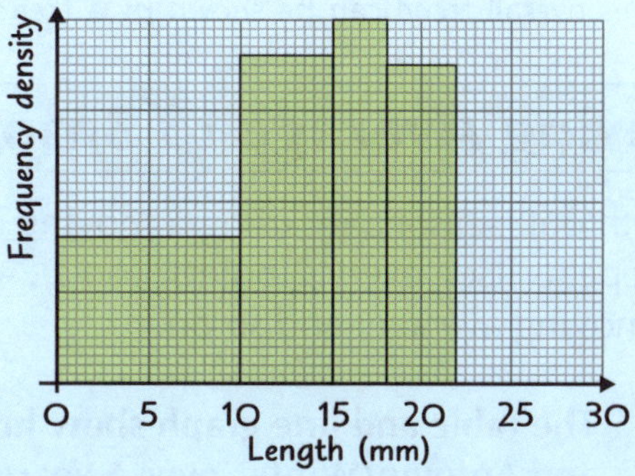

a) Use the histogram to find the missing entry in the table.

1) Add a frequency density column to the table and fill in what you can using the formula.

Frequency density
32 ÷ 10 = 3.2
36 ÷ 5 = 7.2
28 ÷ 4 = 7
16 ÷ 8 = 2

2) Use the frequency densities to label the vertical axis of the graph.

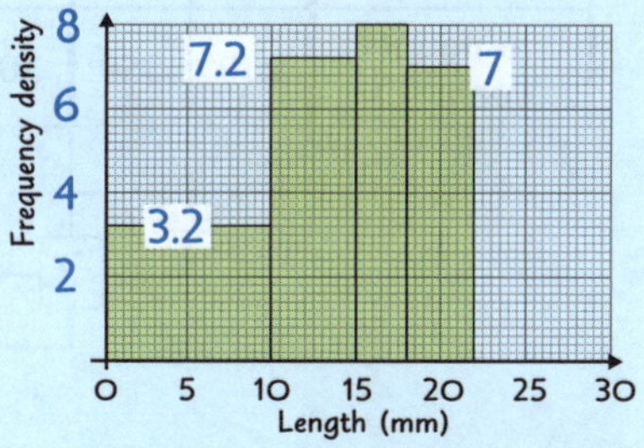

3) Now use the 3rd bar to find the frequency for the class "$15 < x \le 18$".

Frequency density = 8 and class width = 3.

So frequency = frequency density × class width = 8 × 3 = 24

b) Use the table to add the bar for the class "$22 < x \le 30$" to the histogram.

Frequency density = Frequency ÷ Class Width = $\frac{16}{8}$ = 2

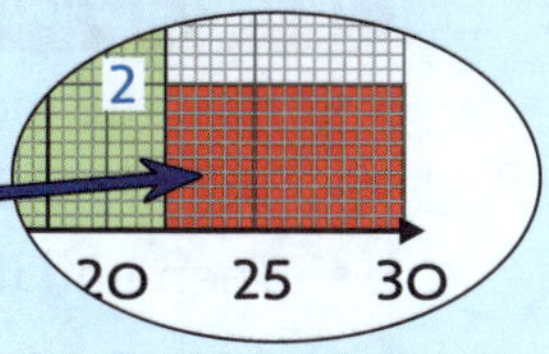

c) Estimate the number of beetles between 7.5 mm and 12.5 mm in length.

Use the formula frequency = frequency density × class width — multiply the frequency density of the class by the width of the part of that class you're interested in.

3.2 × (10 – 7.5) + 7.2 × (12.5 – 10)
= 3.2 × 2.5 + 7.2 × 2.5
= 26

Use the height and width of a bar to find its frequency

Although they look very like harmless bar charts, histograms are actually pretty tricky. Make sure you get your head around the method above...

Q1 a) This table shows information about the lengths of slugs in a garden. Draw a histogram to represent the information. [4 marks]

b) Estimate the number of slugs that are shorter than 70 mm. [3 marks]

Length (mm)	Frequency
$0 < x \le 40$	20
$40 < x \le 60$	45
$60 < x \le 65$	15
$65 < x \le 100$	70

Q1 Video Solution

GRADE 8

Time Series

*A **time series** is what you get if you measure the **same thing** at a number of **different times**.*

Line Graphs can show Time Series

GRADE 3

1) With time series, a basic pattern often repeats itself — this is called seasonality (though it doesn't have to match the seasons).
   The time series plotted in red has a definite repeating pattern.
2) The time taken for the pattern to repeat itself (measured peak-to-peak or trough-to-trough) is called the period.
   This pattern repeats itself every four points.
3) You can also look at the overall trend — i.e. whether the values are generally getting bigger or generally getting smaller (ignoring any repeating pattern). Look at the peaks and troughs — here they're going up slightly each time, which shows a slight upward trend.
   This overall trend can be shown by a trend line — drawn here in blue.

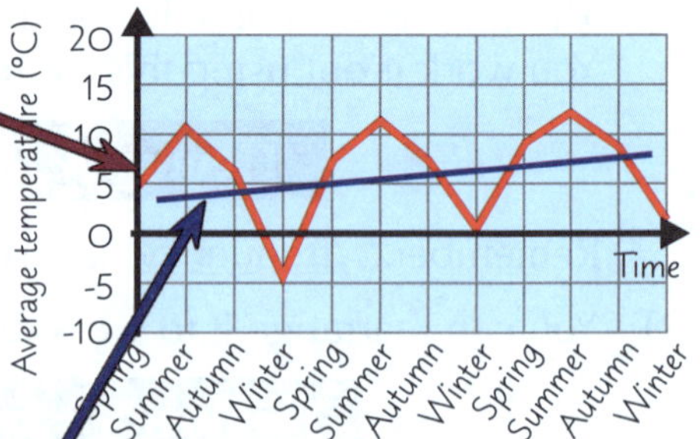

Use a Moving Average to Smooth Out a Time Series

GRADE 6

You can 'smooth out' seasonality using a moving average. This can help you see trends in the data. If the pattern repeats itself every 2 points, use a 2-point moving average, if it repeats every 3 points, use a 3-point moving average, and so on.

EXAMPLE

The table and line graph show how much Jamie spent on heating during Spring/Summer and Autumn/Winter over 4 years. Plot the values for a 2-point moving average and describe the trend in the data.

6-month period	Spr / Sum	Aut / Win	Spr / Sum	Aut / Win	Spr / Sum	Aut / Win	Spr / Sum	Aut / Win
Amount spent (£)	250	350	300	375	290	400	300	425

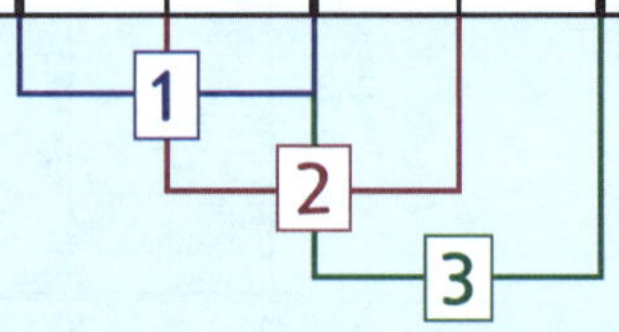

Use a 2-point moving average because the seasonal pattern repeats itself every 2 points.

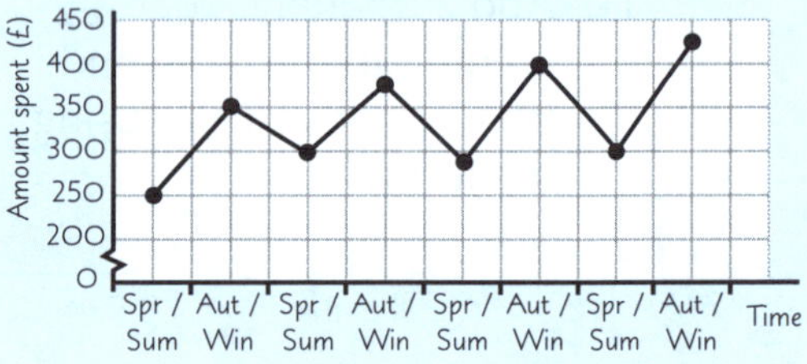

To find a 2-point MOVING AVERAGE:

- Find the mean of the 1st and 2nd values.
- Then find the mean of the 2nd and 3rd values.
- Then find the mean of the 3rd and 4th values.
- And so on, until you reach the last pair.

1: $\frac{250 + 350}{2} = £300$

2: $\frac{350 + 300}{2} = £325$

3: $\frac{300 + 375}{2} = £337.50$

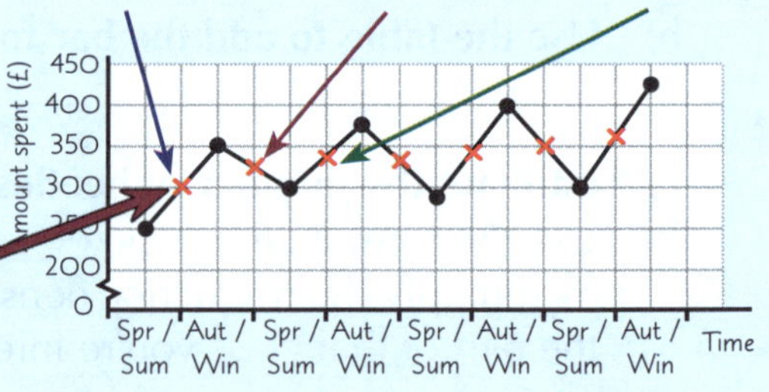

Plot the moving average point in the middle of each pair.

Look at the trend in the points you've plotted.

The data shows an upward trend in the amount Jamie spent.

Use the peaks or the troughs to work out the period

Quite a lot to take in on this page. Here's a question to check you've got it all.

GRADE 6

Q1 This data shows how many times Khalid goes rock-climbing in different quarters over 2 years.

a) Describe the repeating pattern in the data. [1 mark]

b) Find the values of a suitable moving average. [3 marks]

Quarter	1	2	3	4	1	2	3	4
Climbing	2	4	10	6	1	4	11	7

Scatter Graphs

*A **scatter graph** tells you **how closely** two things are **related** — the fancy word is **CORRELATION**.*

Scatter Graphs Show **Correlation**

1) If you can draw a line of best fit pretty close to most of your data points, the two things are correlated. If the points are randomly scattered, and you can't draw a line of best fit, then there's no correlation.
2) Strong correlation is when your points make a fairly straight line — this means the two things are closely related to each other. Weak correlation is when your points don't line up quite so nicely, but you can still draw a line of best fit through them.
3) If the points form a line sloping uphill from left to right, then there is positive correlation — both things increase or decrease together. If the line slopes downhill from left to right, then there is negative correlation — as one thing increases the other decreases.

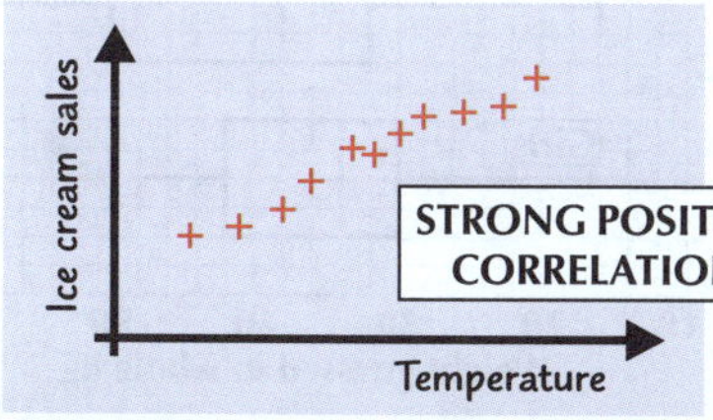

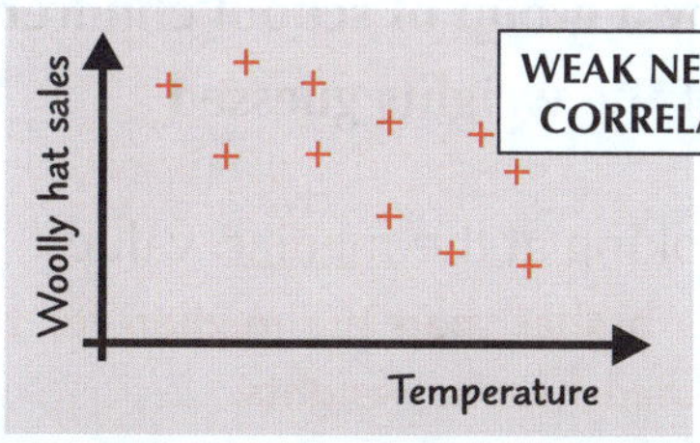

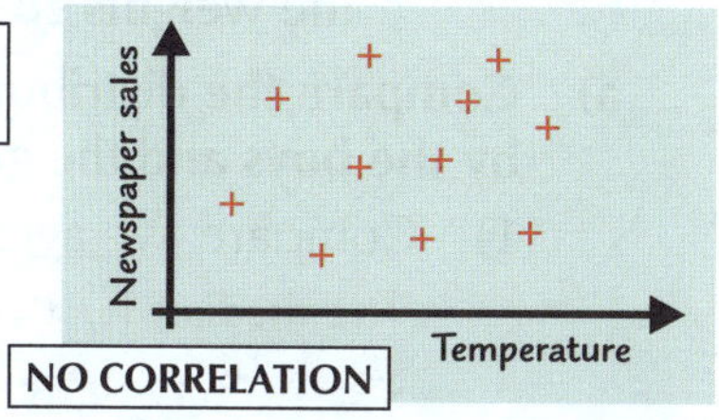

Use a **Line of Best Fit** to Make **Predictions**

1) You can use a line of best fit to make estimates. Predicting a value within the range of data you have should be fairly reliable, since you can see the pattern within this range. If you extend your line outside the range of data your prediction might be unreliable, since you're just assuming the pattern continues.
2) You also need to watch out for outliers — data points that don't fit the general pattern. These might be errors, but aren't necessarily. Outliers can drag your line of best fit away from the other values, so it's best to ignore them when you're drawing the line.

This graph shows the number of zoo visitors plotted against the outside temperature for several Sundays.

Draw a line of best fit to estimate the number of visitors when the temperature is 15 °C. 2250 should be a reliable estimate.

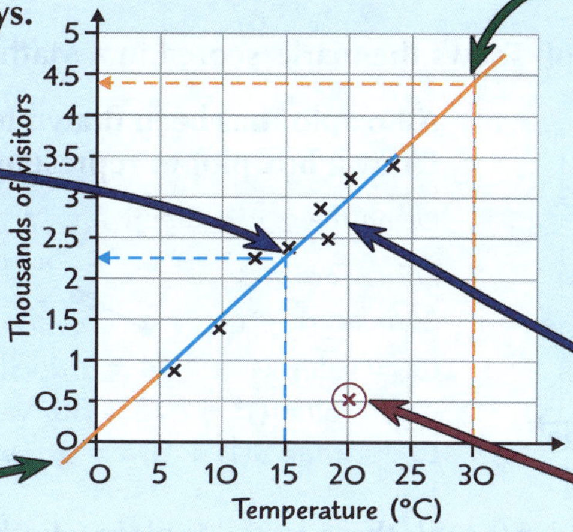

Extending the line you can estimate roughly 4375 visitors for a temperature of 30 °C. But this might be unreliable.

Predicting outside the range of data is called extrapolation.

Predicting within the range of data is called interpolation.

The data shows strong positive correlation — as the temperature increases, so does the number of visitors.

It doesn't make sense to extend the line below zero visitors.

This point is an outlier.

BE CAREFUL with correlation — if two things are correlated it doesn't mean that one causes the other. There could be a third factor affecting both, or it could just be a coincidence.

If you can draw a line of best fit, you have correlation

Time to test how much you can remember from this page with a question...

Q1 This graph shows Sam's average speed on runs of different lengths.

a) Describe the relationship between length of run and average speed. [1 mark]
b) Circle the point that doesn't follow the trend. [1 mark]
c) Estimate Sam's average speed for an 8-mile run. [1 mark]
d) Comment on the reliability of your estimate in part c). [1 mark]

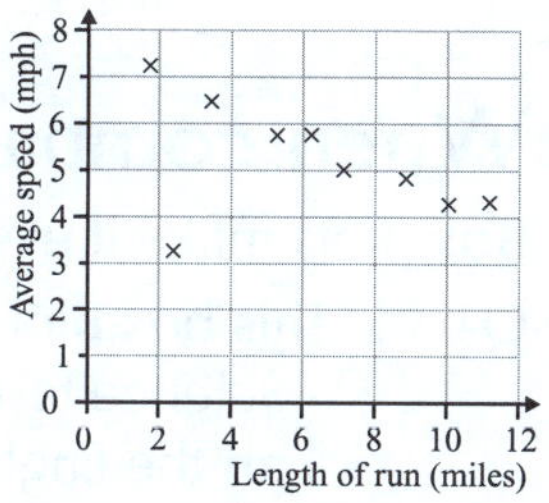

Comparing Data Sets

*You need to be able to **compare the distributions** of two sets of data represented by **graphs and charts**. That might mean comparing the **shapes** of the graphs, or reading off **measures of average** (mean, median or mode), and **spread** (range or interquartile range).*

Compare Data Sets using **Box Plots**

GRADE 6

For a reminder about box plots, see p.179.

From a box plot you can easily read off the median and work out the range and IQR.
Remember to say what these values mean in the context of the data.
A larger spread means the values are less consistent (there is more variation in the data).

EXAMPLE

An animal park is holding a 'guess the weight of the baby hippo' competition. These box plots summarise the weights guessed by a group of school children.

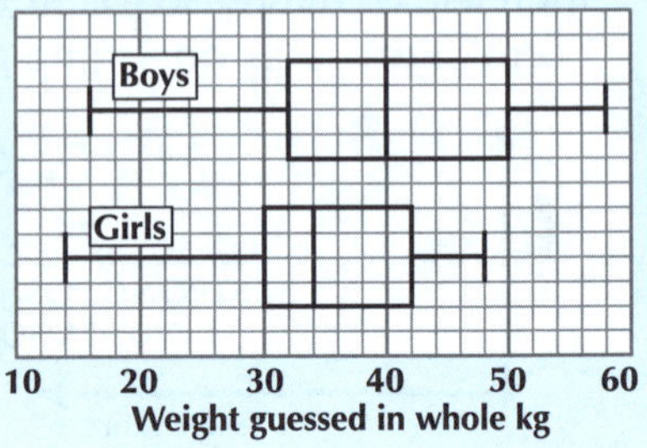

a) Compare the distributions of the weights guessed by the boys and the girls.

1) Compare averages by looking at the median values.

The median for the boys is higher than the median for the girls. So the boys generally guessed heavier weights.

2) Compare the spreads by working out the range and IQR.

Boys' range = 58 – 16 = 42 and IQR = 50 – 32 = 18.
Girls' range = 48 – 14 = 34 and IQR = 42 – 30 = 12.

Both the range and the IQR are smaller for the girls' guesses, so there is less variation in the weights guessed by the girls.

It's important you give your answers in the context of the data.

b) Can you tell from these box plots whether there are more boys or more girls in this group of children? Explain your answer.

The box plots don't show information on the numbers of data values, so you can't tell whether there are more boys or more girls.

EXAMPLE **This scatter graph shows the marks scored in a Maths test and an English test by 11 students.**

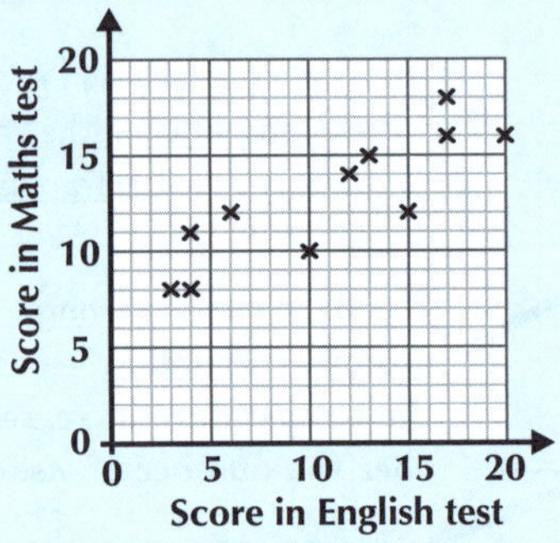

a) A box plot has been drawn to represent the Maths scores. Draw a box plot to represent the English scores.

Using the scatter graph:

Min score = 3
Max score = 20

See p.179 for a reminder.

Q_1 = value (11 + 1)/4 = 3rd value = 4
Q_2 = value (11 + 1)/2 = 6th value = 12
Q_3 = value 3(11 + 1)/4 = 9th value = 17

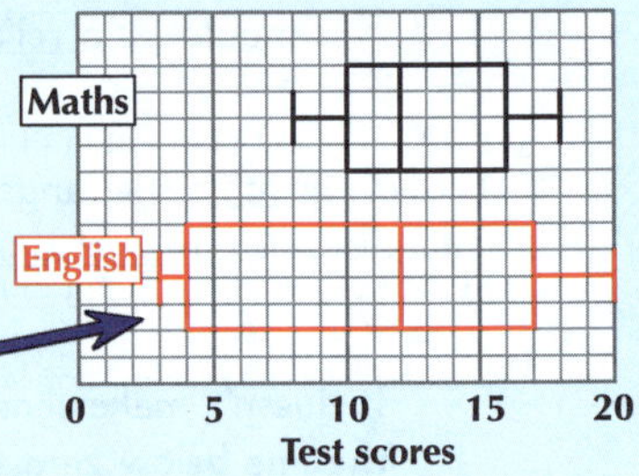

b) A total of 1000 students took these tests. Explain whether you can use the box plots above to compare the English and Maths scores of all the students who took the test.

You can only compare the scores of these 11 students, not all the students, because a sample of 11 isn't big enough to represent the whole population of 1000 students.

When comparing distributions, look at average, range and IQR

Box plots make it easy to see similarities and differences between data sets.

GRADE 5

Q1 This box plot represents the marks scored by the 11 students above in a Science test. Claudia says that these Science scores are more consistent than the English scores. Explain whether Claudia is correct. [2 marks]

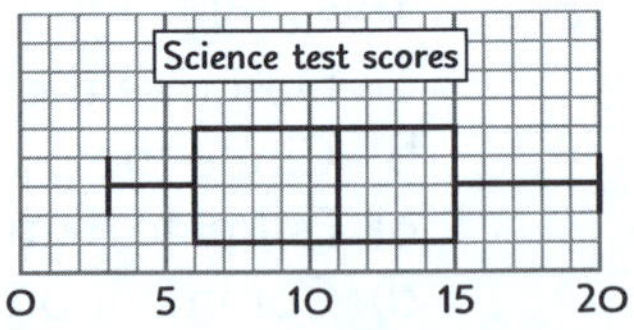

Comparing Data Sets

Compare Data Sets using Histograms

See p.183 for a reminder about histograms.

EXAMPLE

This histogram shows information about the times taken by a large group of children to solve a puzzle.

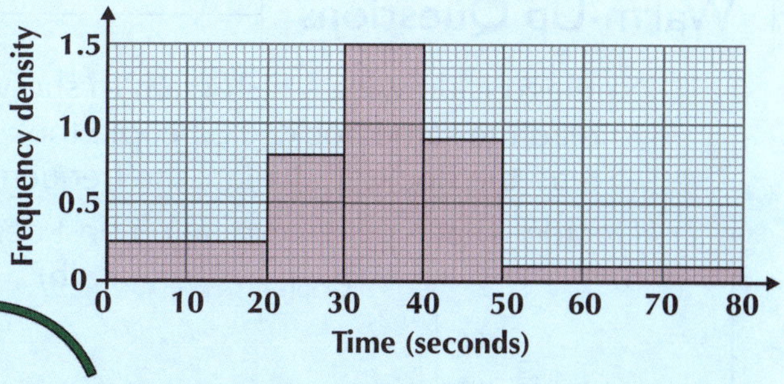

a) **Estimate the mean time taken to solve the puzzle.**

Draw a table and fill in what the graph tells you.

Time (seconds)	Frequency Density	Frequency (f)	x	f × x
0 < t ≤ 20	0.25	0.25 × 20 = 5	10	50
20 < t ≤ 30	0.8	0.8 × 10 = 8	25	200
30 < t ≤ 40	1.5	1.5 × 10 = 15	35	525
40 < t ≤ 50	0.9	0.9 × 10 = 9	45	405
50 < t ≤ 80	0.1	0.1 × 30 = 3	65	195
Total	—	40	—	1375

Find the frequency in each class using:
Frequency = Frequency Density × Class Width

Add a column for the mid-interval values.

Add up the 'Frequency × mid-interval value' column to estimate the total time taken.

Number of children

Mean = $\frac{\text{total time taken}}{\text{number of children}} = \frac{1375}{40}$ = **34.4 seconds (1 d.p.)**

This is just like estimating the mean from a grouped frequency table (see p.178). Now you've found the frequencies, you could also find the class containing the median.

b) **Write down the modal class.**

Modal class is 30 < t ≤ 40

The modal class has the highest frequency density. It's frequency density, not frequency, because the class widths vary.

c) **Estimate the range of times taken.**

Highest class boundary – lowest class boundary = 80 – 0 = 80 seconds

d) **A large group of adults solve the same puzzle with a mean time of 27 seconds. Is there any evidence to support the hypothesis that children take longer to solve the puzzle than adults?**

Yes, there is evidence to support this hypothesis because the mean time for the children is longer.

Large samples mean the results should represent the population.

EXAMPLE **A company makes chocolate bars and decides to change the wrappers they use.**

The histograms show information on the daily sales of the chocolate bar, before and after changing the wrapper. The company claims that changing the wrapper has increased daily sales.

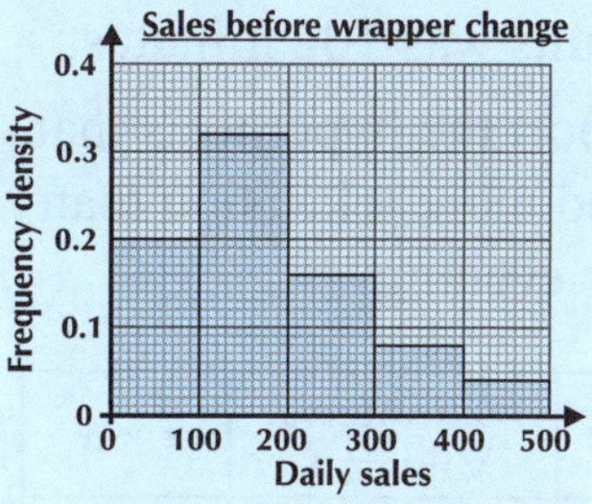

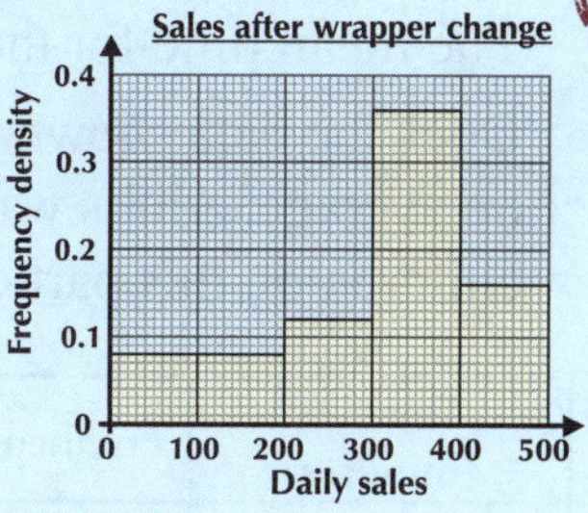

a) **Is there evidence that daily sales have increased since the wrapper change?**

Yes, there is evidence that daily sales have increased because there are more days with high sales values since the wrapper change than before.

Look at the shapes of the histograms — after the change, more data is on the right-hand side (showing more days with high sales).

b) **Comment on the company's claim that changing the wrapper has increased daily sales.**

Changing the wrapper might have caused the increase in sales, but sales could also have been affected by other factors, e.g. pricing, advertising, economic conditions, etc.

You could be asked to use any type of graph

Get comfortable with the different types of graph here and learn what they represent. In the exam you might have to criticise how a graph has been drawn.

Q1 Give three criticisms of the histogram opposite. [3 marks]

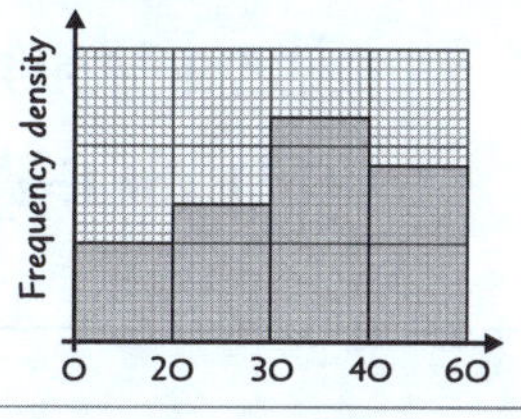

Warm-Up and Worked Exam Questions

There are some lovely warm-up questions here covering stats graphs. Now's the time to go back over any bits you're not sure of — in the exam it'll be too late.

Warm-Up Questions

1) Santiago is drawing a histogram to show how long it took different people to read the same book. 12 people took between 3 and 6 hours to read the book. Calculate the frequency density for this class.
2) Decide what type of correlation best describes the two scatter graphs below.

a)

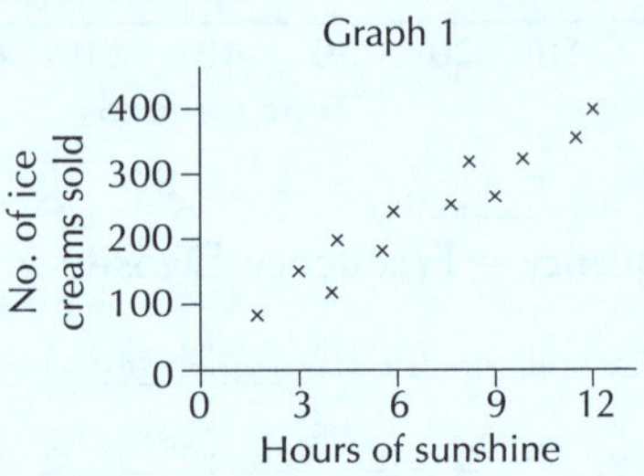

b)

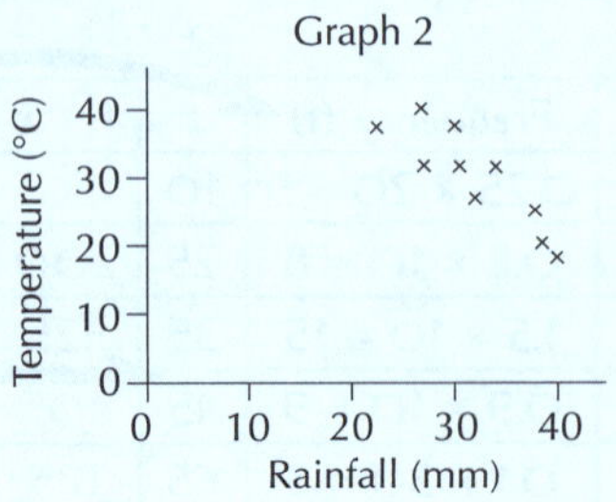

3) The table below shows the number of people who visit a holiday park over three years. Give a suitable moving average that could be used to show the overall trend in the data.

Month	Jan-Apr	May-Aug	Sept-Dec	Jan-Apr	May-Aug	Sept-Dec	Jan-Apr	May-Aug	Sept-Dec
People (thousands)	9.7	14.3	3.2	10.5	15.4	3.8	11.1	15.9	4.5

Worked Exam Question

It's no good learning all the facts in the world if you can't put them into practice in the exam. This worked example shows how to make all those facts into a good answer — and earn yourself marks.

1 The histogram shows the amount of time (in minutes) that 270 children spent watching television one evening.

GRADE 8

A large sample of adults were asked how long they watched television for on the same evening. The mean time for the adults was 102 minutes.

Does the data shown support the hypothesis that, on average, adults watched more television than children on this particular evening?

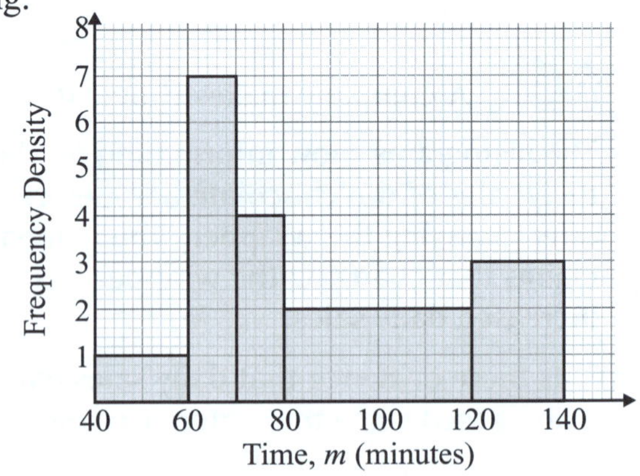

Time, m (minutes)	Frequency (f)	Mid-interval Value (x)	f × x
40 ≤ m < 60	20 × 1 = 20	50	1000
60 ≤ m < 70	10 × 7 = 70	65	4550
70 ≤ m < 80	10 × 4 = 40	75	3000
80 ≤ m < 120	40 × 2 = 80	100	8000
120 ≤ m < 140	20 × 3 = 60	130	7800
Total	270		24 350

Estimate the mean amount of time the children watched TV for, then compare that to the mean time for the adults.

Mean (children) = 24 350 ÷ 270
= 90.185...
= 90.2 minutes (1 d.p.)

The data supports the hypothesis as the mean time for the adults is longer than the mean time for the children, and large samples mean the results should represent the population.

[4 marks]

Exam Questions

2 A furniture company is looking at how effective their advertising is. They are comparing how much they spent on advertising in random months with their total sales value for that month. This information is shown on the graph below.

The table shows the amount spent on advertising and the value of sales for three more months.

Amount spent on advertising (thousands of pounds)	0.75	0.15	1.85
Sales (thousands of pounds)	105	60	170

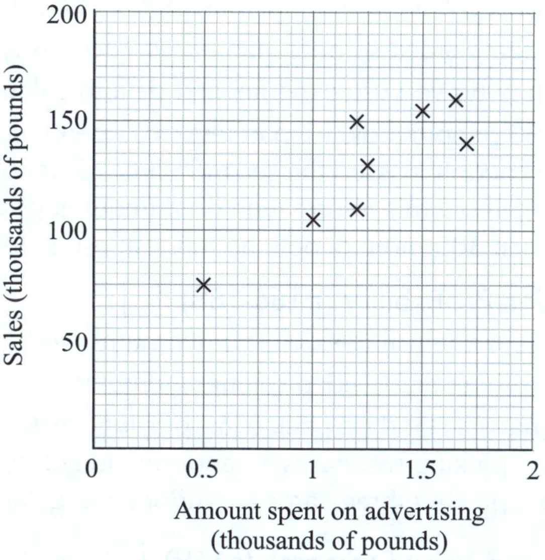

a) Plot the information from the table on the scatter graph.

[1 mark]

b) Describe the relationship between the amount spent on advertising and the value of sales.

....

....

....

[1 mark]

c) Use a line of best fit to estimate how much the company would be likely to spend on advertising in a month where they sold £125 000 worth of furniture.

£ ....

[2 marks]

d) The company plan to increase their monthly spend on advertising to at least £3000. They use the trend in the data above to predict future sales values. Comment on how reliable this prediction is likely to be.

....

....

....

[2 marks]

3 Tom gives a puzzle to a sample of boys and girls. These box plots show information about the time it took the children to finish the puzzle.

Compare the distributions of the times taken by the boys and the times taken by the girls.

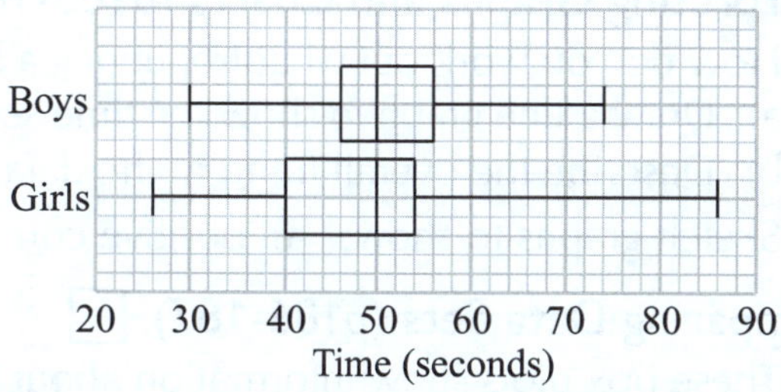

....

....

....

[2 marks]

Revision Questions for Section Seven

Here's the inevitable list of questions to test how much you know.

- Tick off each one when you can get it right without cheating.
- When you're completely happy with a topic, tick that off too.

For even more practice, try the quiz for Section Seven — just scan this QR code!

Section Seven Quiz

Basic Probability (p160-163)

1) I pick a random number between 1 and 50. Find the probability that my number is a multiple of 6.
2) A fair y-sided spinner, numbered 1 to y, is spun twice. What is the probability of getting two 1's?
3) How do you use experimental data to calculate relative frequencies?
4) 160 people took a 2-part test. 105 people passed the first part and of these, 60 people passed the second part. 25 people didn't pass either part.
   a) Show this information on a frequency tree. b) Find the relative frequency of each outcome.
   c) If 300 more people do the test, estimate how many of them would pass both parts.

Harder Probability (p166-168)

5) I spin a fair nine-sided spinner, numbered 1-9, twice. Find P(spinning a 6 then an even number).
6) I spin a fair 20-sided spinner, numbered 1-20. Find P(spinning a factor of 20 or an even number).
7) I have a standard pack of 52 playing cards. Use tree diagrams to find the probability of me:
   a) picking two cards at random and getting two kings if the first card is replaced.
   b) picking three cards at random and getting three kings if no cards are replaced.

Sets and Venn Diagrams (p169)

8) 180 people were asked whether they like tea or coffee. Half the people surveyed said they only like coffee, $2x + 5$ people said they only like tea, x people said they like both and $2x$ people like neither.
   a) Show this information on a Venn diagram. b) Find x.
   c) If one of the 180 people is randomly chosen, find the probability of them liking tea.

Sampling, Data Collection and Averages (p172-175)

9) What is a sample and why does it need to be representative?
10) Is 'eye colour' qualitative, discrete or continuous data?
11) Find the mode, median, mean and range of this data: 2, 8, 11, 15, 22, 24, 27, 30, 31, 31, 41

Grouped Frequency Tables (p178)

12) For this grouped frequency table showing the lengths of some pet alligators:
   a) find the modal class,
   b) find the class containing the median,
   c) estimate the mean.

Length (y, in m)	Frequency
$1.4 \le y < 1.5$	4
$1.5 \le y < 1.6$	8
$1.6 \le y < 1.7$	5
$1.7 \le y < 1.8$	2

Graphs and Charts (p179-180)

13) Draw a cumulative frequency graph for the data in the grouped frequency table in Q12 above.

Interpreting Graphs and Charts (p183-185)

14) How do you work out what frequency a bar on a histogram represents?
15) a) Draw a line graph to show the time series data in this table.
    b) Describe the overall trend in the data.

Quarter	1	2	3	4	1	2	3	4
Sales (1000's)	1	1.5	1.7	3	0.7	0.9	1.2	2.2

16) Sketch graphs to show: a) positive correlation b) negative correlation c) no correlation

Comparing Data Sets (p186-187)

17) These box plots show information about how long it took someone to get to work in summer and winter one year. Compare the travel times in the two seasons.

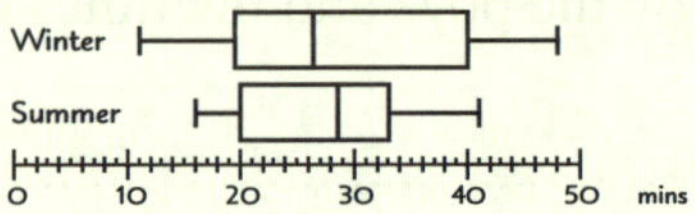

18) An 800 m runner had a mean time of 147 seconds, before she increased her training hours. The histogram shows information about the times she runs after increasing her training hours. Is there any evidence that her running times have improved?

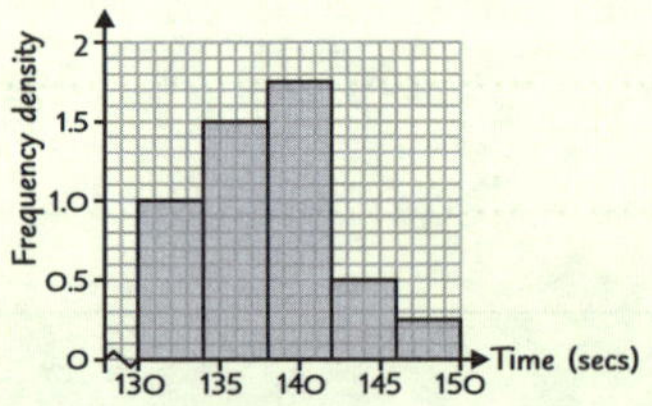

Practice Paper 1: Calculator

As final preparation for the exams, we've included three full practice papers to really put your Maths skills to the test. Paper 1 and Paper 3 (on p.216) require a calculator. Paper 2 (on p.203) is a non-calculator paper. There's a whole page on formulas on p.252. Good luck...

Candidate Surname | Candidate Forename(s)

Centre Number | Candidate Number | Candidate Signature

GCSE

Mathematics

Higher Tier

Paper 1 (Calculator)

Practice Paper

Time allowed: 1 hour 30 minutes

You must have:
Pen, pencil, eraser, ruler, protractor, pair of compasses.
You may use tracing paper.

You **may use** a calculator.

Instructions to candidates

- Use **black** ink to write your answers.
- Write your name and other details in the spaces provided above.
- Answer **all** questions in the spaces provided.
- In calculations, show clearly how you worked out your answers.
- Do all rough work on the paper.
- Unless a question tells you otherwise, take the value of π to be 3.142, or use the π button on your calculator.

Information for candidates

- The marks available are given in brackets at the end of each question.
- You may get marks for method, even if your answer is incorrect.
- There are 23 questions in this paper. There are no blank pages.
- There are 100 marks available for this paper.

Answer ALL the questions.

Write your answers in the spaces provided.

You must show all of your working.

1 The length of a leaf is 11 cm to the nearest centimetre.
Give the lower and upper bounds for the actual length of the leaf.

lower bound = ... cm

upper bound = ... cm

[Total 2 marks]

---

2 The diagram shows a right-angled triangle.

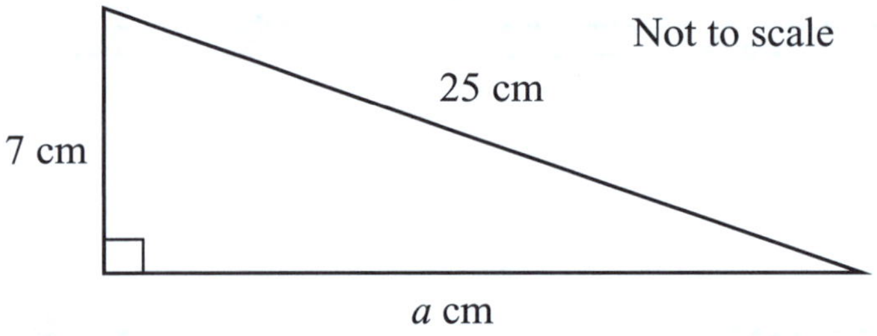

Calculate the value of a.

a = ...

[Total 2 marks]

---

3 Gabi is asked to draw a quadrilateral with these three properties:

- one line of symmetry
- exactly two sides that are equal in length
- two pairs of equal angles

Gabi says, "There is no quadrilateral which has all these properties."

Draw a shape on the grid to show that Gabi is wrong.

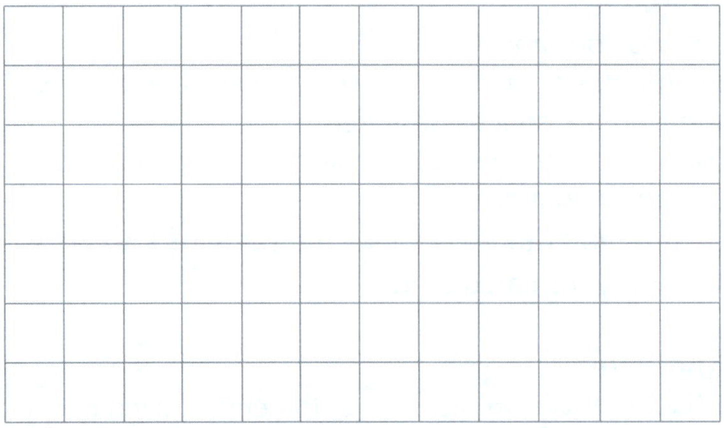

[Total 2 marks]

---

4 Calculate the value of $\frac{18.4 \times 2.56}{\sqrt{21.6 - 4 \times 1.55}}$. Give your answer correct to 3 significant figures.

....

[Total 2 marks]

---

5 A drawer contains ties that are coloured either red or green or white or black.
Zayn picks a tie at random from the drawer. The table shows some of the probabilities.

Colour of tie	Red	Green	White	Black
Probability	0.35	0.20		

The drawer contains exactly twice as many black ties as white ties.

Zayn says, "Half the ties are coloured either red or white."
Is Zayn correct? Tick a box.

Yes ☐ No ☐

Show how you worked out your answer.

[Total 4 marks]

---

6 Amir and Orla each have an expression.

Amir	Orla
$(x + 4)^2 - 1$	$(x + 5)(x + 3)$

Show clearly that Amir's expression is equivalent to Orla's expression.

[Total 3 marks]

---

7 The scatter graph shows the maximum power (in kW) and the maximum speed (in km/h) of a sample of cars.

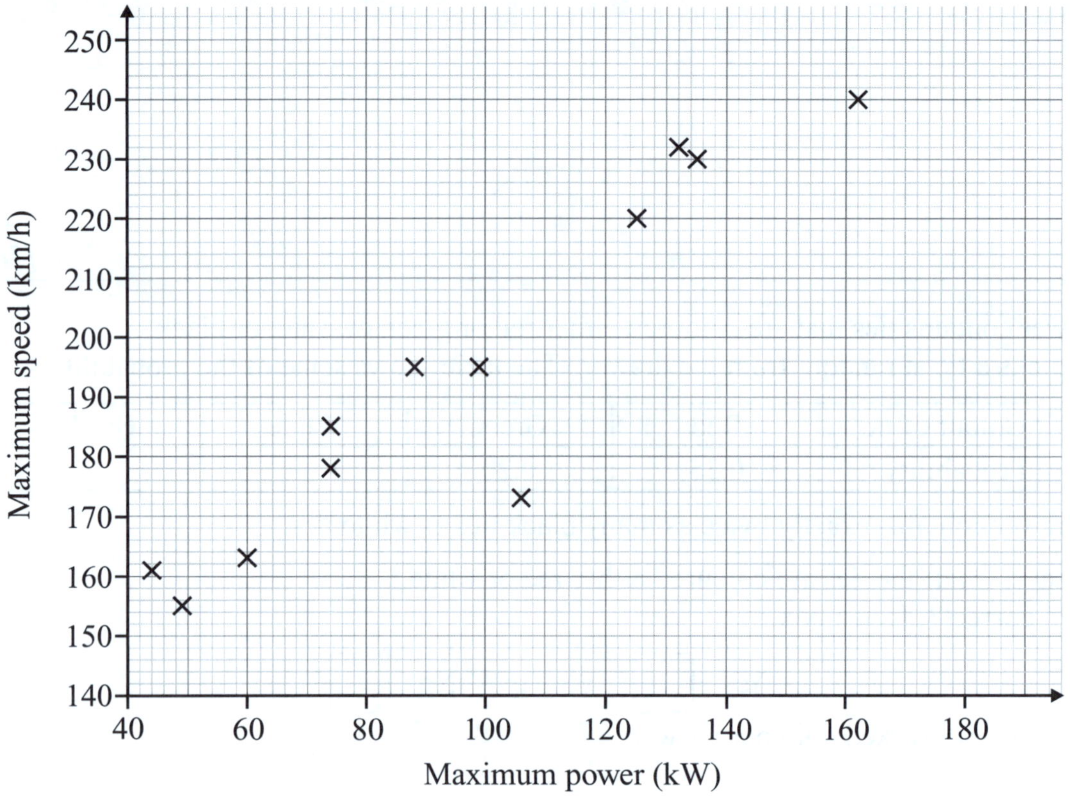

(a) One of the cars has a maximum speed of 220 km/h.
Write down the maximum power of this car.

.... kW
[1]

(b) One of the points is an outlier as it does not fit in with the trend.
Draw a ring around this point on the graph.

[1]

(c) Ignoring the outlier, describe the correlation shown on the scatter graph.

.... correlation
[1]

(d) A different car has a maximum power of 104 kW.
By drawing a suitable line on your scatter graph, estimate the maximum speed of this car.

.... km/h
[2]

(e) Explain why it may not be reliable to use the scatter graph to estimate the maximum speed of a car with a maximum power of 190 kW.

....

....

[1]

[Total 6 marks]

8

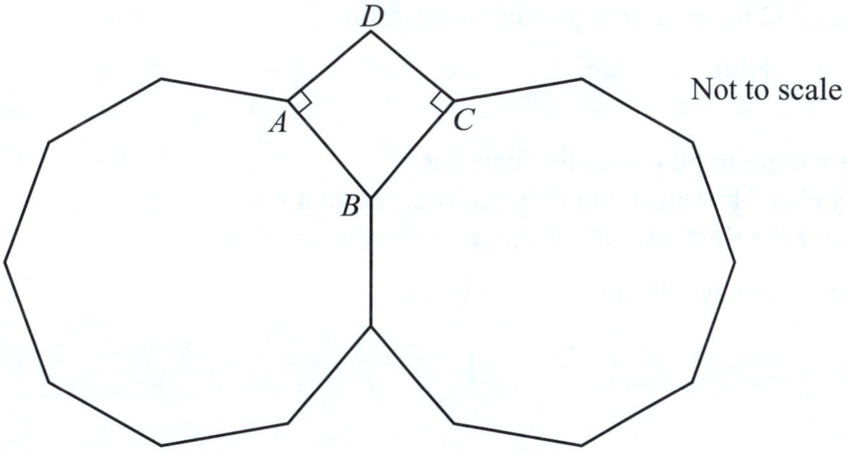

AB and BC are sides of congruent nine-sided regular polygons.
Angle DAB = angle DCB = 90°.

Calculate the size of angle ADC.

.... °

[Total 3 marks]

9 (a) Solve the inequality $5(3 - x) < x - 9$

....

[3]

(b) Make x the subject of the formula $y = \frac{x^3 - 2}{5}$

....

[2]

[Total 5 marks]

10 Beth shares £1200 between four people in the ratio

Alex	:	Hiro	:	Mei	:	Eve
2	:	3	:	4	:	6

Eve uses their share to buy tyres for their car.
Tyres usually cost £150 each, but they are reduced in a sale.
Eve has exactly the right amount of money to buy four tyres.

Calculate the percentage discount on the tyres.

.... %

[Total 5 marks]

---

11 The diagram shows a solid aluminium cylinder and a solid silver cube.

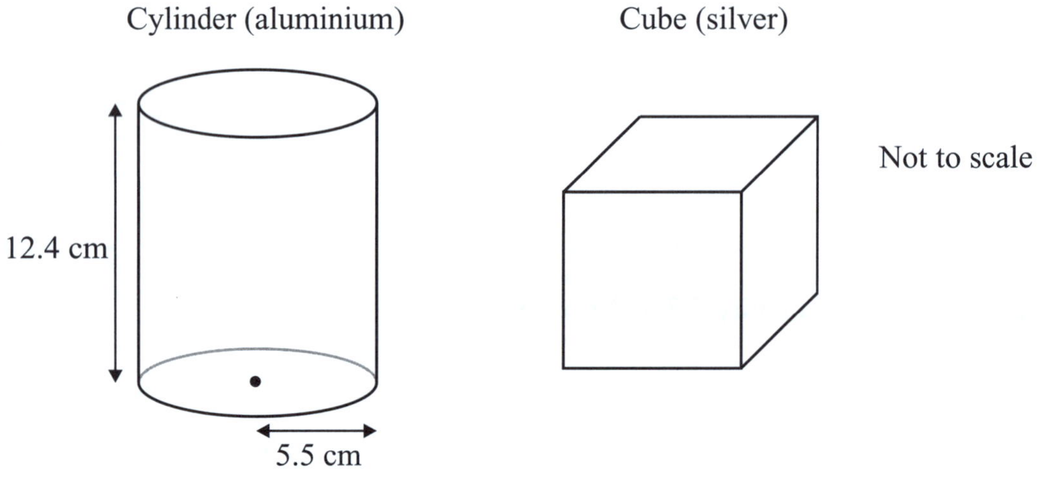

The cylinder and the cube have the same mass.
The density of aluminium is 2.7 g/cm^3 and the density of silver is 10.5 g/cm^3.

Calculate the side length of the cube. Give your answer correct to two significant figures.

.... cm

[Total 5 marks]

12 Shape R is drawn on the grid below:

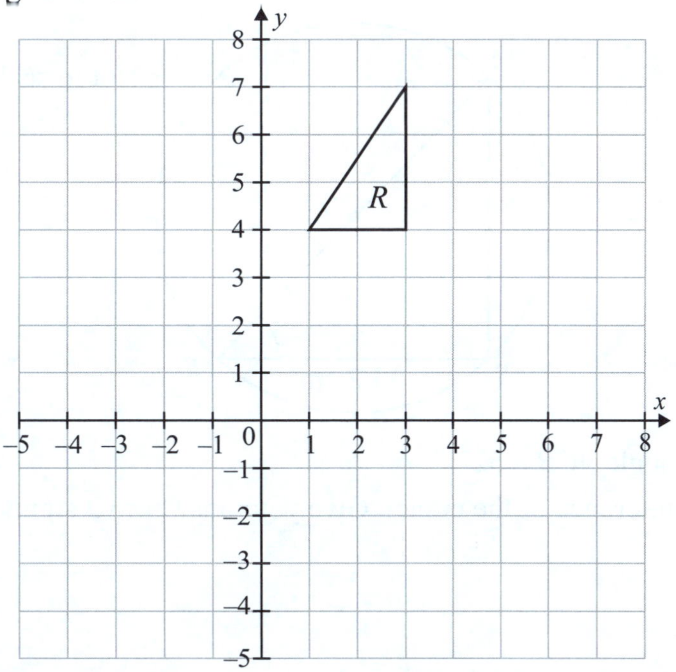

(a) Shape S is produced by a reflection of shape R in the line $y = x$, followed by a reflection in the line $y = 0$. Draw and label the transformed shape S.

[2]

(b) Fully describe the single transformation that would take shape R onto shape S.

....

....

[3]

[Total 5 marks]

---

13 Jamila has two plant pots that are mathematically similar.

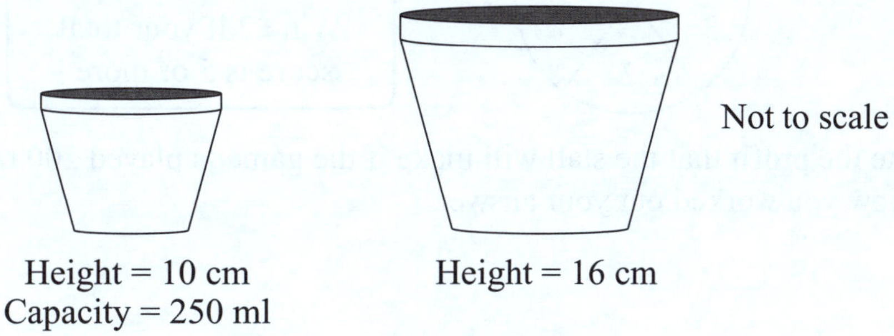

Will the large plant pot hold one litre of compost? Tick a box.

Yes ☐ No ☐

Show how you worked out your answer.

[Total 3 marks]

---

14 *A*, *B* and *C* are points on the circumference of a circle with centre *O*.

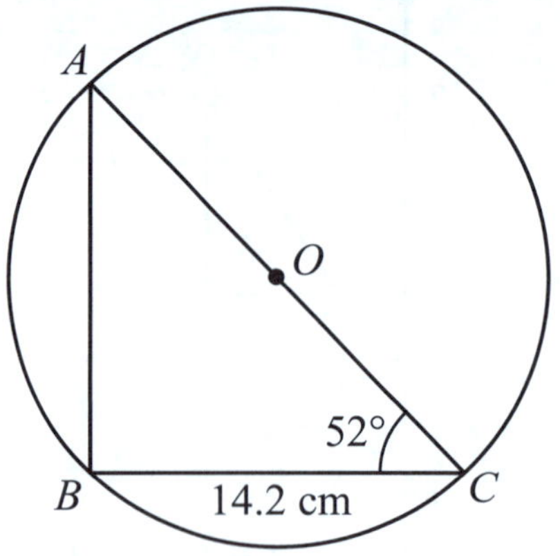

$BC = 14.2$ cm and angle $ACB = 52°$.

Calculate the circumference of the circle. Give your answer to 3 significant figures.

.... cm

[Total 5 marks]

---

15 A funfair stall runs a game played using this spinner. The rules of the game are shown below.

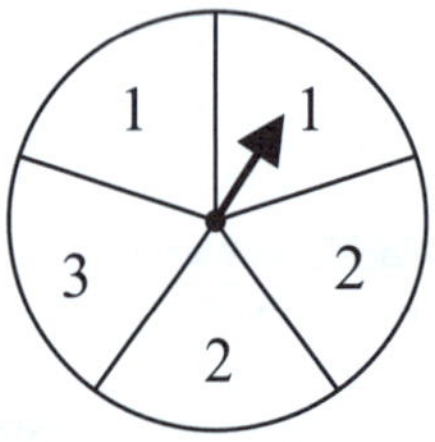

50p a go
Spin the spinner twice.
Win £2 if your total score is 5 or more.

(a) Estimate the profit that the stall will make if the game is played 200 times.
Show how you worked out your answer.

£ ....

[5]

(b) Write down one assumption that you made about the spinner when you answered part (a).

....

....

[1]

[Total 6 marks]

16 (a) Dev has £6000, which they want to invest for three years.
Dev is choosing between two savings accounts which each pay compound interest.

Account 1	**Account 2**
2.5% per annum Fixed for 3 years	Year 1: Interest rate 1.0% Year 2: Interest rate 1.5% Year 3: Interest rate 5.0%

Which account should Dev choose if they want to receive the greatest possible amount of interest? Show how you worked out your answer.

....

[4]

(b) Nina invests a sum of money in an account for two years.
The account pays 2% per annum compound interest.
Nina receives a total of £606 interest.

Work out the amount of money Nina invested initially.

£ ....

[3]

[Total 7 marks]

---

17 The table on the right shows some information about the lengths, l cm, of iron nails in a jar.

Length of nail, l cm	Frequency
$0.5 \leq l < 1.0$	8
$1.0 \leq l < 1.5$	23
$1.5 \leq l < 2.0$	40
$2.0 \leq l < 4.0$	21
$4.0 \leq l < 6.0$	18

(a) What percentage of the nails are over 1.5 cm long?
Give your answer to 2 decimal places.

.... %

[2]

(b) Calculate an estimate for the mean length of the nails.
Give your answer to 2 decimal places.

.... cm

[4]

[Total 6 marks]

---

18 Work out the values of a and b so that

$$\frac{ax+b}{2x^2-32}\times(x^2-2x-8)=3x+6$$

$a =$ ....

$b =$ ....

[Total 5 marks]

---

19 The diagram shows a patio made from a rectangle of width x m joined to a semicircle of radius x m.

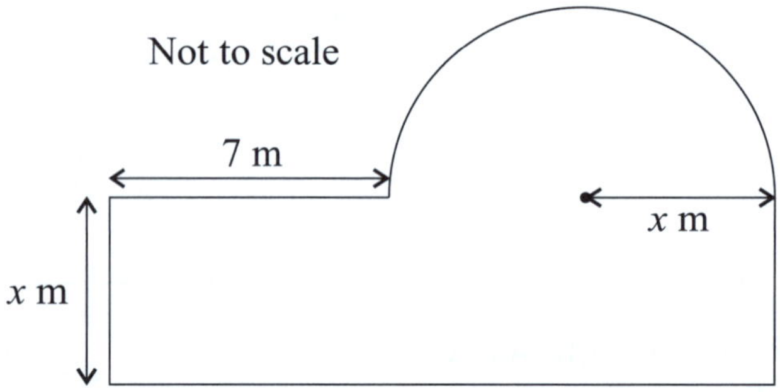

The area of the patio is 40 m^2.

(a) Show that x satisfies the equation $(\pi+4)x^2+14x-80=0$.

[4]

(b) Work out the value of x.

$x =$ ....

[2]

[Total 6 marks]

20 For each part, work out a possible equation of the curve shown by the solid line.
The curve shown by a dotted line on each grid is $y = \cos x$.

(a)

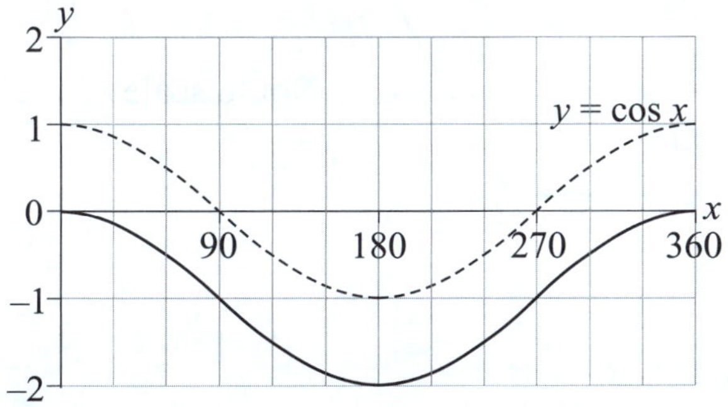

$y =$ ....

[1]

(b)

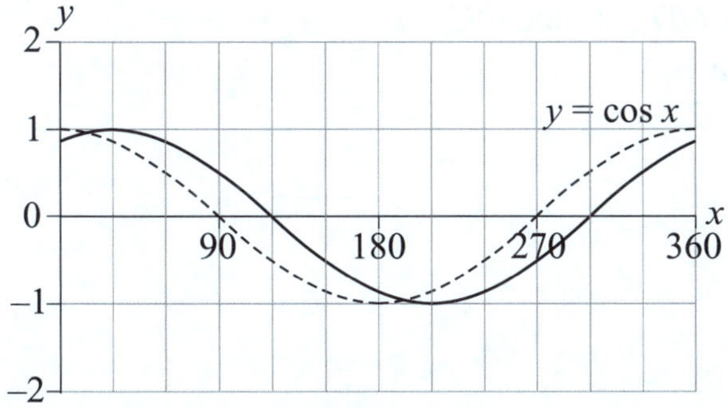

$y =$ ....

[1]

(c)

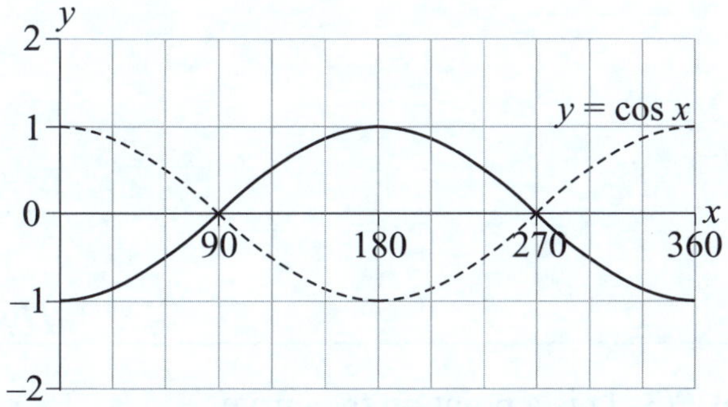

$y =$ ....

[1]

[Total 3 marks]

---

21 Here is a function machine.

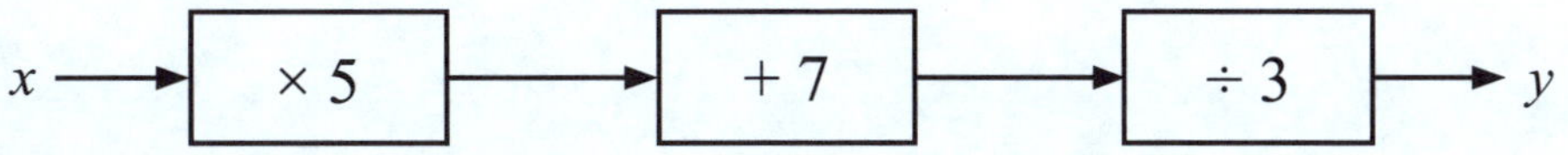

(a) What is the value of x when $y = 154$?

....

[2]

(b) A value of a is put into the function machine to give an output of b.
b is then put back into the function machine to give an output of c.
Write an expression in terms of a for the value of c. Fully simplify your answer.

....

[4]

[Total 6 marks]

22 *OABC* is a parallelogram.

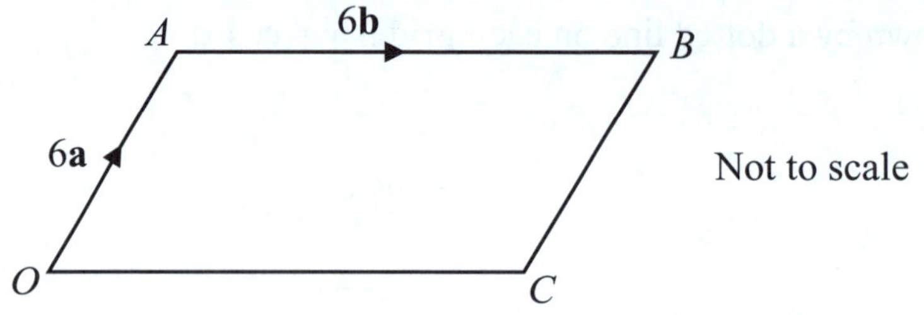

(a) Find the vector $\overrightarrow{AC}$.

....
[1]

When *OC* is extended, *D* is a point such that *OD* is twice *OC*.
M is the point with position vector $\overrightarrow{OM} = 4\mathbf{a}$.
N is the point on *BD* such that $BN : ND = 1 : 2$.

(b) Prove that *MN* is parallel to *AB*.

....
[4]
[Total 5 marks]

23 A curve has equation $x^2 + y^2 = 10$. The point $P(3, 1)$ is a point on the curve.

Work out the equation of the tangent to the curve at *P*.

....
[Total 4 marks]

[TOTAL FOR PAPER = 100 MARKS]

Practice Paper 2: Non-calculator
Right, here's Practice Paper 2 — no calculators allowed for this one. Don't forget there's a page on formulas (p.252) — it tells you which formulas you'll be given in your exam and which you'll need to learn.

Candidate Surname	Candidate Forename(s)	
Centre Number	Candidate Number	Candidate Signature

GCSE

Mathematics — Higher Tier

Paper 2 (Non-Calculator)

Practice Paper

Time allowed: 1 hour 30 minutes

You must have:
Pen, pencil, eraser, ruler, protractor, pair of compasses.
You may use tracing paper.

You are **not allowed** to use a calculator.

Instructions to candidates
- Use **black** ink to write your answers.
- Write your name and other details in the spaces provided above.
- Answer **all** questions in the spaces provided.
- In calculations, show clearly how you worked out your answers.
- Do all rough work on the paper.

Information for candidates
- The marks available are given in brackets at the end of each question.
- You may get marks for method, even if your answer is incorrect.
- There are 22 questions in this paper. There are no blank pages.
- There are 100 marks available for this paper.

Answer ALL the questions.

Write your answers in the spaces provided.

You must show all of your working.

1 Circle the graphs that match the following descriptions.

(a) A straight line has equation $y = mx + c$ where $m > 0$ and $c < 0$.

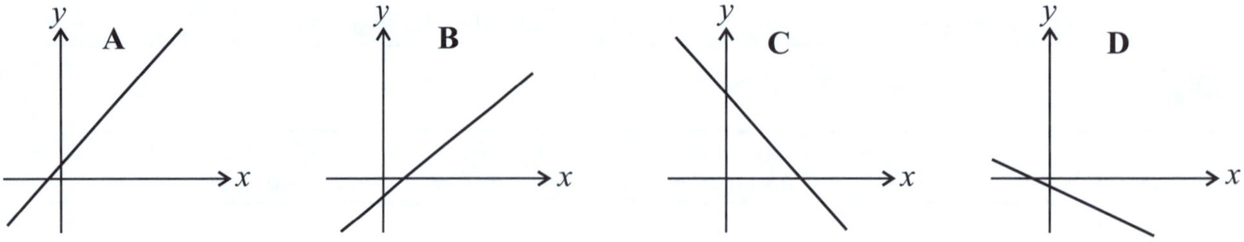

[1]

(b) y is inversely proportional to x.

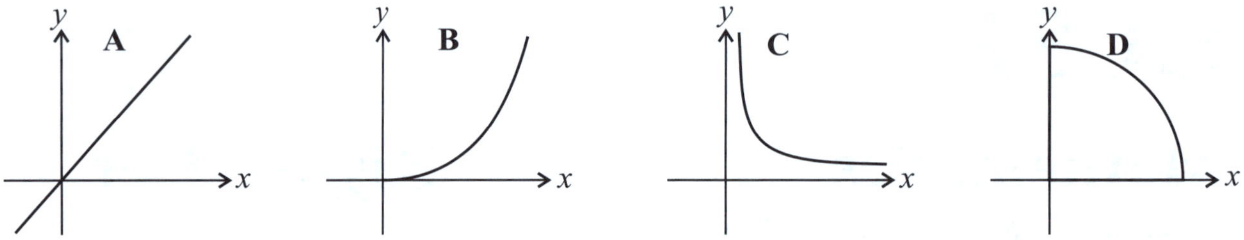

[1]

[Total 2 marks]

---

2 A is 60% of B.
B is 30% of C.

(a) What fraction of C is A? Give your answer in its simplest form.

....

[3]

(b) If $A = 9 \times 10^8$, work out the value of C. Give your answer in standard form.

....

[2]

[Total 5 marks]

---

3 *ABCD* is a rectangular piece of card.

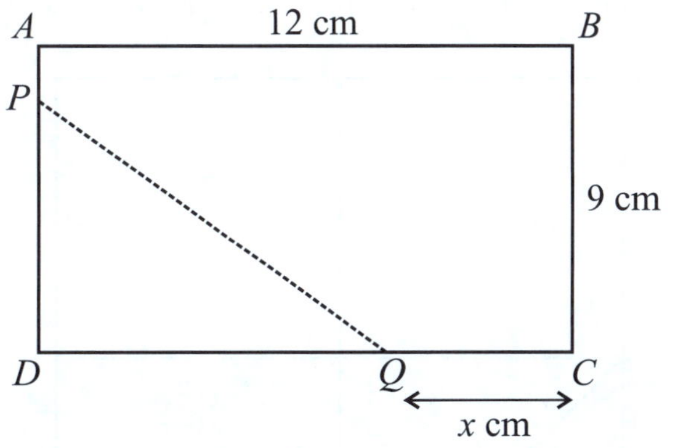

Not to scale

P is the point two-thirds of the way along *DA*.
Q is *x* cm from *C*.

The card is cut into two pieces along the line *PQ*.
The ratio of the areas of the smaller piece to the larger piece is 1 : 3

(a) Find the area of triangle *DPQ*.

.... cm^2

[2]

(b) Work out the value of *x*.

x = ...

[3]

[Total 5 marks]

4 Estimate the value of

$$\sqrt[3]{\frac{785.3 \times 2.156}{0.1972}}$$

Show the numbers you used to work out your estimate.

....

[Total 3 marks]

5 The diagram shows a square *EFGH*.
The square has been divided into smaller squares and isosceles triangles.

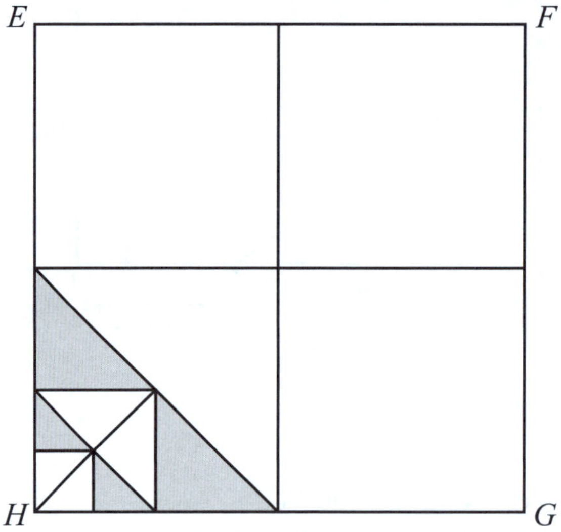

What fraction of the square *EFGH* has been shaded?

....

[Total 3 marks]

6 The *n*th term of a sequence is given by the formula $n^2 + 2n + 5$

(a) Charlie says "The 4th term in the sequence is a prime number."
Is Charlie correct? Tick a box.

Yes ☐ No ☐

Show how you worked out your answer.

....

....

[2]

(b) A different sequence begins 2, 5, 7, 12, 19, ...
Write down the next two terms in the sequence.

.... and ....

[2]

[Total 4 marks]

7 $\xi = \{1, 2, 3, \ldots, 10\}$
$A = \{x : 2 < x \leq 6\}$
$B = \{x : x \text{ is a factor of } 12\}$

(a) Complete the Venn diagram to show the number of elements in each set.

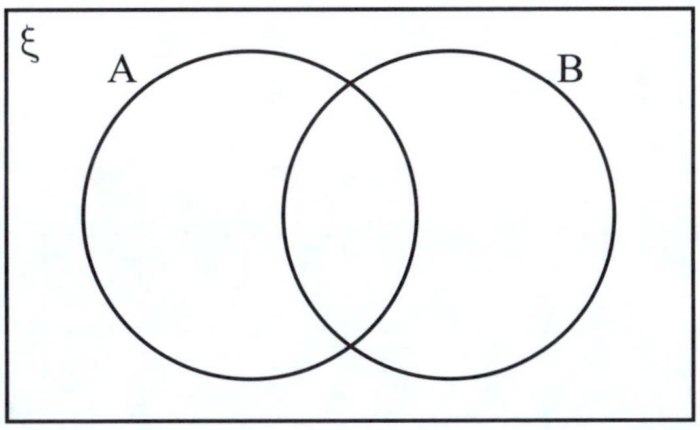

[3]

(b) A random number generator is used to select a number from 1-10.
What is the probability that the number is in set A, given that it is not in set B?

....

[1]

[Total 4 marks]

---

8 (a) Find values of a and b such that

$$3a + 2b = 17$$
$$2a + b = 10$$

$a =$ ....

$b =$ ....

[3]

(b) Hence, work out $a\begin{pmatrix}2\\1\end{pmatrix} - b\begin{pmatrix}3\\2\end{pmatrix}$

....

[2]

[Total 5 marks]

---

9 The scale drawing shows the position of three hospitals, A, B and C, on an island.

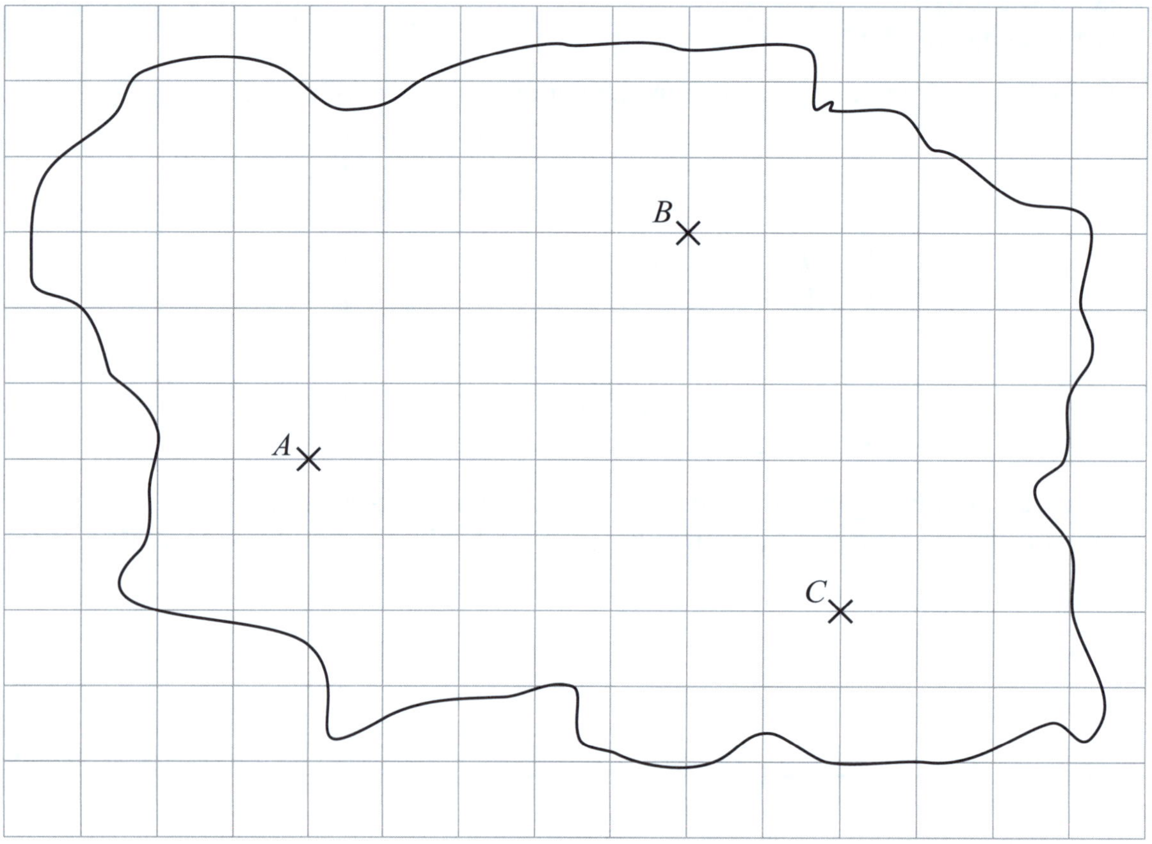

Scale: 1 cm = 10 miles

An ambulance takes a patient to hospital A if they are within a 30 mile radius of this hospital.
Otherwise it will take the patient to hospital B or C depending on which is closer.

Shanti calls from their home for an ambulance.
The ambulance takes Shanti to hospital B.

Show on the map the region where Shanti could live.

[Total 3 marks]

---

10 $p = 2^3 \times 3^2 \times 5 \times 7 \qquad q = \frac{6}{7}p \qquad r = \frac{4}{15}p$

(a) Work out the lowest common multiple of p and q.
Give your answer as a product of prime numbers.

....

[3]

(b) Work out the highest common factor of q and r.
Give your answer as a whole number.

....

[3]

[Total 6 marks]

---

11 The population of Stone City is 4.2×10^5

(a) 7.5% of the population is aged 75 or above.
There are 20 500 more right-handed people than left-handed people aged 75 or above.

Work out the number of left-handed people aged 75 or above.
Give your answer as an ordinary number.

....

[4]

(b) The population of Deepland is 27 times that of Stone City.

Calculate the population of Deepland.
Give your answer in standard form correct to 2 significant figures.

....

[2]

[Total 6 marks]

12 Simplify $\dfrac{3x^2 + 14x + 8}{9x^2 - 4}$

....

[Total 3 marks]

13 The diagram shows a circle A and a sector B.

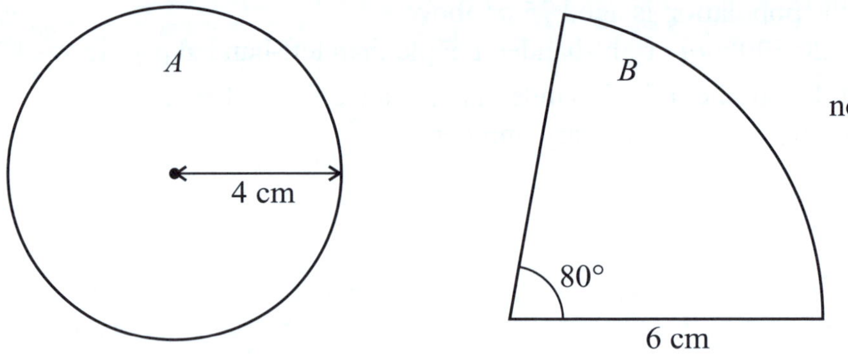

not to scale

Show that the area of A is twice the area of B.

[Total 3 marks]

14 Calculate $\left(2\frac{1}{4}\right)^{-\frac{1}{2}} \div \frac{2}{9}$

....

[Total 4 marks]

15 Sixty teams took part in a charity pram race in 2022.
The cumulative frequency graph shows the times that the teams took to complete the course.

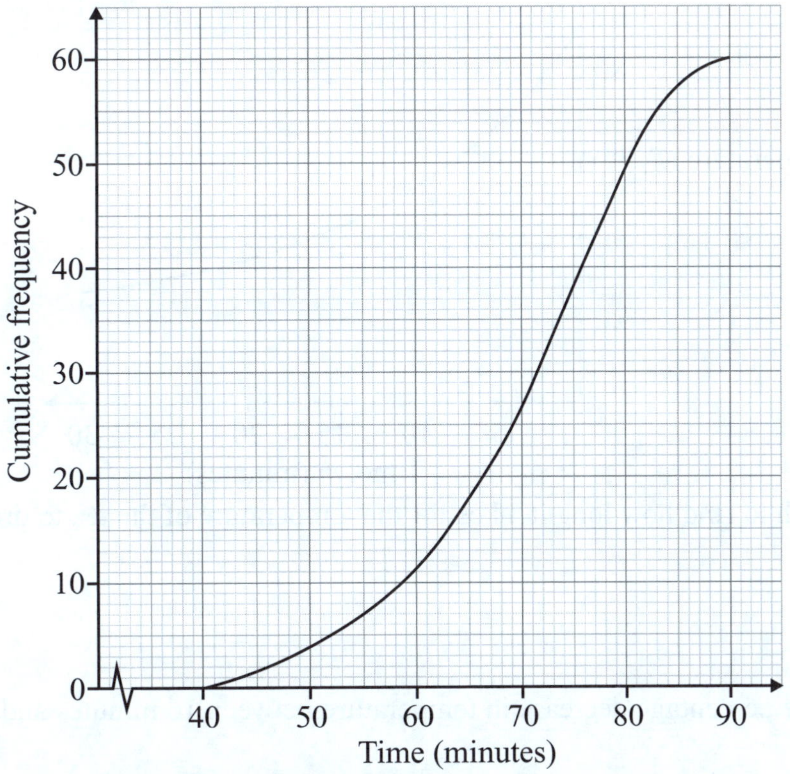

The table below summarises the times that teams took to complete the pram race in 2021.

2021 Pram Race Times	
Median	76 minutes
Interquartile range	18 minutes
Winning time	37 minutes

(a) What is the smallest possible difference between the winning times in 2021 and 2022?

.... minutes
[2]

(b) On average were the teams faster in 2021 or 2022? Explain your answer.

....

....

....
[2]

(c) Were the times more consistent in 2021 or 2022? Explain your answer.

....

....

....
[2]

[Total 6 marks]

16 The diagram shows the temperature, T °C, of a cup of tea m minutes after it is made.

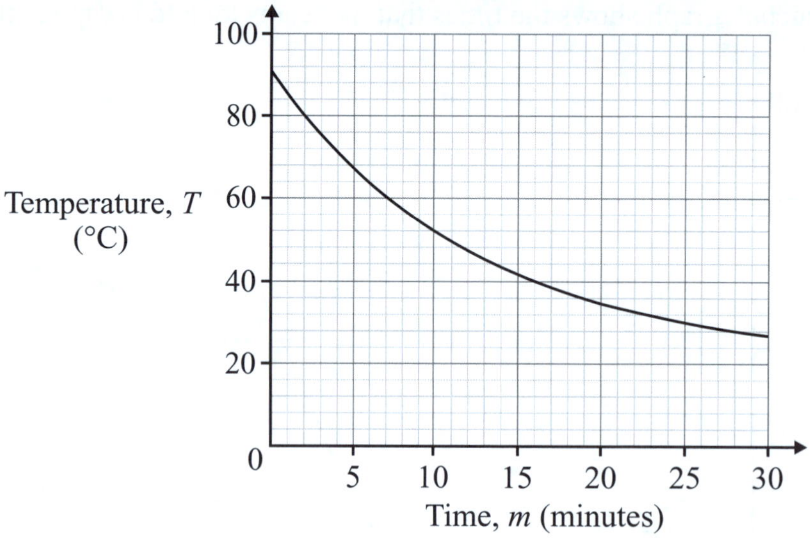

(a) Use the graph to find how long it takes for the temperature of the tea to drop to 36 °C.

.... minutes

[1]

(b) What was the percentage decrease in temperature between 16 minutes and 28 minutes?

.... %

[3]

(c) Estimate the rate at which the temperature of the tea is decreasing 10 minutes after it is made.

.... °C/minute

[3]

[Total 7 marks]

17 Write $\frac{6}{\sqrt{3}} + \sqrt{27}$ in the form $k\sqrt{3}$.

....

[Total 3 marks]

18

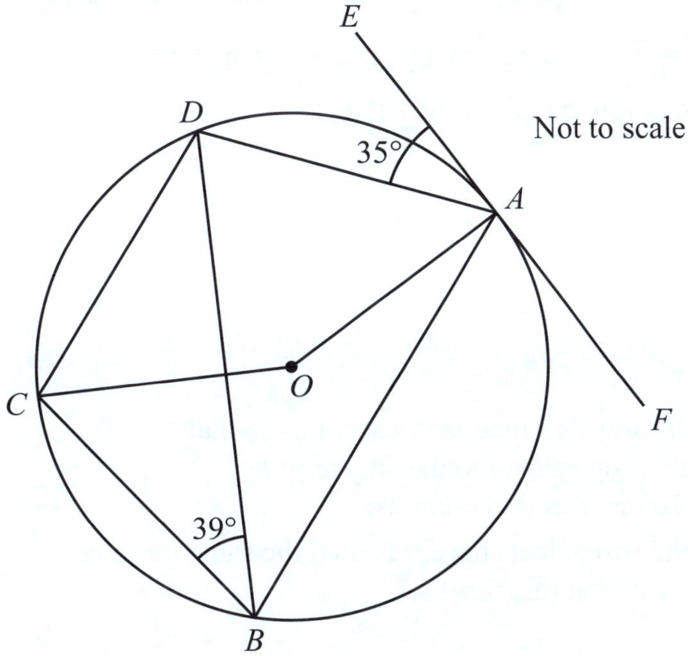

A, *B*, *C* and *D* are points on the circumference of a circle centre *O*.
EF is a tangent to the circle at *A*.
Angle *EAD* = 35° and angle *DBC* = 39°

Work out the size of angle *DCO*.
Show all of your working, some of which may be on the diagram.

[Total 5 marks]

19 Object A is accelerating. The speed, x m/s, of object A after t seconds ($t > 0$) can be found using the equation $x = \frac{18(s + 2t^2)}{5t}$, where s metres is the distance travelled after t seconds.

(a) Rearrange this equation to make s the subject.

$s =$ ....

[2]

(b) A second object, B, travels s metres in t seconds so that:

- s is directly proportional to the square of t.
- it travels 160 metres in 8 seconds.

After 6 seconds the two objects have travelled the same distance.
Calculate the value of x at this time.

$x =$ ....

[5]

[Total 7 marks]

20 (a) Expand $(n + 1)(n - 1)(n + 4)$.

....

[3]

(b) n is a positive integer.
Prove that the value of $n(n + 3)(n + 1) - (n + 1)(n - 1)(n + 4)$ is a multiple of 4.

....

[3]

[Total 6 marks]

21 The diagram on the right shows a frustum made by removing the top part of a cone.
The frustum has a base diameter of 12 cm and a slant height of 10 cm.
The original cone had a slant height of 15 cm.

Calculate the exact surface area of the frustum.

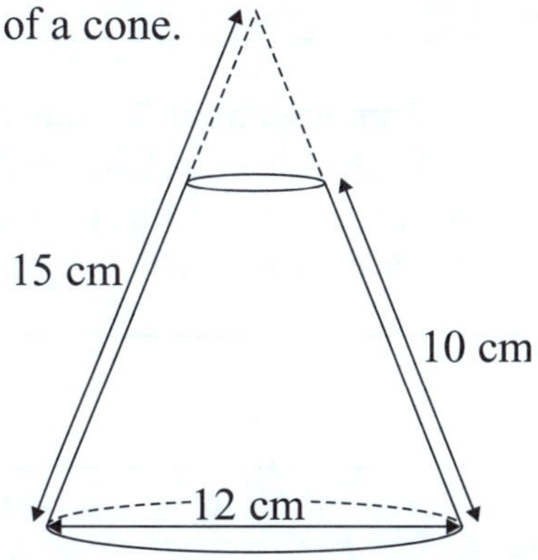

.... cm^2

[Total 6 marks]

22 The diagram shows a quadratic graph.

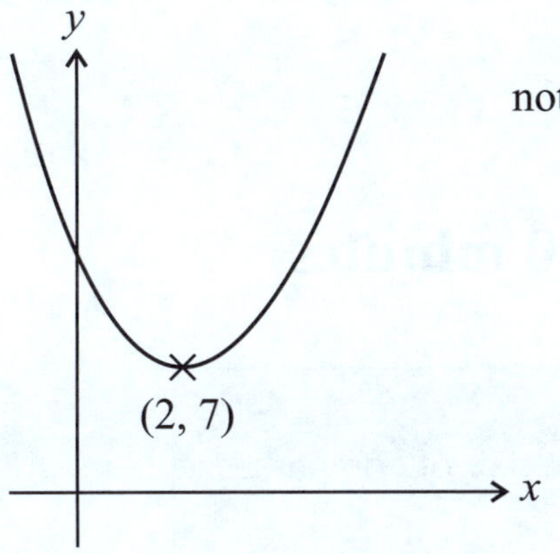

not to scale

The equation of the graph can be written in the form $y = (x - a)^2 + b$
The turning point of the quadratic has coordinates (2, 7).
A point on the graph has coordinates $(k, 128)$

Calculate the two possible values of k.

k = ... or k = ...

[Total 4 marks]

[TOTAL FOR PAPER = 100 MARKS]

Practice Paper 3: Calculator
Finally, here's Practice Paper 3 — you'll be pleased to know that you can use your calculator for this one too. Take a look at the formulas page on p.252 if you need a reminder of which formulas you need to learn and which you'll be given in the exam.

Candidate Surname	Candidate Forename(s)	
Centre Number	Candidate Number	Candidate Signature

GCSE

Mathematics

Higher Tier

Paper 3 (Calculator)

Practice Paper

Time allowed: 1 hour 30 minutes

You must have:
Pen, pencil, eraser, ruler, protractor, pair of compasses.
You may use tracing paper.

You **may use** a calculator.

Instructions to candidates

- Use **black** ink to write your answers.
- Write your name and other details in the spaces provided above.
- Answer **all** questions in the spaces provided.
- In calculations, show clearly how you worked out your answers.
- Do all rough work on the paper.
- Unless a question tells you otherwise, take the value of π to be 3.142, or use the π button on your calculator.

Information for candidates

- The marks available are given in brackets at the end of each question.
- You may get marks for method, even if your answer is incorrect.
- There are 23 questions in this paper. There are no blank pages.
- There are 100 marks available for this paper.

Answer ALL the questions.

Write your answers in the spaces provided.

You must show all of your working.

1 Ivan and Umi each throw an ordinary six-sided dice once.

What is the probability that they both throw a number greater than 4?
Give your answer as a fraction in its simplest form.

....

[Total 2 marks]

---

2 Darcie thinks of a whole number between 10 and 30.
Darcie's number is not a prime number and when they square it, the final digit is 1.

What number did Darcie think of?

....

[Total 2 marks]

---

3 Jack has four number cards.

[7] [12] [?] [?]

Jack's four numbers have a median value of 12 and a mean of 13.
Work out the range of Jack's four numbers. Show how you worked out your answer.

....

[Total 3 marks]

---

4 A function is represented by this number machine:

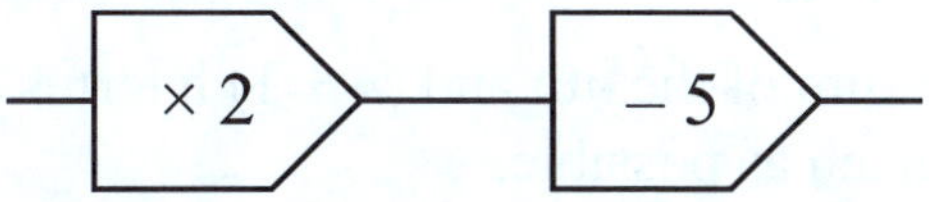

What is the ratio of the outputs when the inputs are 15 and 25?
Give your answer in the form $1 : n$.

....

[Total 3 marks]

---

5 The diagram shows an object made from 8 centimetre cubes.

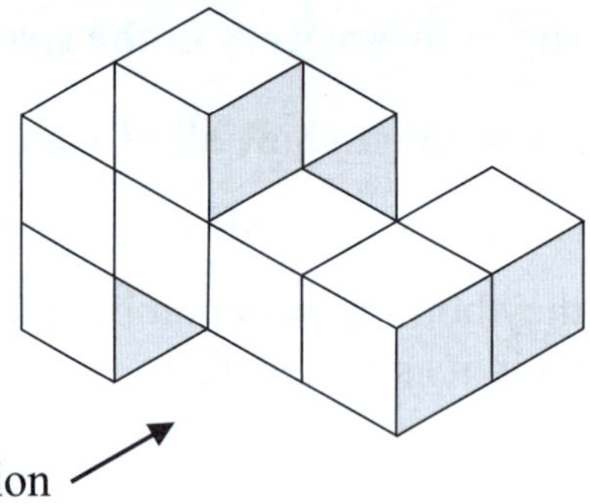

Front elevation

Draw on the grid below the plan view and the front elevation of the object.

Plan view | Front elevation

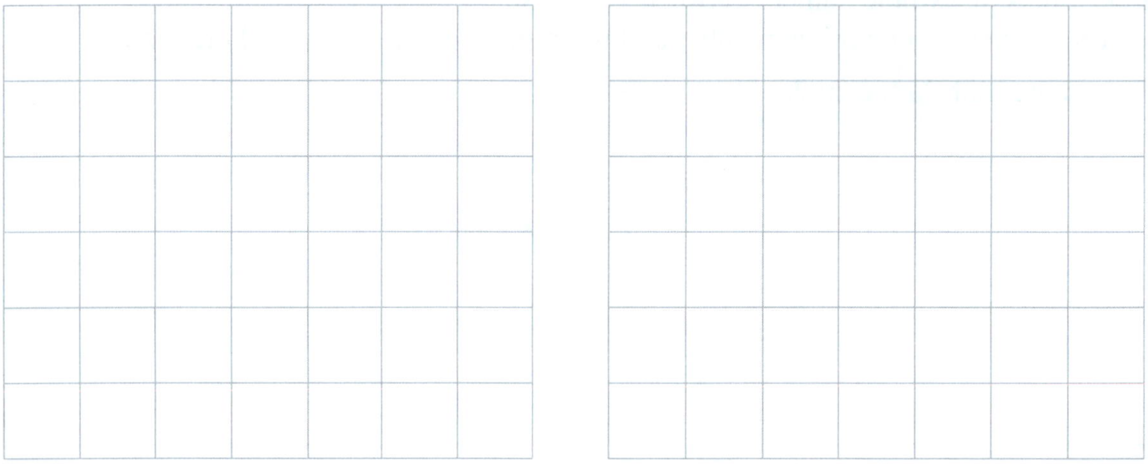

[Total 3 marks]

---

6 Amaya and Leo both think of a sequence of numbers.

Amaya's sequence
4th term = 17
Term-to-term rule is:
Add 3

Leo's sequence
Term-to-term rule is:
Add 6

The 1st term of Amaya's sequence is twice the 1st term of Leo's sequence.

(a) Work out the 5th term of Leo's sequence.

....

[3]

(b) Find an expression for the sum of the nth and $(n + 1)$th terms of Leo's sequence.
Simplify your answer as much as possible.

....

[3]

[Total 6 marks]

---

7 Here are the equations of five straight lines.

$y = 2$ $\quad$ $2y = x$ $\quad$ $y = 2x + 1$ $\quad$ $y - 2x = -3$ $\quad$ $3y = 2x + 2$

(a) Write each of the equations in the correct position in this table.
The first equation has been put in for you.

	Gradient equal to 2	Gradient not equal to 2
Passes though the point (2, 1)		
Does not pass though the point (2, 1)		$y = 2$

[2]

(b) Which of the equations shows that y is directly proportional to x?

....

[1]

[Total 3 marks]

8 A chocolate manufacturer makes boxes of chocolates in four different sizes.

Box A Box B Box C Box D

Box A contains c chocolates.
Box B contains 4 more chocolates than Box A.
Box C contains twice as many chocolates as Box B.
Box D contains 44 more chocolates than Box B.

The number of chocolates in Box D is the same as the total number of chocolates in the other three boxes combined.

Form and solve an equation to find the number of chocolates in Box A.

.... chocolates

[Total 5 marks]

9 The diagram below shows a trapezium. *FG* is parallel to *EH*.
EH is 15.5 cm, *EF* is 4.8 cm and *FG* is 10 cm.

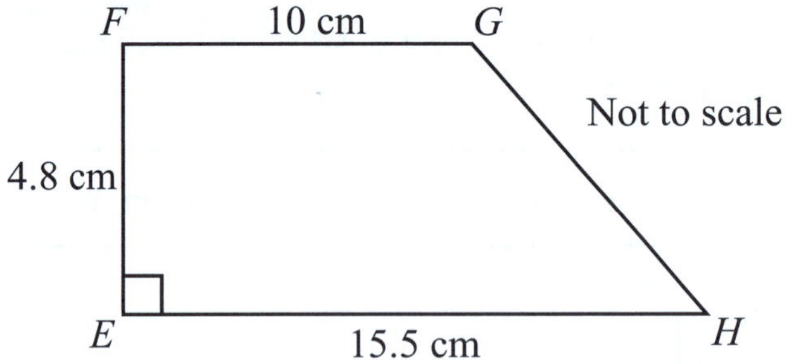

(a) Calculate the length of *GH*.

.... cm
[3]

(b) Calculate the size of angle *EHG*.
Give your answer correct to 1 decimal place.

.... °
[3]

[Total 6 marks]

---

10 The distance between Harlow and Bath is 180 kilometres.
Taylor leaves Harlow at 8.30 am and arrives in Bath at 11.30 am.
Sasha leaves Harlow at 9.00 am.
Sasha's average speed is 12 kilometres per hour faster than Taylor's.

Show that Sasha arrives in Bath at the same time as Taylor.

[Total 5 marks]

11 A bank interviews a sample of 500 of its customers to find out whether they are satisfied with the service the bank provides. The bank has both savings and mortgage customers.

The frequency tree summarises the responses.

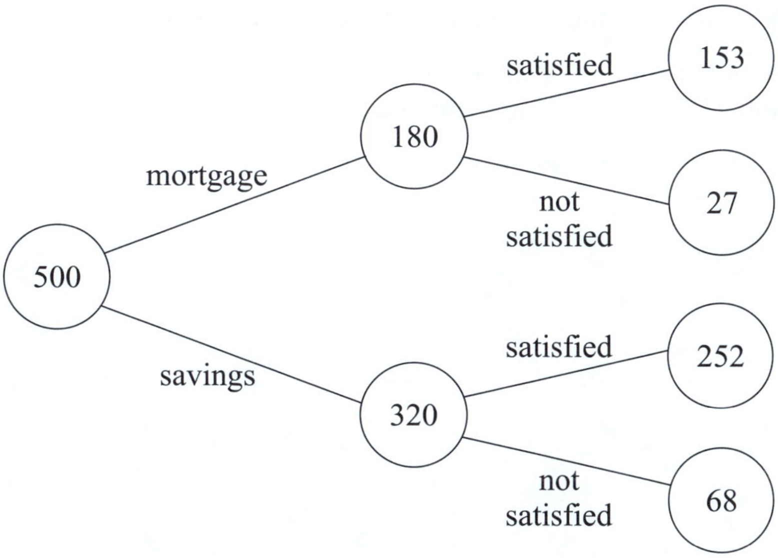

Comment on how satisfied the savings customers are in comparison with the mortgage customers. Give calculations to support your answer.

....

....

....

....

....

....

[Total 3 marks]

---

12 Simplify

(a) a^0

....

[1]

(b) $3a^2b \times 2a^3b^2$

....

[2]

(c) $(3a^3)^2 \div \frac{9a^3b^{\frac{1}{2}}}{4b^{\frac{3}{2}}a^2}$

....

[3]

[Total 6 marks]

---

13 Kareem is planning to paint a large hall with a shade of green paint.
The total area they have to paint is 840 m^2. They need 2 litres of green paint to cover an area of 30 m^2.

Kareem makes 10 litres of the green paint by mixing:
3 litres of yellow paint, 5 litres of blue paint and 2 litres of white paint.

The total amount of paint Kareem has available is:
18 litres of yellow paint, 27.5 litres of blue paint and 12 litres of white paint.

Show that Kareem does not have enough paint to finish the hall.

[Total 4 marks]

14 x and y satisfy these inequalities.

$$x \geq 1 \qquad y \geq \frac{x}{2} \qquad x + 2y \leq 8$$

(a) Show the region on the grid which satisfies these inequalities.

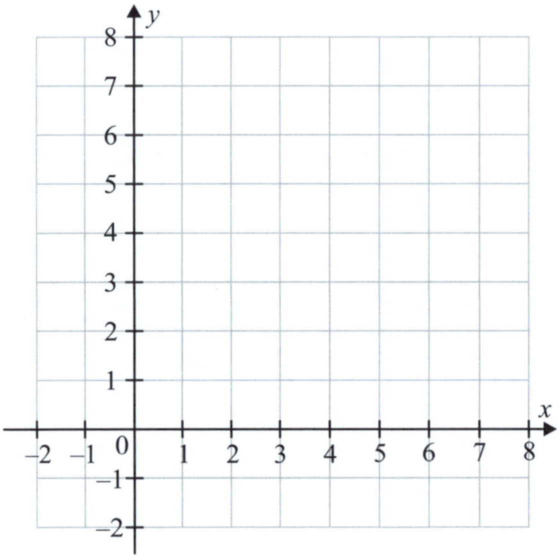

[4]

(b) How many different pairs of integer solutions satisfy all three inequalities?
Explain your answer.

....

....

....

[2]

[Total 6 marks]

15 *AB* and *BC* are perpendicular lines.

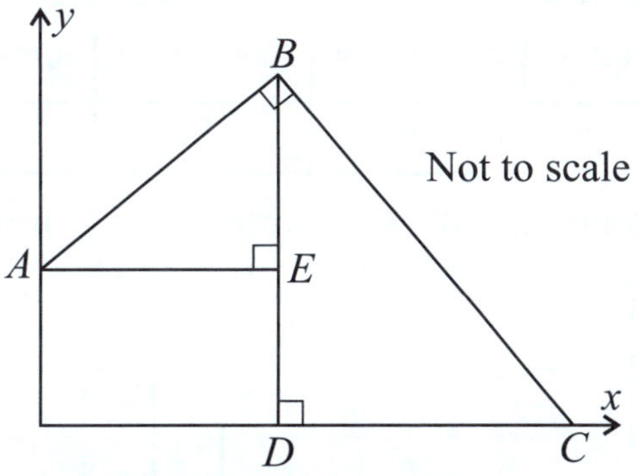

B has coordinates (12, 18). *C* has coordinates (27, 0).
A is a point on the *y*-axis.

(a) Explain why triangles *AEB* and *BDC* are similar.

....

....

....

[2]

(b) Write the ratio *AE* : *BD* in its simplest form.

.... : ....

[2]

(c) Work out the coordinates of *E*.

(............. , ...)

[3]

[Total 7 marks]

---

16 A company is testing an air pressure gauge that is based on a gas-filled ball.

The air pressure is inversely proportional to the cube of the diameter of the ball.
When the air pressure, *p*, is 60 bars, the diameter of the ball, *d*, is 2 cm.

(a) Write down a formula connecting *p* and *d*.

....

[3]

(b) What diameter will the ball be when the pressure is 100 bars?
Give your answer correct to 2 decimal places.

.... cm

[2]

[Total 5 marks]

---

17 The table shows some information about the ages of the adult members of a gym.

Age, A years	$18 < A \leq 20$	$20 < A \leq 25$	$25 < A \leq 30$	$30 < A \leq 40$	$40 < A \leq 60$	$60 < A \leq 70$	$70 < A \leq 90$
Frequency	18	35	40	45	50	75	40

The gym manager draws a histogram to show this information. It is incorrect.

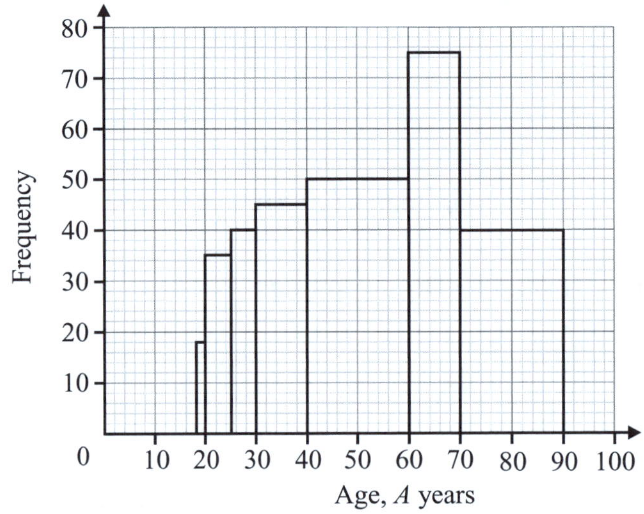

(a) Explain what error the manager has made when drawing the histogram.

....

....

[1]

(b) Draw the histogram correctly on the grid below.

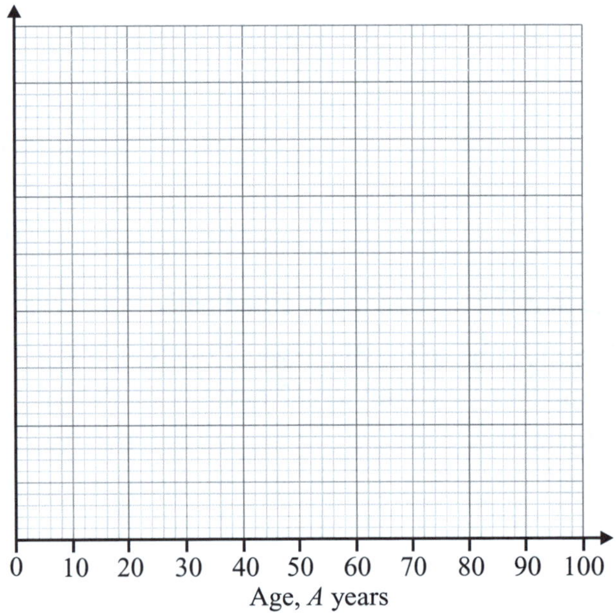

[3]

(c) The mean age of the adult members of the gym is 47 years.
Explain why the mean does not give a very typical age for the members of this gym.

....

....

[1]

[Total 5 marks]

18 The grid shows a quadrilateral Q.

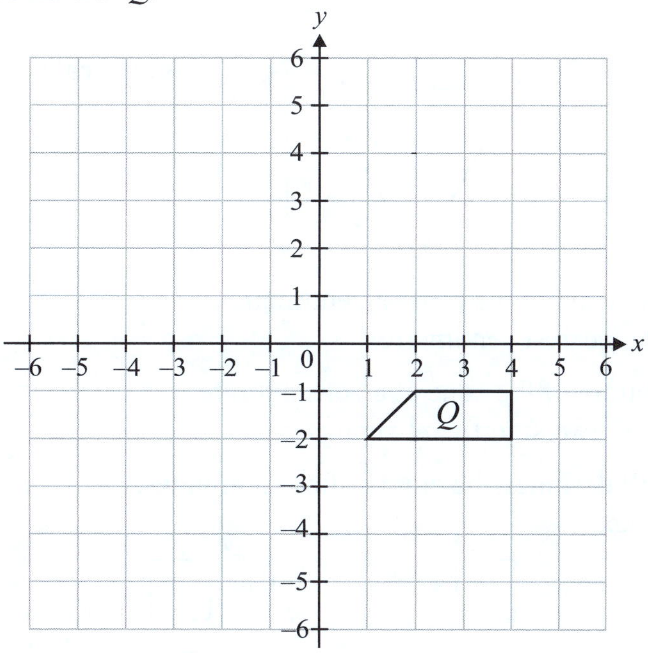

Draw the enlargement of Q using scale factor -2 and centre $(2, 0)$. Label the image R.

[Total 3 marks]

19 A museum has a collection of 200 thimbles.
The two-way table and the Venn diagram show some information about the thimbles.

	Made in Europe	**Made outside Europe**	**Total**
Antique			
Not antique			
Total		80	200

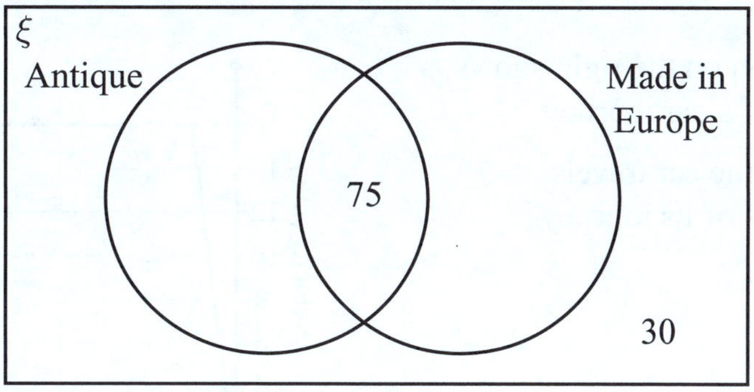

(a) Complete the table and the Venn diagram.

[3]

(b) Find the probability that a randomly chosen thimble is antique, given that it is made in Europe.

....

[2]

[Total 5 marks]

20 This cone is filled with water.

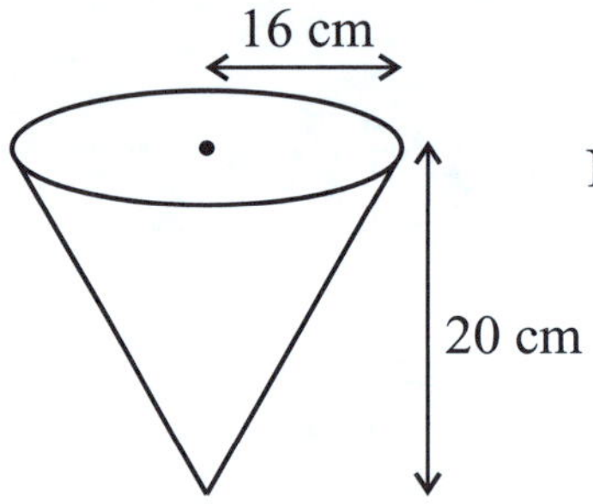

The radius of the cone is 16 cm to the nearest centimetre.
The height is 20 cm to the nearest centimetre.

Water leaks out of the bottom of the cone at a constant rate of 0.39 litres per minute, to two significant figures.

Ariel says, "The cone will definitely be empty after 15 minutes."
Is Ariel correct? Tick a box.

Yes ☐ No ☐

Explain your answer.

[Total 5 marks]

---

21 The velocity-time graph on the right shows the first two minutes of a car journey.

Calculate the distance the car travels in the first two minutes of its journey.

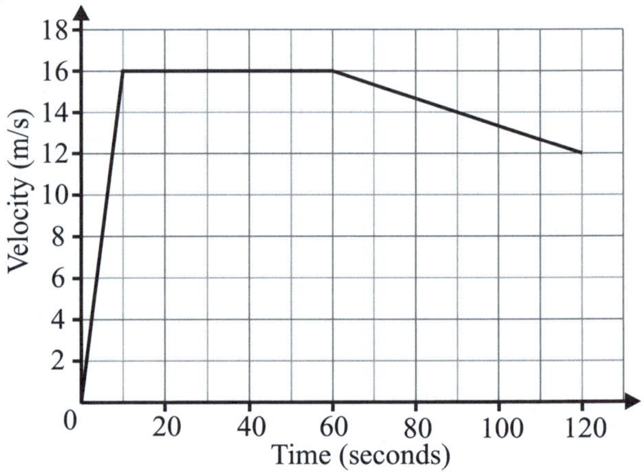

.... m

[Total 4 marks]

---

22 ABC is an isosceles triangle with $AB = AC = 9$ cm.
D is the point on AC such that $AD = 5$ cm and $BD = 7$ cm.

Calculate length BC.

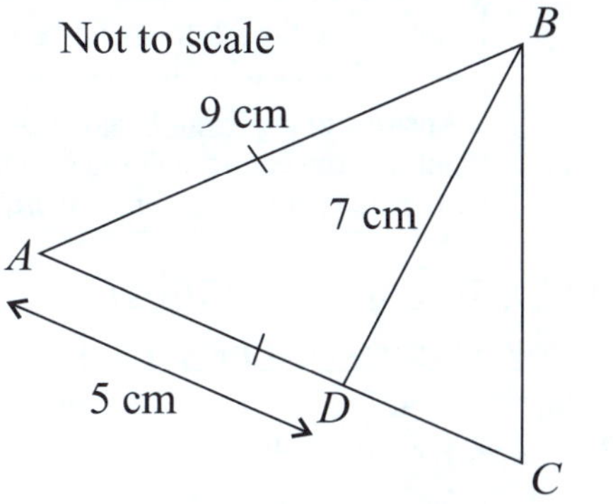

.... cm

[Total 4 marks]

---

23 Emma and Felix both think of a number.
Emma's number is negative. Felix's number is one more than Emma's.

They each take the reciprocal of their numbers. The sum of the reciprocals is $\frac{5}{6}$.

Use algebra to work out Emma's original number.

....

[Total 5 marks]

[TOTAL FOR PAPER = 100 MARKS]

Answers

Answers to the Warm-Up Questions, Exam Questions and Revision Questions (which appear on the green and yellow pages of the book) and answers to the Practice Papers are given on the pages that follow.

Answers to the Exam Practice Questions that appear at the bottom of most other pages can be found online. You can scan this QR code to access them as a printable PDF, or you can find them as part of your free Online Edition — there's more info about how to get hold of this at the front of this book.

Online Answers

Section One — Number

Page 4 (Warm-Up Questions)

1 a) 1, 10, –4, 7 b) $\sqrt{2}$, 6π

2 1, 2, 4, 5, 8, 10, 20, 40

3 34, 51, 68

4 $231 \div 3 = 77$ or $231 \div 7 = 33$ or $231 \div 11 = 21$.
So 231 has more than 2 factors.

5 $2 \times 2 \times 2 \times 5$ (or $2^3 \times 5$)

6 a) 42 b) 6

Page 5 (Exam Questions)

3 a) LCM = $3^7 \times 7^3 \times 11^2$ ***[1 mark]***
b) HCF = $3^4 \times 11$ ***[1 mark]***

4 E.g. $6 = \sqrt{3x + 2y}$, so $36 = 3x + 2y$
Try different values of x and see what y-value each one gives:
$x = 2$: $3x = 6$, so $2y = 36 - 6 = 30$, so $y = 15$
$x = 4$: $3x = 12$, so $2y = 36 - 12 = 24$, so $y = 12$
[2 marks available — 1 mark for each correct pair of x and y values]
The only other possible solution is $x = 6$, $y = 9$ (you'll get full marks if you got this solution instead).

5 Common multiples of 6 and 7: 42, 84, 126, 168, 210, 252, ... ***[1 mark]***
Factors of 252: 1, 2, 3, 4, 6, 7, 9, 12, 14, 18, 21, 28, 36, 42, 63, 84, 126, 252
Factors of 420: 1, 2, 3, 4, 5, 6, 7, 10, 12, 14, 15, 20, 21, 28, 30, 35, 42, 60, 70, 84, 105, 140, 210, 420
[1 mark for both sets of factors]
Common factors of 252 and 420:
1, 2, 3, 4, 6, 7, 12, 14, 21, 28, 42, 84 ***[1 mark]***
So $x = 84$ ***[1 mark]***
[4 marks available in total — as above]

6 Prime factorisation of $A = A$
Prime factorisation of $B = B$ (as A and B are prime)
So LCM = $A \times B$ (or AB)
[2 marks available — 2 marks for the correct answer, otherwise 1 mark for stating that the prime factorisations of A and B are A and B]

Page 12 (Warm-Up Questions)

1 a) $\frac{4}{15}$ b) $\frac{2}{5} \times \frac{3}{2} = \frac{6}{10} = \frac{3}{5}$
c) $\frac{6}{15} + \frac{10}{15} = \frac{16}{15} = 1\frac{1}{15}$ d) $\frac{10}{15} - \frac{6}{15} = \frac{4}{15}$

2 40%

3 66.66666…%

4 a) $\frac{11}{25}$ b) $\frac{4}{9}$ c) $\frac{5}{11}$

5 a) 0.7 b) $0.7777777... = 0.\dot{7}$

Page 13 (Exam Questions)

3 $\frac{2}{5} + \left(1 - \frac{6}{7}\right) = \frac{2}{5} + \frac{1}{7} = \frac{14}{35} + \frac{5}{35} = \frac{19}{35}$
[3 marks available — 1 mark for finding the unshaded region of shape Y, 1 mark for writing over a common denominator and 1 mark for the correct answer]

4 $a = \frac{3}{4}, b = \frac{5}{2}$, so $\frac{1}{a} + \frac{1}{b} = \frac{4}{3} + \frac{2}{5} = \frac{20}{15} + \frac{6}{15} = \frac{26}{15}$ or $1\frac{11}{15}$
[3 marks available — 1 mark for reciprocal fractions, 1 mark for rewriting over a common denominator, 1 mark for the correct answer]

5 $17\frac{1}{2} \times \frac{1}{5} = \frac{35}{2} \times \frac{1}{5} = \frac{35}{10} = \frac{7}{2}$ tonnes of flour used to make cheese scones. ***[1 mark]***
Then $\frac{7}{2}$ out of $25 = \frac{7}{2} \div 25 = \frac{7}{50}$ ***[1 mark]***.
[2 marks available in total — as above]

6 Convert to an equivalent fraction with all nines on the bottom
$\frac{7}{33} = \frac{21}{99}$ ***[1 mark]***
Then the number on the top tells you the recurring part,
so $\frac{7}{33} = 0.\dot{2}\dot{1}$ ***[1 mark]***
[2 marks available in total — as above]
The first mark could also be obtained by using a division method.

7 Let $r = 0.\dot{5}9\dot{0}$, then $10r = 5.\dot{9}\dot{0}$, so $1000r = 590.\dot{9}\dot{0}$ ***[1 mark]***
$990r = 585$ ***[1 mark]***
$r = \frac{585}{990} = \frac{13}{22}$ ***[1 mark]***
[3 marks available in total — as above]

Page 17 (Warm-Up Questions)

1 a) 40.22 b) 39.9 c) 28

2 $\frac{94 \times 1.9}{0.328 + 0.201} \approx \frac{90 \times 2}{0.3 + 0.2} = \frac{180}{0.5} = 360$

3 Upper bound = 14.5 km. Lower bound = 13.5 km

4 a) Maximum = 9.3, minimum = 9.1
b) Maximum = 3.5, minimum = 3.3
c) Maximum = 18.7325, minimum = 17.8125
d) Maximum = 2.23 (to 2 d.p.), minimum = 2.12 (to 2 d.p.)

Page 18 (Exam Questions)

3 a) $\frac{215.7 \times 44.8}{460} \approx \frac{200 \times 40}{500} = \frac{8000}{500} = 16$
[3 marks available — 1 mark for correctly rounding 1 number to 1 significant figure, 1 mark for correctly rounding the other 2 numbers to 1 significant figure, 1 mark for correct answer]
b) The answer to a) will be smaller than the exact answer, because in the rounded fraction the numerator is smaller and denominator is larger compared to the exact calculation.
[2 marks available — 1 mark for 'smaller than the exact answer', 1 mark for correct reasoning]

4 upper bound for $x = 57.5$ mm ***[1 mark]***
upper bound for $y = 32.5$ mm ***[1 mark]***
upper bound for area = 57.5 mm × 32.5 mm = 1868.75 mm^2
= 1870 mm^2 to 3 s.f. ***[1 mark]***
[3 marks available in total — as above]

5 Upper bound of $x = 2.25$ ***[1 mark]***
So upper bound of $4x + 3 = 4 \times 2.25 + 3 = 12$
Lower bound of $x = 2.15$ ***[1 mark]***
So lower bound of $4x + 3 = 4 \times 2.15 + 3 = 11.6$
Written as an interval, this is $11.6 \leq 4x + 3 < 12$
[1 mark for both bounds correct, 1 mark for expressing as an inequality correctly]
[4 marks available in total — as above]

6 lower bound for volume = 0.935 × 0.605 × 0.205 = 0.11596... m^3
upper bound for volume = 0.945 × 0.615 × 0.215 = 0.12495... m^3
Both the upper bound and lower bound round to 0.12 m^3 to 2 d.p. (or 2 s.f.) so the volume to 2 d.p. is 0.12 m^3.
[4 marks available — 1 mark for the correct upper and lower bounds for the dimensions, 1 mark for the correct lower bound for volume, 1 mark for the correct upper bound for volume, 1 mark for rounding to a suitable number of decimal places (or significant figures) to obtain the final answer]

Page 21 (Warm-Up Questions)

1 2.4×10^5 miles

2 2.7×10^{-6} seconds

3 0.0000000000000000000000027 g
(i.e. 22 zeros after the decimal point.)

4 1.2×10^8

5 a) 2.4×10^{11} b) 2×10^4
c) 7.797×10^6 d) 3.2088×10^{12}

Page 22 (Exam Questions)

2 time (s) = distance (miles) ÷ speed (miles/s)
$= (9.3 \times 10^7) \div (1.86 \times 10^5)$ seconds ***[1 mark]***
$= 5 \times 10^2$ seconds ***[1 mark]***
[2 marks available in total — as above]

3 a) number of tablets = dose (grams) ÷ dose per tablet (grams)
$= (4 \times 10^{-4}) \div (8 \times 10^{-5})$ ***[1 mark]***
$= (4 \div 8) \times (10^{-4} \div 10^{-5})$
$= 0.5 \times 10^1$ ***[1 mark]***
$= 5$ ***[1 mark]***
[3 marks available in total — as above]

b) new dose $= 4 \times 10^{-4}$ grams $+ 6 \times 10^{-5}$ grams ***[1 mark]***
$= 4 \times 10^{-4}$ grams $+ 0.6 \times 10^{-4}$ grams ***[1 mark]***
$= (4 + 0.6) \times 10^{-4}$ grams
$= 4.6 \times 10^{-4}$ grams per day ***[1 mark]***
[3 marks available in total — as above]
You could have done this one by turning 4×10^{-4} into 40×10^{-5} and adding it to 6×10^{-5} instead.

4 $\frac{3^2}{2^{122} \times 5^{120}} = \frac{9}{2^2(2^{120} \times 5^{120})} = \frac{9}{2^2 \times 10^{120}} = \frac{9}{4} \times \frac{1}{10^{120}} = 2.25 \times 10^{-120}$
[2 marks available — 1 mark for writing the denominator as a multiple of a power of 10, 1 mark for the correct answer]

Page 23 (Revision Questions)

1 a) Whole numbers — either positive or negative, or zero
b) Numbers that can be written as fractions
c) Numbers which will only divide by themselves or 1 (excluding 1)

2 a) 11 b) 0.5 c) 169

3 8 packs of buns, 3 packs of cheese slices, 4 packs of hot dogs.

4 a) 14 b) 40

5 a) $320 = 2 \times 2 \times 2 \times 2 \times 2 \times 2 \times 5 = 2^6 \times 5$
$880 = 2 \times 2 \times 2 \times 2 \times 5 \times 11 = 2^4 \times 5 \times 11$
b) LCM $= 2^6 \times 5 \times 11 = 3520$
HCF $= 2^4 \times 5 = 80$

6 Divide top and bottom by the same number till they won't go any further.

7 a) $8\frac{2}{9}$ b) $\frac{33}{7}$

8 Multiplying: Multiply top and bottom numbers separately.
Dividing: Turn the second fraction upside down, then multiply.
Adding/subtracting: Put fractions over a common denominator, then add/subtract the numerators.

9 a) $\frac{14}{99}$ b) $3\frac{1}{7}$ or $\frac{22}{7}$
c) $\frac{11}{24}$ d) $7\frac{11}{20}$ or $\frac{151}{20}$

10 a) 210 kg b) $\frac{11}{7}$

11 $\frac{3}{4} = \frac{30}{40}$, $\frac{5}{8} = \frac{25}{40}$, $\frac{7}{10} = \frac{28}{40}$
So $\frac{7}{10}$ is closer to $\frac{3}{4}$ than $\frac{5}{8}$.

12 a) Divide the top by the bottom.
b) Put the digits after the decimal point on the top, and a power of 10 with the same number of zeros as there were decimal places on the bottom.

13 a) (i) $\frac{4}{100} = \frac{1}{25}$ (ii) 4%
b) (i) $\frac{65}{100} = \frac{13}{20}$ (ii) 0.65

14 orange juice = 12.5 litres, lemonade = 10 litres, cranberry juice = 2.5 litres

15 Let $r = 0.\dot{5}\dot{1}$.
Then $100r - r = 51.\dot{5}\dot{1} - 0.\dot{5}\dot{1}$
$\Rightarrow 99r = 51 \Rightarrow r = \frac{51}{99} = \frac{17}{33}$

16 a) 427.96 b) 428.0
c) 430 d) 428.0

17 Estimates should be around 16-20.

18 Estimates should be between 6.6 and 6.8.

19 The upper and lower bounds of a rounded measurement are half a unit either side of the rounded value. The upper and lower bounds of a truncated measurement are the truncated value itself and a whole unit above the truncated value.

20 2.35 litres $\leq V <$ 2.45 litres

21 132.2425 m^2

22 1. The front number must always be between 1 and 10.
2. The power of 10, n, is how far the decimal point moves.
3. n is positive for big numbers, and negative for small numbers.

23 a) 9.7×10^5 b) 3.56×10^9
c) 2.75×10^{-6}

24 0.00456, 270 000

25 a) 2×10^3 b) 2.739×10^{12}

26 2.48×10^9

Section Two — Algebra

Page 30 (Warm-Up Questions)

1 a) $2a - 5c$ b) $7r^2 - 5r - 1$

2 a) $4^3 = 64$ b) $1\frac{32}{49}$
c) 9 d) $\frac{9}{4}$

3 a) $8p + 28$ b) $8x^2 - 2$ c) $5a^2 - 3a$

4 a) $2(3p - 6q + 2)$ b) $2cd(2d - 1 + 5cd^2)$

5 $(x + 2y)(x - 2y)$

6 $\sqrt{30}$

Page 31 (Exam Questions)

3 Height $= 7 \times (f + g) + 9 \times (h - g) + 5 \times 2h$ ***[1 mark]***
$= 7f + 7g + 9h - 9g + 10h$
$= 7f - 2g + 19h$ cm ***[1 mark]***
[2 marks available in total — as above]

4 a) $8^2 = 64$ and $9^2 = 81$, so $x = \sqrt{70} \approx 8.4$
[2 marks available — 2 marks for 8.3, 8.4 or 8.5, otherwise 1 mark for any answer between 8 and 9 with 1 d.p.]
b) $3^2 = 9$ and $3^3 = 27$, so $3^{2.7} \approx 20$, $y \approx 2.7$
[2 marks available — 2 marks for 2.6, 2.7 or 2.8, otherwise 1 mark for any answer between 2 and 3 with 1 d.p.]

5 $(x - 1)(2x + 3)(2x - 3) = (x - 1)(4x^2 - 6x + 6x - 9)$
$= (x - 1)(4x^2 - 9)$
$= 4x^3 - 4x^2 - 9x + 9$
[3 marks available — 3 marks for the correct answer, otherwise 1 mark for correctly multiplying two sets of brackets together, 1 mark for attempting to multiply this product by the third set of brackets]
The trick here is spotting that the second pair of brackets multiply out to give just two terms (a difference of two squares), which makes the second multiplication much easier.

6 $x^3 - 25x = x(x^2 - 25) = x(x + 5)(x - 5)$
[3 marks available — 1 mark for each correct factor]

7 $\frac{1+\sqrt{7}}{3-\sqrt{7}} = \frac{(1+\sqrt{7})(3+\sqrt{7})}{(3-\sqrt{7})(3+\sqrt{7})}$ ***[1 mark]***

$= \frac{3+\sqrt{7}+3\sqrt{7}+7}{9-7}$ ***[1 mark]***

$= \frac{10+4\sqrt{7}}{2}$ ***[1 mark]***

$= 5+2\sqrt{7}$ ***[1 mark]***

[4 marks available in total — as above]

Page 36 (Warm-Up Questions)

1 a) $x = 3$ b) $x = -3$ c) $x = 5$

2 a) p b) t

3 $n = 4(5 - m)$ or $n = 20 - 4m$

4 $z = \frac{3x - y}{2}$

Page 37 (Exam Questions)

3 Let the number of tickets Felix sells be x.
Then Poppy sells $2x$ tickets and Aleena sells $(2x + 25)$ tickets,
so $x + 2x + (2x + 25) = 700$
$5x + 25 = 700$
$5x = 675$ and $x = 135$
So Felix sells 135 tickets, Poppy sells $2 \times 135 = 270$ tickets and Aleena sells $(2 \times 135) + 25$ tickets $= 295$ tickets.
[5 marks available — 1 mark for finding expressions for the number of tickets each person sells, 1 mark for forming an equation to solve, 1 mark for solving to find the number of tickets Felix sells, 1 mark for the number of tickets Poppy sells, 1 mark for the number of tickets Aleena sells]

4 If one number is x, the other number is $3x$.
$3x^2 = 147$ ***[1 mark]***, so $x^2 = 49$, which means that $x = 7$ (as $x > 0$) ***[1 mark]*** and $3x = 21$ ***[1 mark]***,
so Hassan is thinking of 7 and 21.
[3 marks available in total — as above]

5 a) $a + y = \frac{b - y}{a}$, so...
$a(a + y) = b - y$ ***[1 mark]***,
$a^2 + ay = b - y$,
$ay + y = b - a^2$ ***[1 mark]***,
$y(a + 1) = b - a^2$ ***[1 mark]***,
$y = \frac{b - a^2}{a + 1}$ ***[1 mark]***
[4 marks available in total — as above]

b) When $a = 3$ and $b = 6$, $y = \frac{6 - 3^2}{3 + 1} = -\frac{3}{4}$ or -0.75 ***[1 mark]***

6 $x = \sqrt{\frac{(1+n)}{(1-n)}}$,
so $x^2 = \frac{(1+n)}{(1-n)}$ ***[1 mark]***,
$x^2(1 - n) = 1 + n$,
$x^2 - x^2n = 1 + n$ ***[1 mark]***,
$x^2 - 1 = n + x^2n$ ***[1 mark]***,
$x^2 - 1 = n(1 + x^2)$ ***[1 mark]***,
$n = \frac{x^2 - 1}{1 + x^2}$ ***[1 mark]***
[5 marks available in total — as above]

Page 43 (Warm-Up Questions)

1 a) $(x + 4)(x + 7)$ b) $(x + 14)(x + 2)$
c) $(x + 14)(x - 2)$.

2 a) $x = -3$ or $x = -5$ (it factorises to $(x + 3)(x + 5) = 0$)
b) $x = 2$ or $x = -7$ (it factorises to $(x - 2)(x + 7) = 0$)
c) $x = 3$ or $x = 4$
(Rearrange to give $x^2 - 7x + 12 = 0$, then factorise to give $(x - 3)(x - 4) = 0$, so $x = 3$ or $x = 4$.)

3 $(4x - 3)(x - 6)$

4 $x = \frac{1}{3}$ or $x = -7$ (it factorises to $(3x - 1)(x + 7) = 0$)

5 $x = 1.46$ or $x = -0.46$
(use the quadratic formula, with $a = 3$, $b = -3$ and $c = -2$).

6 $(x + 4)^2 - 21$, so $x = -4 + \sqrt{21}$ or $x = -4 - \sqrt{21}$
($(x + 4)^2$ gives $x^2 + 8x + 16$ so complete the square by subtracting 21:
$(x + 4)^2 - 21 = 0$
$(x + 4)^2 = 21$
$(x + 4) = \pm\sqrt{21}$
$x = -4 + \sqrt{21}$ or $x = -4 - \sqrt{21}$.)

7 $2(x + 4)^2 + 7$
($2(x + 4)^2 = 2x^2 + 16x + 32$, so $+7$ to complete the square)

Page 44 (Exam Questions)

3 $(x + 2)^2 - 9 = x^2 + 4x + 4 - 9$ ***[1 mark]*** $= x^2 + 4x - 5$
$a = 4$ and $b = -5$ ***[1 mark]***
[2 marks available in total — as above]

4 The area of the square is $(x + 3)(x + 3) = x^2 + 6x + 9$. ***[1 mark]***
The area of the triangle is $\frac{1}{2}(2x + 2)(x + 3)$
$= \frac{1}{2}(2x^2 + 6x + 2x + 6) = \frac{1}{2}(2x^2 + 8x + 6)$
$= x^2 + 4x + 3$ ***[1 mark]***
So the area of the whole shape is $x^2 + 6x + 9 + x^2 + 4x + 3$
$= 2x^2 + 10x + 12$ ***[1 mark]***
$2x^2 + 10x + 12 = 60$, so $2x^2 + 10x - 48 = 0$ ***[1 mark]***
So $x^2 + 5x - 24 = 0$
$(x - 3)(x + 8) = 0$ ***[1 mark]***
$x - 3 = 0$ or $x + 8 = 0$
$x = 3$ or $x = -8$ ***[1 mark for both solutions]***
A length can't have a negative value so the answer must be $x = 3$ ***[1 mark]***
[7 marks available in total — as above]

5 $a = 3$, $b = -2$ and $c = -4$

$$x = \frac{-(-2) \pm \sqrt{(-2)^2 - 4 \times 3 \times -4}}{2 \times 3} = \frac{2 \pm \sqrt{52}}{6} = \frac{2 \pm 2\sqrt{13}}{6}$$

$x = \frac{1+\sqrt{13}}{3}$ or $x = \frac{1-\sqrt{13}}{3}$

[3 marks available — 1 mark for correct substitution, 1 mark for each correct solution. Lose a mark if answers aren't simplified]

6 a) $2(x^2 - 4x) + 19$
$4 \div 2 = 2$, so the first bit is $2[(x - 2)^2]$
Expanding the brackets: $2(x^2 - 4x + 4) = 2x^2 - 8x + 8$
To complete the square: $19 - 8 = 11$
So $2x^2 - 8x + 19 = 2(x - 2)^2 + 11$
[4 marks available — 1 mark for dividing the first two terms by 2, 1 mark dividing the x-term by 2 to find the value of b, 1 mark for finding the value of c, 1 mark for the full correct answer]

b) Minimum value = 11, which occurs at $x = 2$, so the coordinates of the minimum point are (2, 11) ***[1 mark]***

c) This quadratic is u-shaped and its minimum value is 11, so it's always greater than 0. This means it never crosses the x-axis. ***[1 mark]***

Page 48 (Warm-Up Questions)

1 $\frac{4ab}{c^2}$

2 $\frac{y}{2y + x}$

3 a) 8, 13, 18, 23, 28, 33
b) 8, 11, 14, 17, 20, 23

4 a) $5n$ b) $3n + 4$

5 There is always one cross in the centre, and the number of other crosses is 4 times the pattern number (because there are 4 "arms" coming from the centre). So in the nth pattern there will be a total of $4n + 1$ crosses.

Page 49 (Exam Questions)

3 a) $u_1 = 2,\quad u_2 = \frac{-1}{2(2)} = -0.25,$

$u_3 = \frac{-1}{2(-0.25)} = 2,\quad u_4 = \frac{-1}{2(2)} = -0.25$

[2 marks available — 2 marks for all three values correct, otherwise 1 mark for one or two values correct]

b) $u_{50} = -0.25$ ***[1 mark]***

4 a) Number of grey squares as a sequence: 1, 5, 9, 13, ...
Common difference = 4 so $4n$ is in the formula. To get from $4n$ to each term, you have to subtract 3, so the expression for the nth term is $4n - 3$. ***[2 marks available — 2 marks for correct expression, otherwise 1 mark for finding 4n]***

b) Assume Giles makes the nth and $(n + 1)$th patterns.
He uses $4n - 3$ grey squares in the nth pattern and
$4(n + 1) - 3 = 4n + 4 - 3 = 4n + 1$ grey squares in the $(n + 1)$th pattern ***[1 mark]***.
He uses 414 grey squares in total, so
$(4n - 3) + (4n + 1) = 414$ ***[1 mark]***
$8n - 2 = 414$
$8n = 416$
$n = 52$
So Giles has made the 52nd and 53rd patterns ***[1 mark]***.
[3 marks available in total — as above]

c) Total number of squares:
1 7 17 31
First difference: 6 10 14
Second difference: 4 4 ***[1 mark]***
The second differences are constant so the sequence is quadratic. Coefficient of $n^2 = 4 \div 2 = 2$ ***[1 mark]***.
Actual sequence – $2n^2$ sequence: –1, –1, –1, –1
So the nth term of the sequence is $2n^2 - 1$ ***[1 mark]***.
[3 marks available in total — as above]

5 $\frac{2a-8}{a^2-9} \div \frac{a^2-2a-8}{a^2+5a+6} \times (2a^2 - a - 15)$

$= \frac{2a-8}{a^2-9} \times \frac{a^2+5a+6}{a^2-2a-8} \times (2a^2 - a - 15)$

$= \frac{2(a-4)}{(a+3)(a-3)} \times \frac{(a+3)(a+2)}{(a+2)(a-4)} \times (2a+5)(a-3)$

$= 2(2a + 5)$ (or $4a + 10$)

[5 marks available — 1 mark for converting to a multiplication, 1 mark for factorising the first fraction, 1 mark for factorising the second fraction, 1 mark for factorising the quadratic term, 1 mark for cancelling to reach correct answer]

Page 54 (Warm-Up Questions)

1 $x = 13, 14, 15, 16$

2 $n = -3, -2, -1, 0, 1, 2, 3$

3 Dividing by 4 gives $2 < x < 5$,
but x must be an integer so $x = 3, 4$

4 $2q + 2 \le 12 \Rightarrow 2q \le 10 \Rightarrow q \le 5$

5 $4p + 12 > 30 \Rightarrow 4p > 18 \Rightarrow p > 4.5$

6

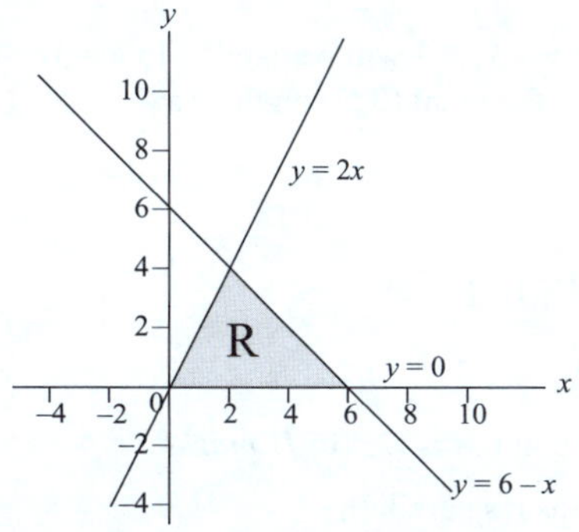

7 $x^3 + 6x = 69$, so $x^3 + 6x - 69 = 0$
Sub in $x = 3$: $(3)^3 + 6(3) - 69 = -24$
Sub in $x = 4$: $(4)^3 + 6(4) - 69 = 19$
There's a sign change between $x = 3$ and $x = 4$, so a solution lies in this interval.

8 $x^3 - 11x = 100$, so $x^3 - 11x - 100 = 0$

x	$x^3 - 11x - 100$	
5.0	–30	–ve
5.1	–23.449	–ve
5.2	–16.592	–ve
5.3	–9.423	–ve
5.4	–1.936	–ve
5.5	5.875	+ve
5.41	–1.169579	–ve
5.42	–0.399912	–ve
5.43	0.373007	+ve

$x = 5.4$ to 1 d.p.

Page 55 (Exam Questions)

3 $2n + (2n + 2) + (2n + 4) < 1000$ ***[1 mark]***
$6n + 6 < 1000$
$n < 165.666...$ ***[1 mark]***
So the largest possible values of the numbers are obtained when $n = 165$, which gives 330, 332 and 334 ***[1 mark]***.
[3 marks available in total — as above]

4 $5n - 3 \le 17$, so $5n \le 20$, so $n \le 4$ ***[1 mark]***
$2n + 6 > 8$, so $2n > 2$, so $n > 1$ ***[1 mark]***
Putting these together gives $1 < n \le 4$, so $n = 2, 3$ or 4 ***[1 mark]***.
[3 marks available in total — as above]

5 a) Find the value of the expression at $x = 1.6$ and $x = 2$:
$x = 1.6$: $3(1.6) - 2(1.6)^3 + 5 = 1.608$
$x = 2$: $3(2) - 2(2)^3 + 5 = -5$
There is a sign change between $x = 1.6$ and $x = 2$, so there is a solution in this interval.
[2 marks available — 1 mark for finding the values when x = 1.6 and x = 2, 1 mark for stating that a sign change means there's a solution]

b)

x	$3x - 2x^3 + 5$	
1.6	1.608	Positive
2.0	– 5	Negative
1.8	–1.264	Negative
1.7	0.274	Positive
1.75	–0.46875	Negative

The sign change shows that there's a solution between $x = 1.7$ and $x = 1.75$, and all values in the range $1.7 < x < 1.75$ round to 1.7 to 1 d.p., so $x = 1.7$ to 1 d.p.
[4 marks available — 1 mark each for rows showing a sign change between x = 1.6 and x = 1.8, x = 1.7 and x = 1.8, and x = 1.7 and x = 1.75, 1 mark for the correct value of x]

6 a) $5x^2 < 80$, so $x^2 < 16$. The solutions of $x^2 = 16$ are $x = 4$ and $x = -4$, so $x^2 < 16$ when $-4 < x < 4$.
[3 marks available — 1 mark for finding the square roots of 16, 1 mark for –4 < x, 1 mark for x < 4]

b) $x^2 + 1 = x + 7$ rearranges to give $x^2 - x - 6 = 0$.
$x^2 - x - 6 = 0$ factorises to give $(x + 2)(x - 3) = 0$.
The graph of $y = x^2 - x - 6$ is a u-shaped quadratic that crosses the x-axis at $x = -2$ and $x = 3$:

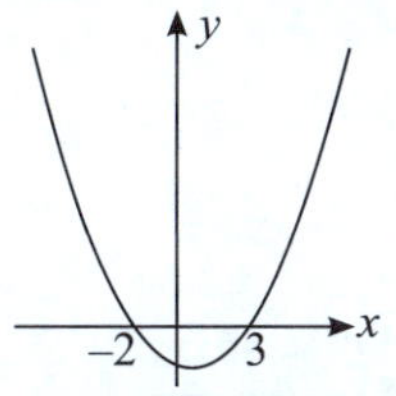

The graph is below 0 when x is greater than –2 and less than 3. So $-2 < x < 3$.
[3 marks available — 1 mark for rearranging and factorising the quadratic to find the solutions, 1 mark for –2 < x, 1 mark for x < 3]

Page 61 (Warm-Up Questions)

1 $x = 3, y = -2$

2 $x = 2, y = 5$

3 $x = -1, y = 3$ and $x = 3, y = -5$

4 Take two odd numbers, $2n + 1$ and $2m + 1$, where n and m are integers. Then $(2n + 1)(2m + 1) = 4nm + 2n + 2m + 1$ $= 2(2nm + n + m) + 1$. This is in the form $2k + 1$, where $k = (2nm + n + m)$, so the product of two odd numbers is odd.

5 $(2x - 5)^2 - (x - 1)^2 = [4x^2 - 20x + 25] - [x^2 - 2x + 1] =$ $3x^2 - 18x + 24 = 3(x^2 - 6x + 8)$. This can be written as $3k$, where $k = (x^2 - 6x + 8)$, so it is a multiple of 3 for all integer values of x.

6 $x = 36$

Page 62 (Exam Questions)

3 $2x + 3y = 12 \xrightarrow{\times 5} 10x + 15y = 60$ ***[1 mark]***

$5x + 4y = 9 \xrightarrow{\times 2} 10x + 8y = 18$ ***[1 mark]***

$$\begin{array}{r} 10x + 15y = 60 \\ -\ 10x + 8y = 18 \\ \hline 7y = 42 \\ y = 6 \end{array}$$ ***[1 mark]***

$2x + 3y = 12$

$2x = 12 - (3 \times 6)$

$2x = -6$

$x = -3$ ***[1 mark]***

[4 marks available in total — as above]

4 E.g. When $a = 1$, $b = 2$, $c = 3$ and $d = 10$ then $a < b < c < d$

$\frac{a}{b} = \frac{1}{2}$ and $\frac{c}{d} = \frac{3}{10}$ but $\frac{1}{2} > \frac{3}{10}$ so $\frac{a}{b} > \frac{c}{d}$

which contradicts what Meena says, so she is not correct.

[3 marks available — 1 mark for finding values of a, b, c and d, 1 mark for showing that these values form a counter example, 1 mark for stating that Meena is wrong]

5 a) Inputting x into machine A gives an output of $y = 8x - 6$

Inputting $8x - 6$ into machine B gives an output of

$z = \frac{8x - 6}{2} + 9 = 4x - 3 + 9 = 4x + 6$

So a possible function machine would be, e.g.

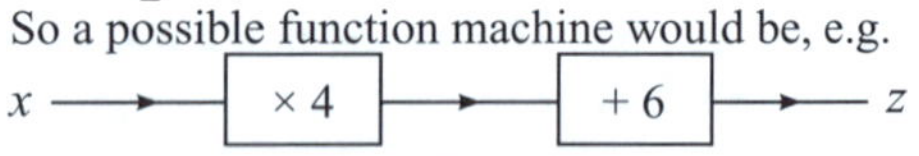

[3 marks available — 3 marks for a correct function machine, otherwise 2 marks for finding the correct output of machine B in terms of x, or 1 mark for finding the output from machine A in terms of x]

There are other correct function machines, e.g. +3, ×2.

b) $x = \frac{z - 6}{4}$

[2 marks available — 2 marks for the correct answer, otherwise 1 mark for the correct numerator or denominator]

6 If $2^{64} - 1$ is prime then its only factors are 1 and itself

$2^{64} - 1 = (2^{32})^2 - 1^2 = (2^{32} + 1)(2^{32} - 1)$ ***[1 mark]***

So $(2^{32} + 1)$ and $(2^{32} - 1)$ are factors of $2^{64} - 1$ ***[1 mark]***

But neither $(2^{32} + 1)$ or $(2^{32} - 1)$ are equal to 1 or $2^{64} - 1$ so $2^{64} - 1$ cannot be prime. ***[1 mark]***

[3 marks available in total — as above]

Page 63 (Revision Questions)

1 $5x - 4y - 5$

2 a) x^9 b) y^2 c) z^{12}

3 a) $6x + 3$ b) $x^2 - x - 6$ c) $x^3 + 7x^2 + 7x - 15$

4 a) $2(2x + y)(2x - y)$ b) $(7 + 9pq)(7 - 9pq)$

c) $12(x + 2y)(x - 2y)$

5 a) $3\sqrt{3}$ b) 5

6 $3\sqrt{2}$

7 a) $x = 2$ b) $x = \pm 3$

8 a) $p = -\frac{4y}{3}$ b) $p = \frac{qr}{q + r}$

9 a) $x = -3$ or $x = -6$ b) $x = 4$ or $x = -\frac{3}{5}$

10 $x = \frac{-b \pm \sqrt{b^2 - 4ac}}{2a}$

11 a) $x = 1.56$ or $x = -2.56$ b) $x = 0.27$ or $x = -1.47$

c) $x = 0.44$ or $x = -3.44$

12 a) $x = -6 \pm \sqrt{21}$ b) $x = 3 \pm \sqrt{11}$

13 $y = x^2 - 4x + 9$, so $p = -4$, $q = 9$

14 $\frac{3x + 1}{(x + 3)(x - 1)}$

15 a) $2n + 5$ b) $-3n + 14$

c) $n^2 + n + 3$

16 Yes, it's the 5th term.

17 a) $x \geq -2$ b) $x < -6$ or $x > 6$

18

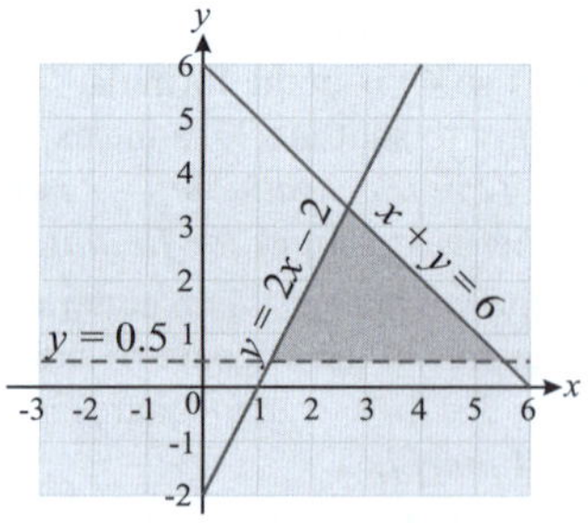

19 $x = 3$ gives a value of -6. $x = 4$ gives a value of 5. There is a sign change so there is a solution between 3 and 4.

20 $x = 2, y = 3$

21 $x = -2, y = -2$ and $x = -4, y = -8$

22 Take an even number, $2p$, and an odd number, $2q + 1$. Their product is $2p \times (2q + 1) = 4pq + 2p = 2(2pq + p)$, which is even as $(2pq + p)$ is an integer (sums and products of integers are also integers).

23 41

Section Three — Graphs

Page 70 (Warm-Up Questions)

1 a) $y = x$

b) Horizontal line ($y = 4$)

c) Vertical line ($x = -1$)

d) $y = -x$

2 a)

x	0	2	3
y	-4	2	5

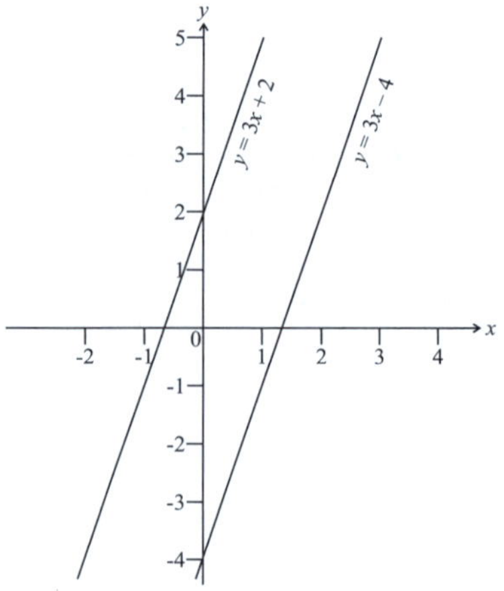

b) See diagram above. $y = 3x + 2$ will be parallel to $y = 3x - 4$ and will pass through the point (0, 2) on the y axis.

3 a) $y = 2x + 4$

b) $-\frac{1}{2}$

Page 71 (Exam Questions)

2 $3x + 4 = 2x + 6$ ***[1 mark]***

$x = 2$ ***[1 mark]***

so $y = 3(2) + 4 = 10$ and point M is (2, 10) ***[1 mark]***

Gradient of line **N** $= \frac{1}{2}$ (as it's parallel),

so $y = \frac{1}{2}x + c$ ***[1 mark]***

$10 = \frac{1}{2} \times 2 + c$, so $c = 10 - 1 = 9$

$y = \frac{1}{2}x + 9$ ***[1 mark]***

[5 marks available in total — as above]

3 a) $2a + 4 = 2c$, so $a + 2 = c$
Substitute values $a + 2 = c$ and $b - 6 = d$ into point (c, d):
$(c, d) = (a + 2, b - 6)$
Gradient of **S** $= \frac{b - 6 - b}{a + 2 - a} = \frac{-6}{2} = -3$
[3 marks available — 1 mark for correctly substituting values into a point, 1 mark for finding change in y over change in x, 1 mark for correct answer]

b) Gradient $= \frac{1}{3}$ ***[1 mark]***
So $y = \frac{1}{3}x + c$.
Substitute (6, 3) into the equation:
$3 = \frac{1}{3} \times 6 + c$
$c = 1$
Line **R**: $y = \frac{1}{3}x + 1$ ***[1 mark]***
[2 marks available in total — as above]

4 Midpoint of line AB: $\left(\frac{5+1}{2}, \frac{7-1}{2}\right) = (3, 3)$
Midpoint of line CD: $\left(\frac{13+3}{2}, \frac{4-2}{2}\right) = (8, 1)$
Gradient of line AB: $\frac{7-(-1)}{5-1} = \frac{8}{4} = 2$
Gradient of the line joining the midpoints of AB and CD:
Gradient $= \frac{1-3}{8-3} = \frac{-2}{5}$
$\frac{-1}{2} \neq \frac{-2}{5}$, therefore Eshe is incorrect.
[4 marks available — 1 mark for saying Eshe is wrong, 1 mark for finding both midpoints, 1 mark for finding both gradients, 1 mark for comparing gradients to show that the lines aren't perpendicular]

Page 79 (Warm-Up Questions)

1 a)

x	–2	–1	0	1	2	3	4	5
y	7	2	–1	–2	–1	2	7	14

b)

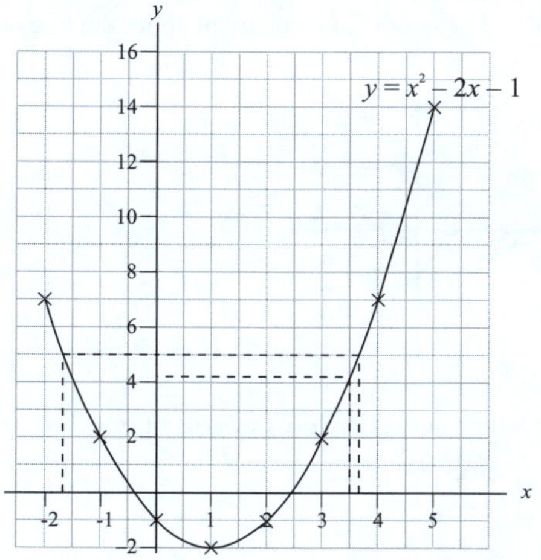

c) 4.25 (a value between 4 and 4.5 is acceptable).

d) $x = -1.65$ and 3.65 (values between –1.6 and –1.7 and between 3.6 and 3.7 are acceptable).

2

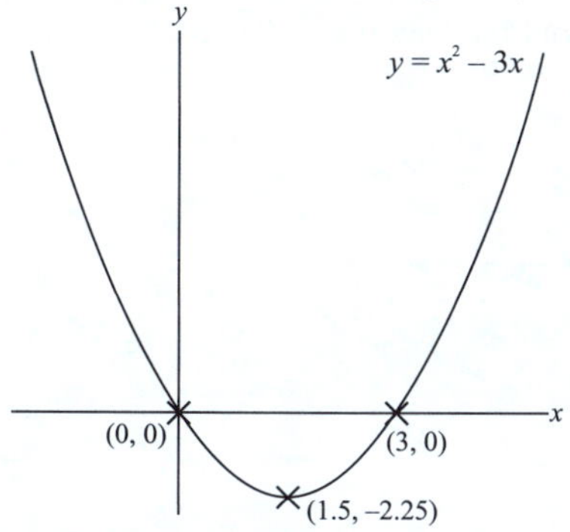

3 $x = 310°$ (answer from 308° – 312° is acceptable)

4 The graphs have the same shape, but the second is shifted up the y-axis by 2 units.

Pages 80-81 (Exam Questions)

2 a) B ***[1 mark]***
b) C ***[1 mark]***
c) A ***[1 mark]***

3 a)

x	2.5	3	3.5	4
y	–5.375	–5	–2.125	4

[2 marks available — 2 marks for all answers correct, otherwise 1 mark for two correct answers]

b)

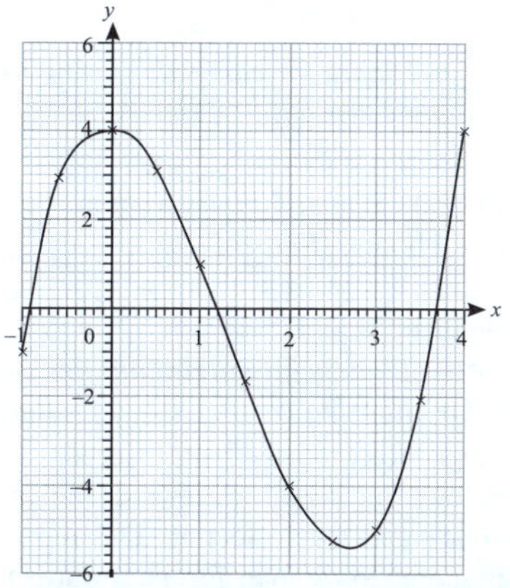

[2 marks available — 1 mark for plotting correct points, 1 mark for joining them with a smooth curve]

c) Reading off the graph, where the curve intersects the x-axis, $x = -0.9$, $x = 1.2$ and $x = 3.7$ ***[1 mark]***
You'll get the mark if your answers are within 0.1 of the answer.

4 Find the equation of the line that should be drawn:
$x^2 + x = 6$
$x^2 + x - 5 = 1$
$x^2 + 2x - 5 = x + 1$
So draw the line $y = x + 1$ to find the solutions. ***[1 mark]***

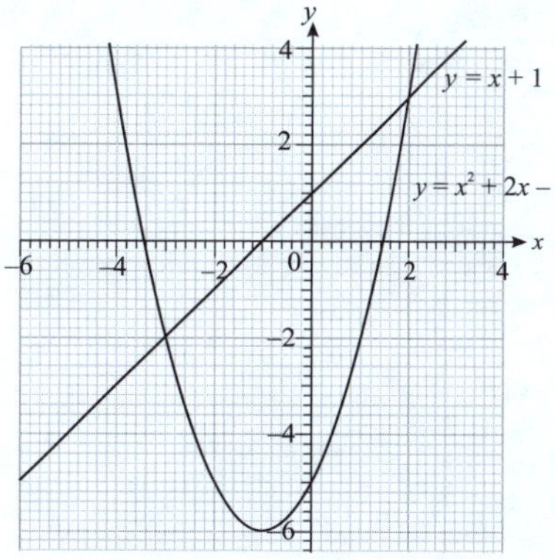

[1 mark]

The solutions to $x^2 + x = 6$ are:
$x = -3$ ***[1 mark]*** $x = 2$ ***[1 mark]***
[4 marks available in total — as above]

5 a)

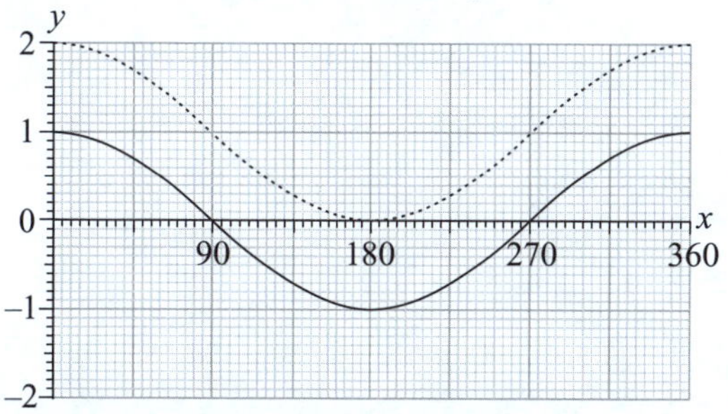

[2 marks available — 1 mark for correct graph shape, 1 mark for correct position]

b) $x = 90° - 30° = 60°$ and $x = 270° - 30° = 240°$
[2 marks available — 1 mark for each correct answer]
cos x crosses the x-axis at x = 90° and x = 270°.
The reflection doesn't change the x-intercepts, but the translation shifts the graph 30° left (or –30° right).

Page 86 (Warm-Up Questions)

1 Graph A and 2, Graph B and 3
Graph C and 4, Graph D and 1

2 –1

Page 87 (Exam Questions)

2 a) Plan A: £25 ***[1 mark]***
Plan B: £28 ***[1 mark]***
[2 marks available in total — as above]

b) Mr Barker should use Plan A because it is cheaper. Using 85 units with Plan A would cost £26.50. 85 units with Plan B would cost £34.
[2 marks available — 1 mark for correctly stating which plan, 1 mark for giving a reason]

3 a) 1 hour ***[1 mark]***

b) Tyrone. He reaches 30 km after 5 hours whereas Selby reaches 30 km after 6 hours. ***[1 mark]***

c) Gradient = $\frac{\text{change in } y}{\text{change in } x} = \frac{25-15}{3-1.5} = \frac{10}{1.5} = 6.67$ km/h (2 d.p)
[2 marks available — 2 marks for correct answer, otherwise 1 mark for choosing correct x and y values]

d) E.g. Selby is the most likely to have been injured.
The gradient of Selby's line decreases towards the end of the race, whereas Tyrone's gets much steeper. This means Selby was moving much more slowly than Tyrone towards the end of the race. ***[2 marks available — 1 mark for stating Selby is the injured runner, 1 mark for a correct explanation referring to gradients or steepness of lines]***

Page 88 (Revision Questions)

1
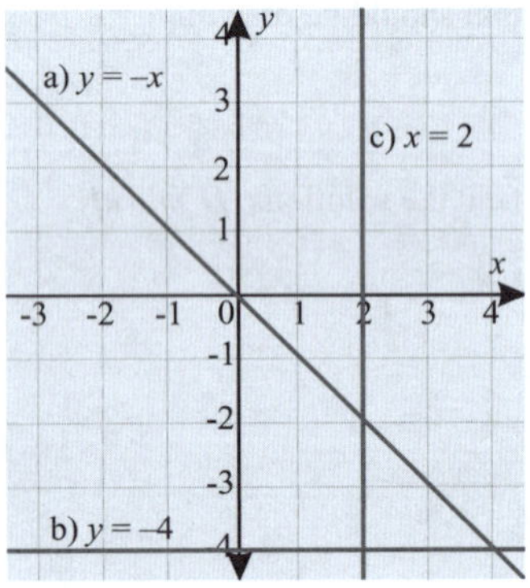

2
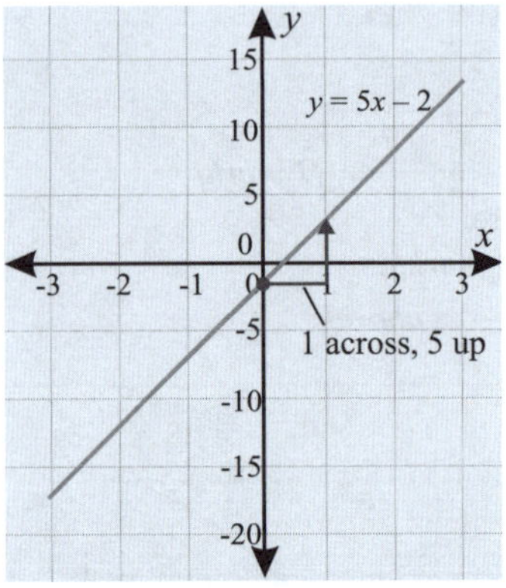

3 $y = 2x + 10$

4 $y = x - 9$

5 $y = -\frac{1}{2}x + 4$

6 a)

x	–3	–2	–1	0	1
y	–7	–9	–9	–7	–3

b)
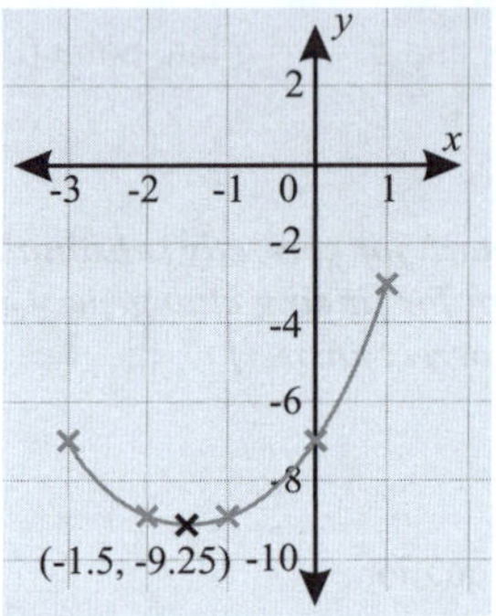

7
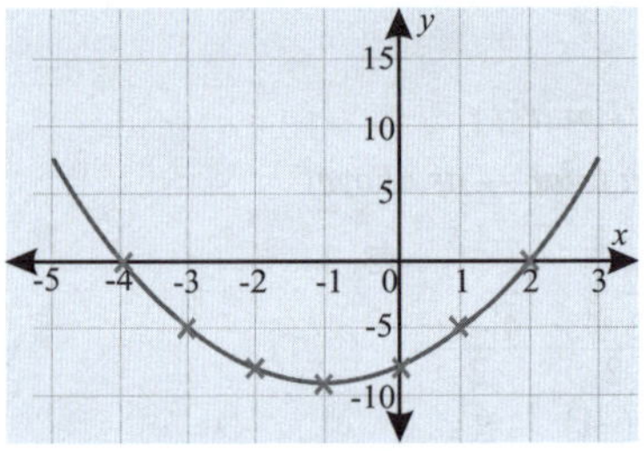

$x = -3.6$ or 1.6 (both ± 0.2)

8 a) A graph with a "wiggle" in the middle. E.g.
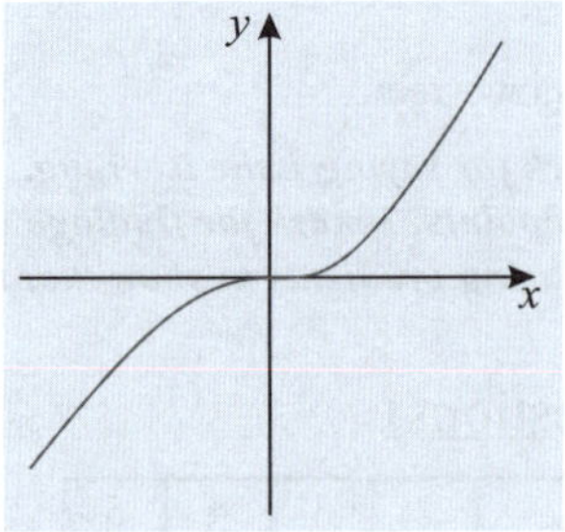

b) A graph made up of two curves in diagonally opposite quadrants. The graph is symmetrical about the lines $y = x$ and $y = -x$.
E.g.
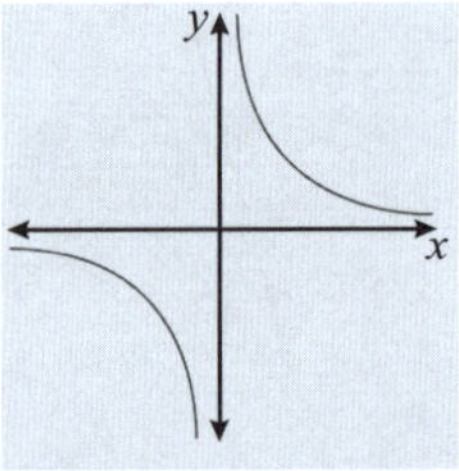

c) A graph which curves rapidly upwards. E.g.
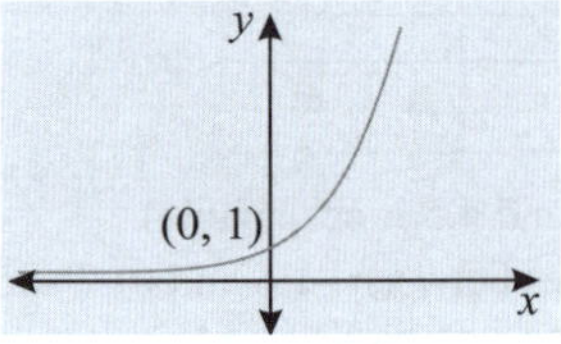

d) A circle with radius r, centre (0, 0). E.g.
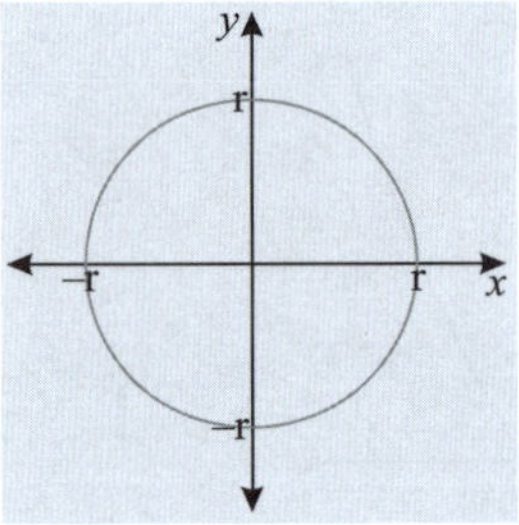

9 b = 0.25, c = 8

10 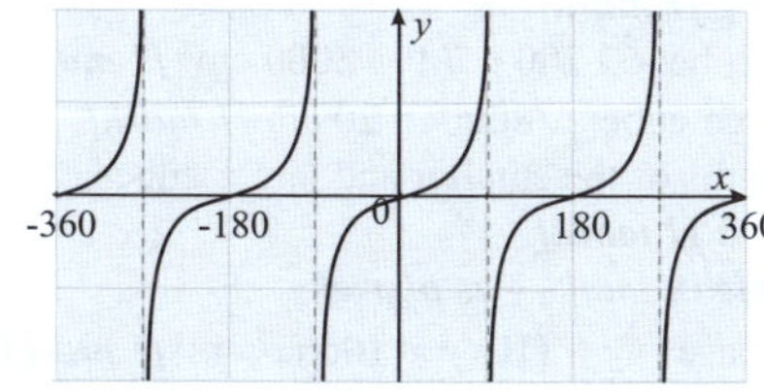

11 $x = 8, y = 12$

12 $y = -x + 6$

13 Translation on y-axis
Translation on x-axis
Reflection in the x-axis or y-axis.

14 a) $y = (-x)^3 + 1$ is the original graph reflected in y-axis.
b) $y = (x + 2)^3 + 1$ is the original graph translated by 2 units in the negative x-direction.
c) $y = x^3 + 4$ is the original graph translated upwards by 3 units.
d) $y = x^3 - 1$ is the original graph translated downwards by 2 units.

15 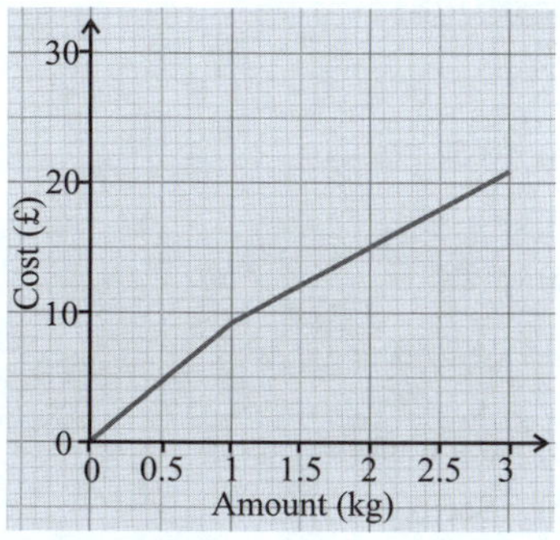

16 a) 480 metres
b) 1.25 m/s²
c) 1.25 m/s²

Section Four — Ratio, Proportion and Rates of Change

Page 92 (Warm-Up Questions)

1 a) 1:2 b) 4:9 c) 2:9
d) 16:7 e) 5:4

2 1:4.4

3 300 g ÷ 2 = 150 g; 150 g × 3 = 450 g of flour

4 20

5 45, 60, 75
(3 + 4 + 5 = 12 parts, so 180 ÷ 12 = 15 per part.)

6 7 parts – 5 parts = 2 parts,
2 parts = 44 cm,
so 1 part = 22 cm,
9 parts = 198 cm
So Owen is 198 cm tall.

Page 93 (Exam Questions)

3 a) $3\frac{3}{4} : 1\frac{1}{2} = 4 \times 3\frac{3}{4} : 4 \times 1\frac{1}{2} = 15:6$ ***[1 mark]***
$= 5:2$ ***[1 mark]***
[2 marks available in total — as above]

b) 5 + 2 = 7 parts
1 part: 2800 ml ÷ 7 = 400 ml
Yellow paint = 400 ml × 5 = 2000 ml
Blue paint = 400 ml × 2 = 800 ml
[2 marks available — 1 mark for finding the amount of 1 part, 1 mark for finding the correct amounts for both yellow and blue paint]
If your answer to part a) was incorrect, but your answers to part b) were correct for your incorrect ratio, you still get the marks for part b).

4 Catrin, Ariana, Nasir and Simone shared the money in the ratio 1:2:4:8 ***[1 mark for 1:2:4:8 or an equivalent ratio (order may be different)]***
660 ÷ (1 + 2 + 4 + 8) = 44 ***[1 mark]***
Simone got £44 × 8 = £352 ***[1 mark]***
[3 marks available in total — as above]
You could answer this question using a formula — if you let x be the amount of money that Catrin gets, then x + 2x + 4x + 8x = £660.

5 Call the number of black olives b and the number of green olives g.
$b:g = 5:11$ and $b - 3 : g - 1 = 3:7$ ***[1 mark for both]***
$\frac{b}{g} = \frac{5}{11}$ and $\frac{b-3}{g-1} = \frac{3}{7}$ ***[1 mark for both]***
$11b = 5g$ and $7(b - 3) = 3(g - 1)$
$11b - 5g = 0$ [1] and $7b - 3g = 18$ [2] ***[1 mark]***
[1] × 3: $33b - 15g = 0$ [3]
[2] × 5: $35b - 15g = 90$ [4] ***[1 mark for both]***
[4] – [3]: $2b = 90$
$b = 45$ ***[1 mark]***
Sub into [1]: $(11 \times 45) - 5g = 0$
$5g = 495$, so $g = 99$ ***[1 mark]***
Solution: $b = 45, g = 99$
[6 marks available in total — as above]
You might have started with $\frac{b}{b+g} = \frac{5}{16}$ *and* $\frac{b-3}{b+g-4} = \frac{3}{10}$ *and used these to form simultaneous equations instead.*

Page 96 (Warm-Up Questions)

1 £1080

2 30 hours

3 12 V

4 a) $A = kr^2$ b) $D = k/R$
c) $H = k/D^3$ d) $V = kS^3$

Page 97 (Exam Questions)

3 a) 1 litre of petrol will keep 8 go-karts going for:
20 ÷ 12 = 1.666... minutes ***[1 mark]***
18 litres of petrol will keep 8 go-karts going for:
1.666... × 18 = 30 minutes ***[1 mark]***
18 litres of petrol will keep 1 go-kart going for:
30 × 8 = 240 minutes ***[1 mark]***
18 litres of petrol will keep 6 go-karts going for:
240 ÷ 6 = 40 minutes ***[1 mark]***
[4 marks available in total — as above]
You might have done these steps in a slightly different order — you'd still get all the marks as long as you got the same answer.

b) In 1 minute, 8 go-karts will use 12 ÷ 20 = 0.6 litres ***[1 mark]***
In 45 minutes, 8 go-karts will use 0.6 × 45 = 27 litres ***[1 mark]***
27 litres of petrol cost: £1.37 × 27 = £36.99 ***[1 mark]***
[3 marks available in total — as above]

4 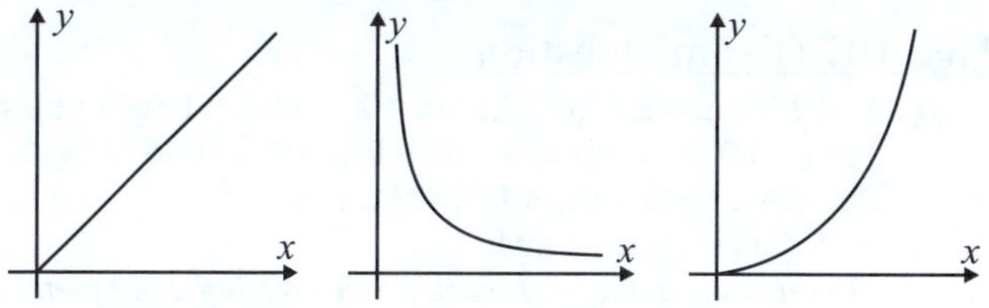

[3 marks available — 1 mark for each correct graph]

5 a) $h \propto S^2$, so $h = kS^2$ ***[1 mark]***
When $h = 35$ and $S = 50$, $35 = k \times 50^2$,
so $k = 35 \div 50^2 = 0.014$ ***[1 mark]***
So $h = 0.014S^2$
$h = 0.014S^2$ so when $S = 40$, $h = 0.014 \times 40^2 = 22.4$ ***[1 mark]***
[3 marks available in total — as above]

b) If h_0 is the initial height difference, S_0 is the initial speed and h_1 is the changed height difference, then:
$h_0 = k(S_0)^2$
$h_1 = k(1.3S_0)^2$ ***[1 mark]***
$= 1.69k(S_0)^2$ ***[1 mark]*** $= 1.69h_0$ ***[1 mark]***
So the height needs to be increased by 69%. ***[1 mark]***
[4 marks available in total — as above]

Page 102 (Warm-Up Questions)

1 £13.50

2 £129.60

3 45%

4 £208

5 £3376.53

Page 103 (Exam Questions)

3 For every 2 grapes there are 5 cherries so there are $\frac{2}{5}$ = 40% as many grapes as cherries. ***[1 mark]***

4 A ratio of 3 : 7 means that 3 out of 10 = 30% of the customers were children.
60% of 30% = 0.6 × 30% = 18% were blond-haired children.
100% – 30% = 70% were adults.
20% of 70% = 0.2 × 70% = 14% were blond-haired adults.
So, 18% + 14% = 32% of the customers had blond hair.
[4 marks available — 1 mark for finding blond-haired children %, 1 mark for finding blond-haired adults %, 1 mark for adding together the two percentages, 1 mark for correct answer]

5 5000 × 0.16 = 800 trees are planted in 2018 ***[1 mark]***
A maximum of 800 × 0.75 = 600 trees are cut down
At the end of 2018 there is a minimum of
5000 + (800 – 600) = 5200 pine trees ***[1 mark]***
5200 × 0.16 = 832 trees are planted in 2019 ***[1 mark]***
A maximum of 832 × 0.75 = 624 trees are cut down
At the end of 2019 there is a minimum of
5200 + (832 – 624) = 5408 pine trees ***[1 mark]***
[4 marks available in total — as above]

6 Multiplier = 1 – 0.25 = 0.75 ***[1 mark]***
$N_0 \times (0.75)^{35-31}$ = 2 000 000 ***[1 mark]***
N_0 = 2 000 000 ÷ 0.75^4 = 6 320 987.654...
= £6 300 000 (to the nearest £100 000) ***[1 mark]***
[3 marks available in total — as above]

Page 106 (Warm-Up Questions)

1 a) 12.7 × 1000 = 12 700 g b) 1430 ÷ 100 = 14.3 m

2 10 kg. (If you can't remember whether to multiply or divide, do both and see which answer is more sensible. Remember to cross out any incorrect working though.)

3 3 000 000 000 mm³

4 11.3 g/cm³ (density = mass ÷ volume)

5 96 g (Volume = 5 × 4 × 6 = 120 cm³. Then use mass = density × volume.)

6 90 km/h (Speed in m/s = 100 ÷ 4 = 25 m/s. Multiply by 3600 to get m/h, then divide by 1000 to get km/h.)

7 9 km (distance = speed × time)

Page 107 (Exam Questions)

3 a) E.g. 2500 m = 2.5 km. 2.5 km = 2.5 ÷ 1.6 = 1.5625 miles.
102 s ÷ 60 = 1.7 minutes ÷ 60 = 0.02833... hours.
Speed = 1.5625 miles ÷ 0.02833... hours
= 55 mph (to nearest mph)
[3 marks available — 1 mark for converting 2500 metres to miles, 1 mark for converting 102 seconds into hours, 1 mark for the correct final answer]
It doesn't matter whether you do the conversion to miles per hour at the start or the end of the calculation — you could find the speed in m/s, km/s or km/h, and then change it to mph. Whichever way, you should get the same answer.

b) E.g. time = 1.5625 miles ÷ 50 mph = 0.03125 hours
0.03125 hours × 60 × 60 = 113 s (to nearest second)
[2 marks available — 1 mark for dividing the distance by the speed limit, 1 mark for the correct answer]

4 a) Volume = 4 cm × 4 cm × 4 cm = 64 cm³ ***[1 mark]***
Mass = 7.9 × 64 ***[1 mark]***
= 505.6 g ***[1 mark]***
[3 marks available in total — as above]

b) 63.2 kg = 63 200 g ***[1 mark]***
Volume of large cube: 63 200 ÷ 7.9 = 8000 cm³ ***[1 mark]***
Side length of large cube: $\sqrt[3]{8000}$ = 20 cm ***[1 mark]***
Ratio of side lengths of the smaller and larger cubes:
4 cm : 20 cm = 1 : 5 ***[1 mark]***
[4 marks available in total — as above]

5 a) Area of circular base = $\pi \times (10x)^2 = 100\pi x^2$ cm² ***[1 mark]***
$100\pi x^2$ cm² = $(100\pi x^2 \div 100 \div 100)$ m²
= $0.01\pi x^2$ m² ***[1 mark]***
Weight = $650 \times 0.01\pi x^2$ ***[1 mark]***
= $6.5\pi x^2$ N ***[1 mark]***
[4 marks available in total — as above]

b) E.g. If the diameter is halved, the area of the circular base becomes: $\pi \times (5x)^2 = 25\pi x^2$ cm² = $0.0025\pi x^2$ m²
Pressure = $6.5\pi x^2 \div 0.0025\pi x^2$ = 2600 N/m²
2600 N/m² ÷ 650 N/m² = 4
If the diameter of the circle is halved the pressure increases and is 4 times greater.
[2 marks available — 1 mark for saying that the pressure increases, 1 mark for saying it's 4 times greater]

Page 108 (Revision Questions)

1 $\frac{13}{8}$ or 1.625

2 a) 3 : 4 b) 3.5 : 1

3 240 blue scarves

4 a) $\frac{1}{5}$ b) 128

5 10

6 $x = 36$, $y = 9$

7 a) 960 flowers b) 3.9 hours

8 a) $y = kx^2$ b) See p.95

9 0.91 Pa (2 d.p.)

10 a) 19 b) 39
c) 21.05% (2 d.p.) d) 475%

11 percentage change = (change ÷ original) × 100

12 35% decrease

13 17.6 m

14 2%

15 $N = N_0 \times (\text{multiplier})^n$

16 a) £157.37 (to the nearest penny)
b) 14 years

17 a) 5600 cm³ b) 240 cm
c) 10.8 km/h d) 12 000 000 cm³
e) 12.8 cm² f) 2750 mm³

18 42 mph

19 12 500 cm³

20 11 m²

Section Five — Geometry and Measures

Page 114 (Warm-Up Questions)

1 $a = 115°$, angles on a straight line add to 180°, so $a = 180° – 65°$.
$b = 115°$, a and b are corresponding angles, so $a = b$.
$c = 65°$, c and 65° are also corresponding angles.
$d = 115°$, angles on a line add to 180°, so $d = 180° – c$.
There are often different ways of going about angle questions. Just keep scribbling down angles as you find them. It can make it easier to get the angle you want.

2 6 sides

3 Pentagon

4 Number of sides = 360° ÷ 24° = 15

5 Kite

Page 115 (Exam Questions)

2 Let a be the third angle in the triangle.
$a = 180° - y - z$ (angles in a triangle add to 180°) ***[1 mark]***
$x = 180° - a$ (angles on a straight line add to 180°) ***[1 mark]***
So $x = 180° - (180° - y - z) = y + z$ ***[1 mark]***
[3 marks available in total — as above]

3 Angle BDC = angle $BCD = x$ (triangle BCD is isosceles)
Angle $CBD = 180° - x - x = 180° - 2x$ ***[1 mark]***
(angles in a triangle add to 180°)
Angle BDE = angle $CBD = 180° - 2x$ ***[1 mark]*** (alternate angles)
Angle AED = angle $BDE = 180° - 2x$ ***[1 mark]*** ($ABDE$ is an isosceles trapezium so has a vertical line of symmetry)
$y = 360°$ – angle AED ***[1 mark]***
(angles round a point add to 360°)
So $y = 360° - (180° - 2x) = 180° + 2x$ ***[1 mark]***
[5 marks available in total — as above]
There's more than one way to do this question — as long as you show your working and explain each step you'll get the marks.

4 Exterior angle = 180° – 150° = 30° ***[1 mark]***
Number of sides = 360° ÷ 30° ***[1 mark]***
= 12 ***[1 mark]***
[3 marks available in total — as above]

5 Interior angle of regular n-sided polygon = 180° – exterior angle
= 180° – (360° ÷ n)

Interior angle of regular octagon = 180° – (360° ÷ 8) = 135°
Interior angle of regular hexagon = 180° – (360° ÷ 6) = 120°
Angle CBK = angle ABC – angle IJK = 135° – 120° = 15°
[2 marks available — 1 mark for using correct method to find interior angle of octagon or hexagon, 1 mark for correct answer]

Page 119 (Warm-Up Question)

1 $M = N = 64°$
(Angles M and N are equal to the 64° angle given, using the alternate segment theorem.)
$L = 52°$ *(Angles in a triangle add to 180°, so 180° – 64° – 64° = 52°.)*

Page 120 (Exam Questions)

2 Angle DBC = 62° ***[1 mark]*** (angles in the same segment are equal)
Angle ABC = 90° ***[1 mark]*** (the angle in a semicircle is 90°)
Angle x = 90° – 62° = 28° ***[1 mark]***
[3 marks available in total — as above]

3 Angle DEG = 53° and angle AEF = 37°
(alternate segment theorem) ***[1 mark]***
Angle AED = 180° – 53° – 37° = 90°
(angles on a straight line) ***[1 mark]***
The chord AD must be a diameter of the circle (angle in a semicircle is 90°), so AD must pass through the centre of the circle. ***[1 mark]***
[3 marks available in total — as above]

4 Opposite angles in a cyclic quadrilateral add up to 180°, so angle ADC = 180° – angle ABC = 180° – 119° = 61°
Angle CDX = angle ADC – angle ADX = 61° – 31° = 30°
If X was the centre of the circle, XD and XC would be radii, so triangle CXD would be isosceles and angles CDX and XCD would be equal. Here angle CDX = 30° and angle XCD = 28° so the angles are not equal, and therefore X is not the centre of the circle.
[3 marks available — 1 mark for finding angle ADC, 1 mark for finding angle CDX, 1 mark for using "two radii from an isosceles triangle" to explain why X cannot be the centre]

Page 125 (Warm-Up Questions)

1 a) A and E b) B and F

2 a) 0.222 or $\frac{2}{9}$
Note that the enlargement scale factor is less than one — so the 'enlargement' actually makes the shape smaller.
b) 2.6 cm

3 $\begin{pmatrix}-3\\3\end{pmatrix}$

4 (–3, 5)

Page 126 (Exam Questions)

2 a) Rotation 90° anticlockwise around the point (0, 0)
[3 marks available — 1 mark for rotation, 1 mark for correct angle and direction of rotation, 1 mark for correct centre of rotation]
b)
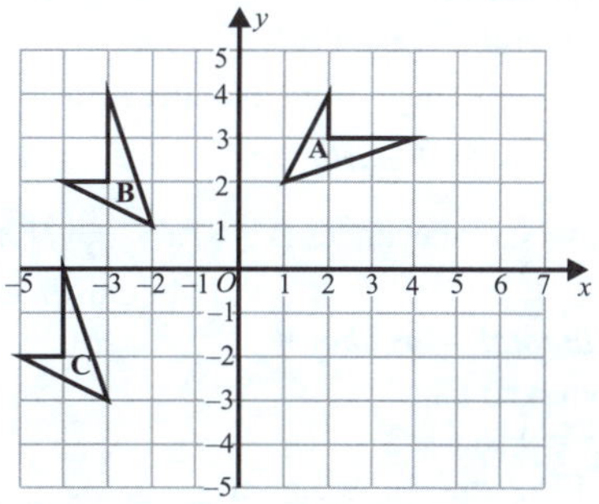

[1 mark for correct translation]

3 63 m = 6300 cm
Scale factor = 6300 ÷ 60 ***[1 mark]***
= 105 ***[1 mark]***
Height of flagpole = 8 cm × 105 = 840 cm = 8.4 m ***[1 mark]***
[3 marks available in total — as above]
The triangles created between Huan's eyes and his finger and his eyes and the flagpole are similar.

4
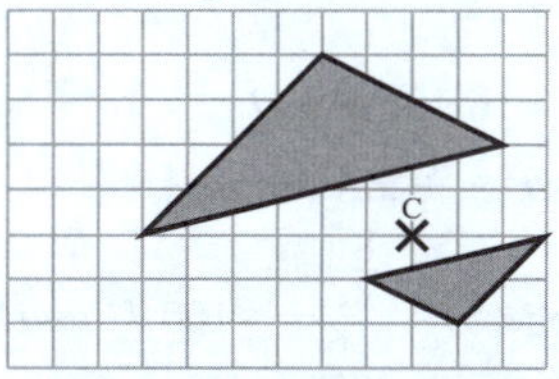

[2 marks available — 2 marks for correct shape in correct position, otherwise 1 mark for one vertex correct]

Page 133 (Warm-Up Questions)

1 a) 1963.5 mm² *($\pi r^2 = 25^2 \times \pi$)*
b) 37 146 mm² *(area of rectangle minus area of four circles)*
c) 148 584 mm³ *(volume = area × thickness)*

2 672 cm³ *(area of triangle × length = ½ × 12 × 8 × 14)*

3 1440 m³ *(volume of a sphere = $\frac{4}{3}\pi r^3$)*

4 a) b) c)
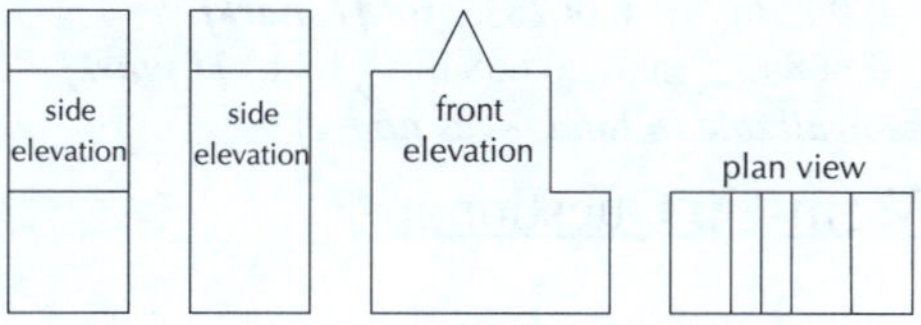

Pages 134-135 (Exam Questions)

2
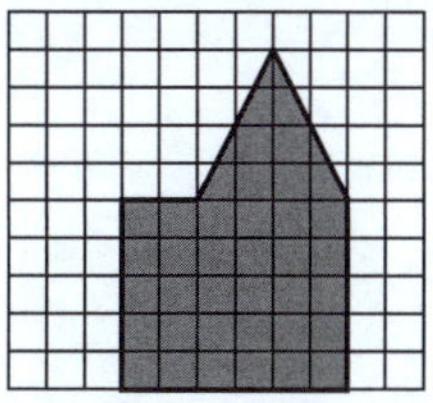
[2 marks available — 1 mark for rectangular part correct, 1 mark for triangular part correct]

3 Shorter parallel side of the trapezium
= 52 cm – 16 cm – 16 cm = 20 cm
To see this, split the shape into two triangles and a rectangle. The two triangles are both isosceles, so the base of each triangle is 16 cm long.
Area of trapezium = 0.5 × (52 + 20) × 16 = 576 cm²
[2 marks available — 1 mark for finding shorter parallel side of trapezium, 1 mark for correct answer]

4 Surface area of curved part of hemisphere =
$\frac{1}{2}$ × surface area of a sphere = $\frac{1}{2} \times 4 \times \pi \times 7^2$ ***[1 mark]***
= 307.876... cm^2
Surface area of curved part of cone = $\pi \times 2 \times 12$ ***[1 mark]***
= 75.398... cm^2
Surface area of flat top of hemisphere = $(\pi \times 7^2) - (\pi \times 2^2)$
= 141.371... cm^2 ***[1 mark]***
Total surface area = 307.876... + 75.398... + 141.371...
= 525 cm^2 (to 3 s.f.) ***[1 mark]***
[4 marks available in total — as above]

5 Let x be the scale factor for length from cylinder A to cylinder B.
$x^3 = \frac{64}{27}$, so $x = \sqrt[3]{\frac{64}{27}} = \frac{4}{3}$ ***[1 mark]***
Then $x^2 = \frac{4^2}{3^2} = \frac{16}{9}$, so s.a. of cylinder B = $81\pi \times \frac{16}{9}$ ***[1 mark]***
= 144π cm^2 ***[1 mark]***
[3 marks available in total — as above]

6 a) Scale factor from **A** to **C**:
$n^2 = 108\pi \div 12\pi = 9$ so $n = 3$
Volume of **A** = 135π cm^3 ÷ 3^3 = 5π cm^3
[4 marks available — 1 mark for finding n^2, 1 mark for finding n, 1 mark for dividing volume of A by n^3, 1 mark for correct answer]

b) Scale factor from **A** to **B**:
$m^2 = 48\pi \div 12\pi = 4$ ***[1 mark]***
$m = 2$ ***[1 mark]***
Perpendicular height of **B** = 4 cm × 2 = 8 cm ***[1 mark]***
[3 marks available in total — as above]

7 Let r be the radius of the spheres.
Volume of cuboid = $2r \times 2r \times 4r = 16r^3$ ***[1 mark]***
Volume of both spheres = $2 \times \frac{4}{3}\pi r^3 = \frac{8}{3}\pi r^3$ ***[1 mark]***
Percentage of box occupied by spheres = $\dfrac{\frac{8}{3}\pi r^3}{16r^3} \times 100$ ***[1 mark]***
= $\frac{800\pi}{48}$ = 52.4% ***[1 mark]***
[4 marks available in total — as above]

8 a) Cross-sectional area of pipe: $0.2^2 \times \pi = 0.12566...$ m^2
Cross-sectional area of water: 0.12566... ÷ 2 = 0.06283... m^2 ***[1 mark]***

b) Rate of flow = 2520 litres per minute
= 2520 ÷ 60 litres per second
= 42 litres per second ***[1 mark]***
= 42 000 cm^3/s
= 0.042 m^3/s ***[1 mark]***
Speed = Rate of flow ÷ cross-sectional area of water
= 0.042 m^3/s ÷ 0.06283... m^2 ***[1 mark]***
= 0.66845... m/s = 0.668 m/s (3 s.f.) ***[1 mark]***
[4 marks available in total — as above]

Page 141 (Warm-Up Questions)

1
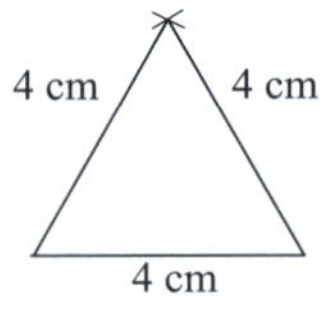

2
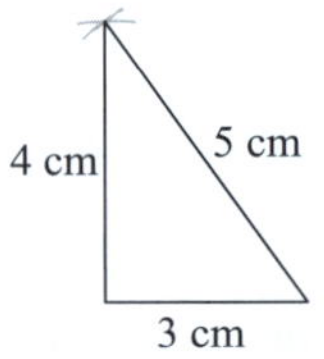

3 Shaded area = where house could be built
(Each square should be 1 cm × 1 cm in your diagram.)
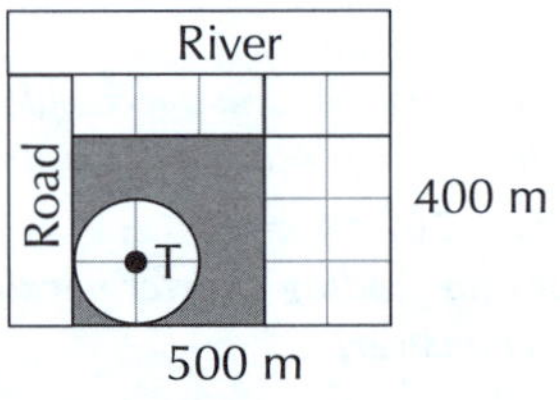

4
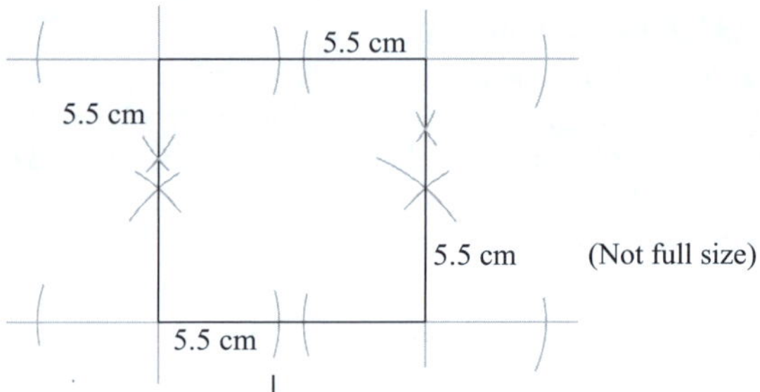

5
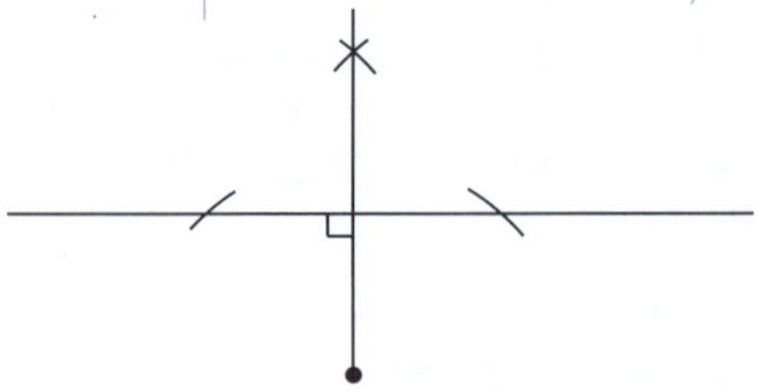

6 Scale: 1 cm represents 1 mile
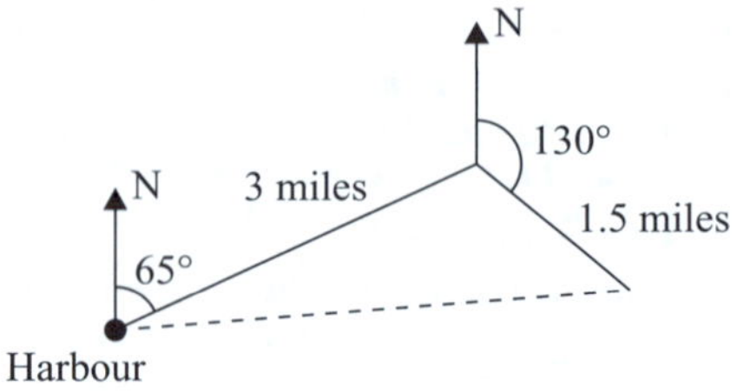

So the distance is 3.9 miles (allow 3.8-4 miles)

Page 142 (Exam Questions)

3 a)
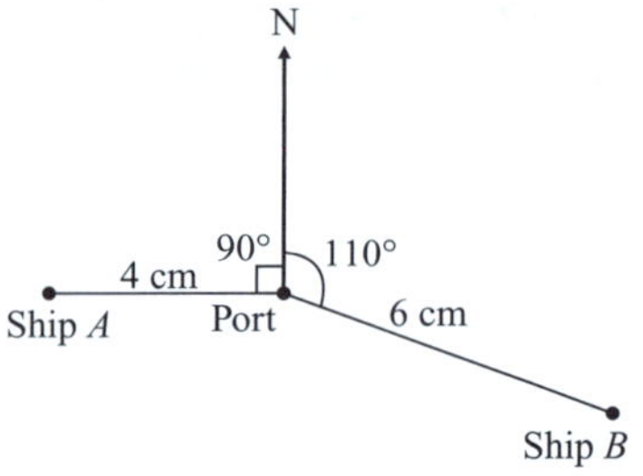

[4 marks available — 1 mark for ship A 4 cm from the port, 1 mark for correct bearing for ship A, 1 mark for ship B 6 cm from the port, 1 mark for correct bearing for Ship B]
This diagram has been drawn a bit smaller to make it fit — your measurements should match the labels given on the diagram here.

b) 102° (accept answers between 100° and 104°) ***[1 mark]***

c) 180° – 102° = 78°
360° – 78° = 282° (accept answers between 280° and 284°)
[2 marks available — 1 mark for correctly using 102°, 1 mark for correct answer]
You could also do this by adding 180° to 102°.

4
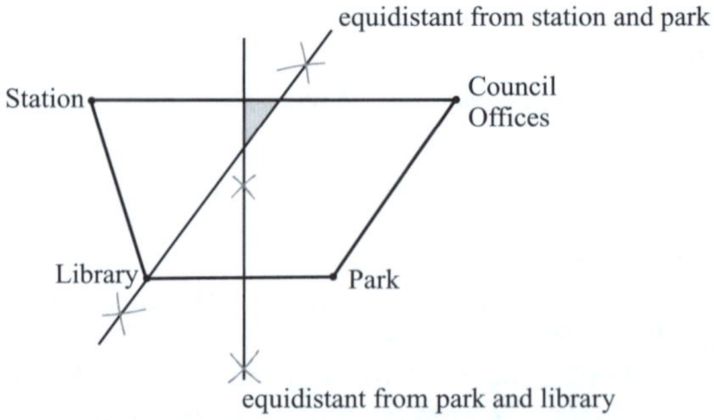

[3 marks available — 1 mark for perpendicular bisector between Library and Park, 1 mark for perpendicular bisector between Station and Park, 1 mark for the correct shaded area]

5

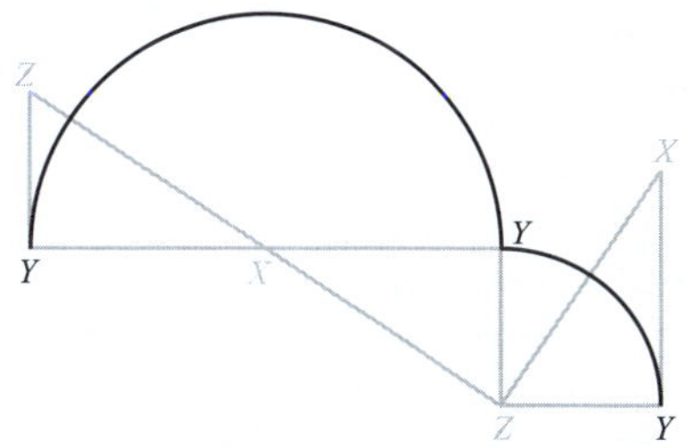

[3 marks available — 1 mark for constructing either the semicircle or quarter circle correctly, 1 mark for the two parts of the locus being joined together, 1 mark for a completely correct diagram]

Pages 143-144 (Revision Questions)

1 See p.109

2 a) $x = 154°$ b) $y = 112°$ c) $z = 58°$

3 Exterior angle = 60°,
sum of interior angles = 720°

4 Equilateral triangle:
lines of symmetry = 3
order of rotational symmetry = 3
Isosceles triangle:
lines of symmetry = 1
order of rotational symmetry = 1
Scalene triangle:
lines of symmetry = 0
order of rotational symmetry = 1

5 E.g. rhombus and parallelogram

6 See p.116-117

7 a) $x = 53°$ b) $y = 69°$ c) $z = 33°$

8 No — opposite angles in a cyclic quadrilateral add up to 180°, but 88° + 95° = 183° ≠ 180°.

9 SSS, AAS, SAS, RHS

10 E.g. angles ACB and ACD are right angles (as it's a perpendicular bisector of a chord)
AB = AD (they're both radii)
CB = CD (as the chord is bisected)
So the condition RHS holds and the triangles are congruent.

11 $x = 2.5$ cm

12 a) Translation by vector $\begin{pmatrix}-7\\-5\end{pmatrix}$,
or rotation 180° about point (1, 2)

b) Enlargement of scale factor $\frac{1}{3}$ and centre of enlargement (0, 0).

13

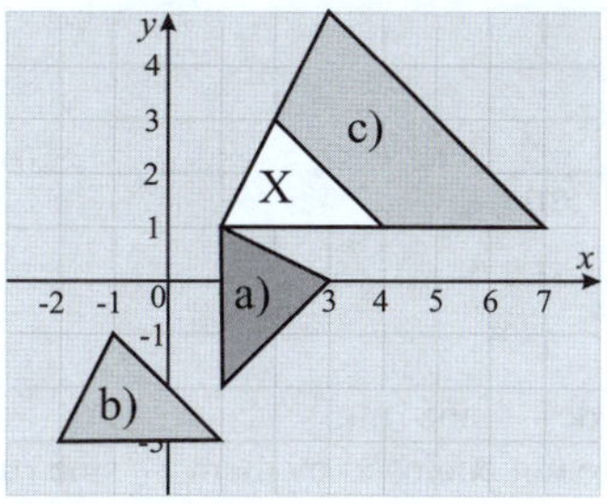

14 $A = \frac{1}{2}(a + b) \times h_v$

15 69 cm²

16 30 cm

17 Circumference = 16π cm, area = 64π cm²

18 39.27 cm²

19 S. A. of a sphere = $4\pi r^2$
S. A. of a cylinder = $2\pi rh + 2\pi r^2$
S. A. of a cone = $\pi rl + \pi r^2$

20 75π cm²

21 1030 cm³ (3 s.f.)

22 a) 129.85 cm³ (2 d.p.)
b) 5.2 s (1 d.p.)

23 80 cm²

24

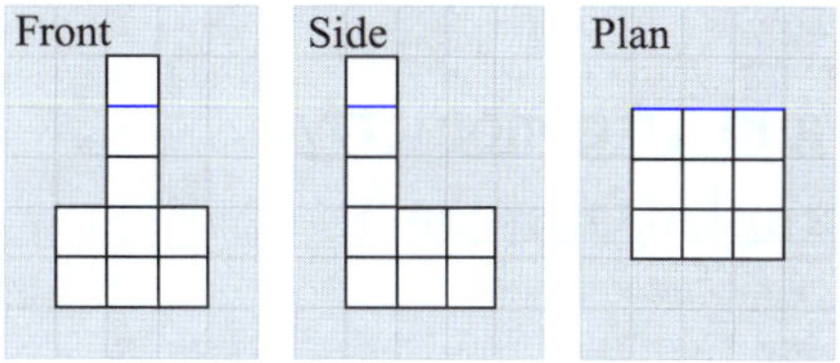

25

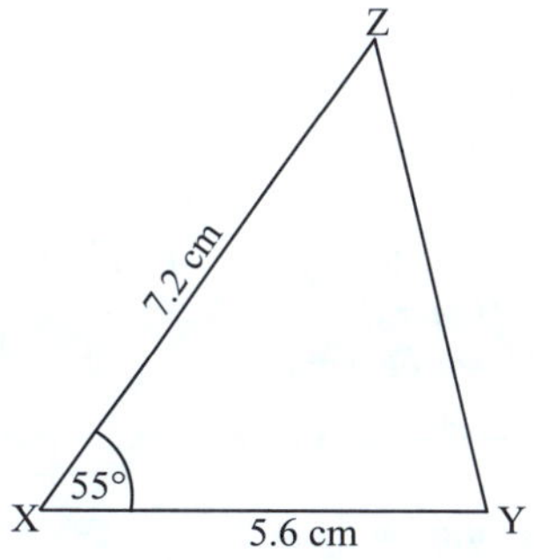

26

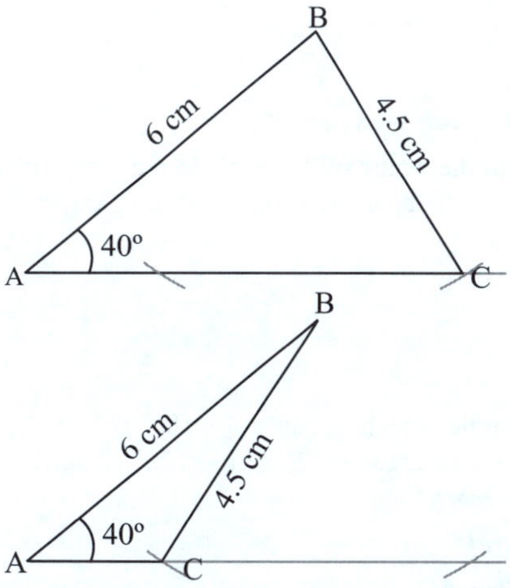

27 A circle

28

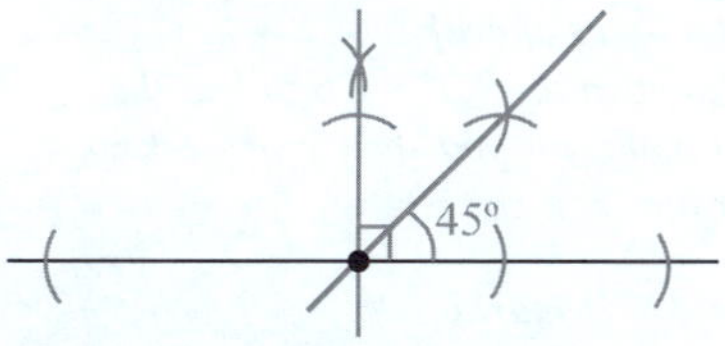

29 See p.137

30

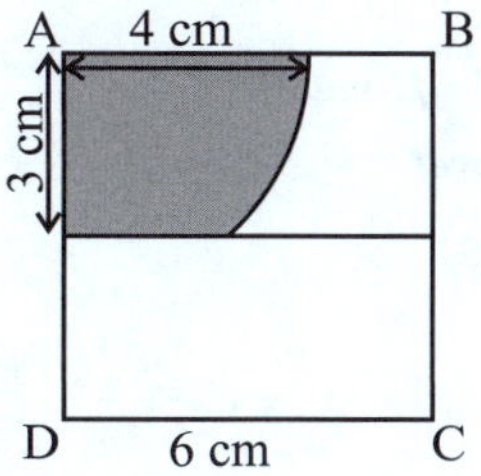

31 Put your pencil on the diagram at the point you're going FROM. Draw a north line at this point. Draw in the angle clockwise from the north line — this is the bearing you want.

32

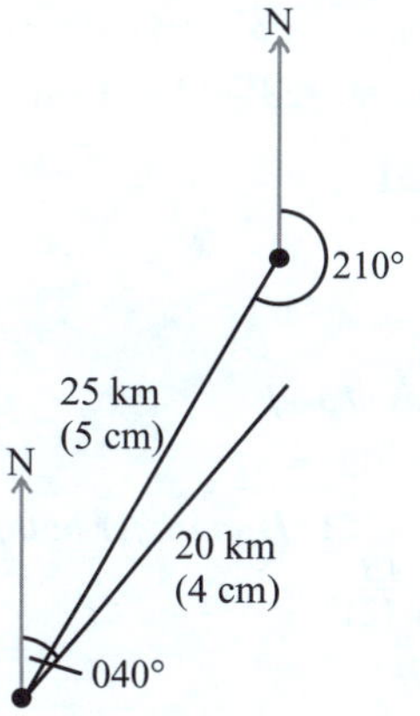

Section Six — Pythagoras and Trigonometry

Page 149 (Warm-Up Questions)

1 a) 13.1 cm *(by Pythagoras)*

b) 40° *($\tan x = 8.4 \div 10$)*

2 3 cm

Page 150 (Exam Questions)

3 Difference in x-coordinates = 8 – 2 = 6
Difference in y-coordinates = 8 – –1 = 9 ***[1 mark for both]***
So length of line segment = $\sqrt{6^2 + 9^2} = \sqrt{36 + 81}$ ***[1 mark]***
$= \sqrt{117} = \sqrt{9 \times 13} = 3\sqrt{13}$ ***[1 mark]***
[3 marks available in total — as above]

4 Let h be the height of the triangle:
$13^2 = 5^2 + h^2$ ***[1 mark]***
$h = \sqrt{169 - 25} = \sqrt{144}$ ***[1 mark]***
$h = 12$ cm ***[1 mark]***
Area, $A = \frac{1}{2} \times 10 \times 12$
$A = 60$ cm² ***[1 mark]***
[4 marks available in total — as above]

5 Call the distance from the centre of the circle to the centre of an edge x. The radius bisects the interior angle forming angle a.

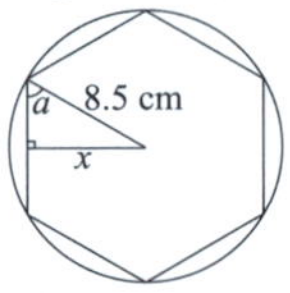

Sum of the interior angles of a hexagon = 4 × 180° = 720°
Each interior angle of a hexagon = 720° ÷ 6 = 120° ***[1 mark]***
$a = 120 \div 2 = 60°$ ***[1 mark]***
$\sin 60° = \frac{x}{8.5}$ ***[1 mark]***
$x = 8.5 \times \sin 60°$ ***[1 mark]***
$x = 7.36$ cm (2 d.p.) ***[1 mark]***
[5 marks available in total — as above]
You could also use the calculation cos 30° × 8.5 to find the value of x. As long as you make sure you show your working, you'll get full marks if your answer is correct.

6 $\sin 60° + \tan 60° = \frac{\sqrt{3}}{2} + \sqrt{3}$ ***[1 mark]***

$= \frac{\sqrt{3}}{2} + \frac{2\sqrt{3}}{2}$ ***[1 mark]***

$= \frac{\sqrt{3} + 2\sqrt{3}}{2} = \frac{3\sqrt{3}}{2}$ ***[1 mark]***

[3 marks available in total — as above]

Page 157 (Warm-Up Questions)

1 $AC = 9.6$ cm

2 104.5°

3 $FDG = 31°$ (2 s.f.)

4 a) $\overrightarrow{CD} = -2\mathbf{a}$ *(since ABCD is a parallelogram, AB = DC)*

b) $\overrightarrow{AC} = 2\mathbf{d} + 2\mathbf{a}$ *(because you know that AD = BC)*

c) $\overrightarrow{BL} = \mathbf{d} - \mathbf{a}$ *(you could find this in a few different ways — for example $\overrightarrow{BL} = \overrightarrow{BC} + \frac{1}{2}\overrightarrow{CA} = 2\mathbf{d} - (\mathbf{d} + \mathbf{a}) = \mathbf{d} - \mathbf{a}$)*

Page 158 (Exam Questions)

2 $BH^2 = 6^2 + 3^2 + 4^2$ ***[1 mark]***
$BH = \sqrt{61}$ ***[1 mark]***
$BH = 7.81$ cm (3 s.f.) ***[1 mark]***
[3 marks available in total — as above]

3 Angle $ABD = 180° - 90° - 31° - 12° = 47°$
Angle $ACB = 180° - 12° - 47° = 121°$ ***[1 mark for both]***
Use the sine rule: $\frac{3.3}{\sin 12°} = \frac{AB}{\sin 121°}$
$AB = \frac{3.3}{\sin 12°} \times \sin 121°$ ***[1 mark]***
$AB = 13.6050...$ m ***[1 mark]***
Find length BD: $\cos 47° = \frac{BD}{13.6050...}$
$BD = \cos 47° \times 13.6050...$ ***[1 mark]***
$BD = 9.2786... = 9.28$ m (3 s.f.) ***[1 mark]***
[5 marks available in total — as above]
There's more than one way of doing this question. As long as you've used a correct method to get the right answer you'll still get the marks.

4 a) $\overrightarrow{BX} = \overrightarrow{BC} + \overrightarrow{CX}$
$\overrightarrow{BC} = 6\overrightarrow{BW} = 6\mathbf{b}$ ***[1 mark]***
As AX = 2XC, CX must be one third of CA, so:
$\overrightarrow{CX} = \frac{1}{3}\overrightarrow{CA}$ ***[1 mark]***
$\overrightarrow{CA} = -3\mathbf{a} - 6\mathbf{b}$ ***[1 mark]***
$\overrightarrow{CX} = \frac{1}{3}(-3\mathbf{a} - 6\mathbf{b}) = -\mathbf{a} - 2\mathbf{b}$
$\overrightarrow{BX} = 6\mathbf{b} - \mathbf{a} - 2\mathbf{b} = 4\mathbf{b} - \mathbf{a}$ ***[1 mark]***
[4 marks available in total — as above]
There are other ways of doing this one — you could start by writing $\overrightarrow{BX}$ as $\overrightarrow{BA} + \overrightarrow{AX}$ instead.

b) From part a) $\overrightarrow{BX} = 4\mathbf{b} - \mathbf{a}$:
$ABCD$ is a parallelogram, so:
$\overrightarrow{CD} = \overrightarrow{BA} = -\overrightarrow{AB} = -3\mathbf{a}$ ***[1 mark]***
$\overrightarrow{CM} = \frac{1}{2}\overrightarrow{CD} = -\frac{3}{2}\mathbf{a}$ ***[1 mark]***
$\overrightarrow{BM} = \overrightarrow{BC} + \overrightarrow{CM}$
$= 6\mathbf{b} - \frac{3}{2}\mathbf{a} = \frac{3}{2}(4\mathbf{b} - \mathbf{a})$ ***[1 mark]***
B, X and M must be three points on a straight line because the lines BM and BX are both scalar multiples of the vector $4\mathbf{b} - \mathbf{a}$. ***[1 mark]***
[4 marks available in total — as above]
There are other ways of doing this bit too — for example, you could show that XM and BX are scalar multiples of each other.

Page 159 (Revision Questions)

1 $a^2 + b^2 = c^2$
You use Pythagoras' theorem to find the missing side of a right-angled triangle.

2 4.72 m

3 7.8

4

5 33.4°

6 See p.148

7 $4\sqrt{3}$ cm

8 Sine rule: $\frac{a}{\sin A} = \frac{b}{\sin B} = \frac{c}{\sin C}$

Cosine rule: $a^2 = b^2 + c^2 - 2bc \cos A$

Area = $\frac{1}{2}ab \sin C$

9 Two angles given plus any side — sine rule.
Two sides given plus an angle not enclosed by them — sine rule.
Two sides given plus the angle enclosed by them — cosine rule.
All three sides given but no angles — cosine rule.

10 56.4° (3 s.f.)

11 6.84 cm (3 s.f.)

12 48.1 cm² (3 s.f.)

13 a) 6.52 m   b) 48.7 m²

14 $a^2 + b^2 + c^2 = d^2$

15 11.9 m (3 s.f.)

16 15.2° (3 s.f.)

17 52.5° (1 d.p.)

18 Multiplying by a scalar changes the size of a vector but not its direction.

19 a) $\begin{pmatrix}-3\\-8\end{pmatrix}$ b) $\begin{pmatrix}20\\-10\end{pmatrix}$

c) $\begin{pmatrix}19\\0\end{pmatrix}$ d) $\begin{pmatrix}-30\\-4\end{pmatrix}$

20 a) $\overrightarrow{AX} = \frac{1}{3}\mathbf{a}$

b) $\overrightarrow{DX} = \frac{4}{3}\mathbf{a} - \mathbf{b}$

$\overrightarrow{XB} = \frac{8}{3}\mathbf{a} - 2\mathbf{b}$

c) $\overrightarrow{XB} = 2\,\overrightarrow{DX}$, so DXB is a straight line.

Section Seven — Probability and Statistics

Page 164 (Warm-Up Questions)

1 $\frac{1}{4}$ or 0.25

2 $x = 0.3$

3

		Dice 1					
		1	2	3	4	5	6
Dice 2	1	2	3	4	5	6	7
	2	3	4	5	6	7	8
	3	4	5	6	7	8	9
	4	5	6	7	8	9	10
	5	6	7	8	9	10	11
	6	7	8	9	10	11	12

P(total = 3) = $\frac{2}{36} = \frac{1}{18}$

4 a)

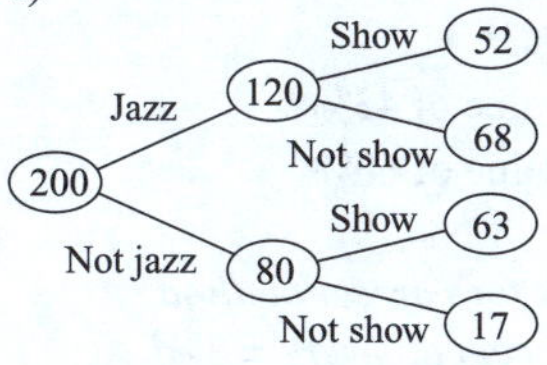

b) Jazz, Show = 0.26
Jazz, Not show = 0.34
Not jazz, Show = 0.315
Not jazz, Not show = 0.085

5 100 times

Page 165 (Exam Questions)

2 P(stripy sock) = $2y$ ***[1 mark]***
$0.4 + y + 2y = 1$ ***[1 mark]***
$3y = 0.6$
$y = 0.2$ ***[1 mark]***
[3 marks available in total — as above]

3

		Dice					
		1	2	3	4	5	6
Cards	2	3	4	5	6	7	8
	4	5	6	7	8	9	10
	6	7	8	9	10	11	12
	8	9	10	11	12	13	14
	10	11	12	13	14	15	16

P(scoring more than 4) = $\frac{\text{number of ways to score more than 4}}{\text{total number of possible outcomes}}$

$= \frac{28}{30} = \frac{14}{15}$

[3 marks available — 1 mark for finding the total number of possible outcomes, 1 mark for finding the number of ways to score more than 4, 1 mark for the correct answer]

4 a)

Number on counter	1	2	3	4	5
Frequency	23	25	22	21	9
Relative Frequency	0.23	0.25	0.22	0.21	0.09

[2 marks available — 2 marks for all correct answers, otherwise 1 mark for any frequency ÷ 100]

b) Elvin is likely to be wrong. The bag seems to contain fewer counters numbered 5. ***[1 mark]***

c) P(odd number) = 0.23 + 0.22 + 0.09 ***[1 mark]***
= 0.54 ***[1 mark]***
[2 marks available in total — as above]

5 Combinations of sandwich and drink = 5 × 8 = 40
Combinations of sandwich and snack = 5 × 4 = 20
Combinations of sandwich, snack and drink = 5 × 4 × 8 = 160
Total number of possible combinations = 40 + 20 + 160 = 220
[3 marks available — 1 mark for the correct number of combinations for one meal deal, 1 mark for the correct number of combinations for the other two meal deals, 1 mark for the correct answer]

Page 170 (Warm-Up Questions)

1 $\frac{1}{216}$ or 0.0046. $(\frac{1}{6} \times \frac{1}{6} \times \frac{1}{6})$

2 0.5

3 a) 0.63
b) 0.37

Page 171 (Exam Questions)

2 a) $(x + 4) + 3x + (2x + 3) + 9 = 40$
$6x + 16 = 40$ ***[1 mark]***
$6x = 24$
$x = 4$ ***[1 mark]***
[2 marks available in total — as above]

b) number of elements in both sets A and $B = 3x = 12$ ***[1 mark]***
P(element in both set A and B) = $\frac{12}{40} = \frac{3}{10}$ ***[1 mark]***
[2 marks available in total — as above]

3 a) P(losing) = 1 – 0.3 = 0.7
P(losing 3 games) = 0.7 × 0.7 × 0.7 = 0.343 ***[1 mark]***

b) P(wins at least 1 prize) = 1 – P(doesn't win a prize)
= 1 – (0.7 × 0.7) ***[1 mark]***
= 1 – 0.49 = 0.51 ***[1 mark]***
[2 marks available in total — as above]

4 a) P(2nd milk given 1st is milk) = $\frac{6}{11}$ ***[1 mark]***
1 milk chocolate has been taken, so there are 6 milk chocolates left out of 11 remaining chocolates.

b) P(at least one milk) = 1 – P(no milk)
= 1 – P(white then white)
$= 1 - \left(\frac{5}{12} \times \frac{4}{11}\right)$ ***[1 mark]***
$= 1 - \frac{20}{132} = \frac{112}{132} = \frac{28}{33}$ ***[1 mark]***
[2 marks available in total — as above]

c) P(milk and white) = P(milk then white) + P(white then milk)
$= \left(\frac{7}{12} \times \frac{5}{11}\right) + \left(\frac{5}{12} \times \frac{7}{11}\right)$ ***[1 mark]***
$= \frac{35}{66}$ ***[1 mark]***
[2 marks available in total — as above]
You might find it helpful to draw a tree diagram for parts b) and c).

Page 176 (Warm-Up Questions)

1 The population is all Birmingham households. The sample is made up of 600 residential addresses in Birmingham.
The results obtained are likely to be biased as the interview timing will exclude households where the occupants are at work during normal working hours.

2 Mode = most common. Median = middle value.
Mean = total of items ÷ number of items
Range = difference between the highest and the lowest values.

3

Number of cars	0	1	2	3	4	5	6	Total
Frequency	1	24	36	31	22	9	1	124
No. of cars × F	0	24	72	93	88	45	6	328

a) Mean = 328 ÷ 124 = 2.645 (3 d.p.)
b) Median is half-way between the 62nd and 63rd values, so median = 3
c) Mode = 2   d) Range = 6 – 0 = 6

Page 177 (Exam Questions)

2 a) E.g.

Number of chocolate bars	Tally	Frequency
0-2		
3-5		
6-8		
9-11		
12 or more		

[2 marks available — 1 mark for a suitable tally table, 1 mark for non-overlapping classes that cover all possible values]

b) E.g. Faye's results are likely to be biased because she hasn't selected her sample at random from all the teenagers in the UK. Also, her sample is too small to represent the whole population. So Faye can't use her results to draw conclusions about teenagers in the UK.
[2 marks available — 1 mark for a correct comment based on bias or sample size, 1 mark for stating that Faye can't draw conclusions about teenagers in the UK with reasoning]

3 a) 23, 26, 36 (in any order)
range = 13, median = 26
[2 marks available — 1 mark for all three weights correct, 1 mark for both range and median correct]

b) 32 + 23 + 31 + 28 + 36 + 26 = 176
4 × 27.25 = 109 ***[1 mark]***
176 – 109 = 67 ***[1 mark]***
so, goats weighing 31 kg and 36 kg ***[1 mark]***
[3 marks available in total — as above]

4 Call the five consecutive numbers n, $n + 1$, $n + 2$, $n + 3$ and $n + 4$
Median = middle value = $n + 2$

$$\text{Mean} = \frac{n+(n+1)+(n+2)+(n+3)+(n+4)}{5} = \frac{5n+10}{5} = n+2$$

Difference between mean and median = $(n + 2) – (n + 2) = 0$
[3 marks available — 1 mark for writing correct expressions for five consecutive numbers, 1 mark for a correct expression for the mean, 1 mark for showing that the difference between the expression for the mean and the expression for the median is zero]

Page 181 (Warm-Up Questions)

1

Height (cm)	$145 \le x < 155$	$155 \le x < 165$	$165 \le x < 175$	$175 \le x < 185$	Total
Frequency	18	22	24	15	79
Midpoint	150	160	170	180	
Midpoint × F	2700	3520	4080	2700	13 000

a) Mean = 13000 ÷ 79 = 164.56 (2 d.p.)
b) Median is in the group containing the 40th value, so the median group is $155 \le x < 165$.
c) Modal Group = $165 \le x < 175$

2 Range = 51, IQR = 21

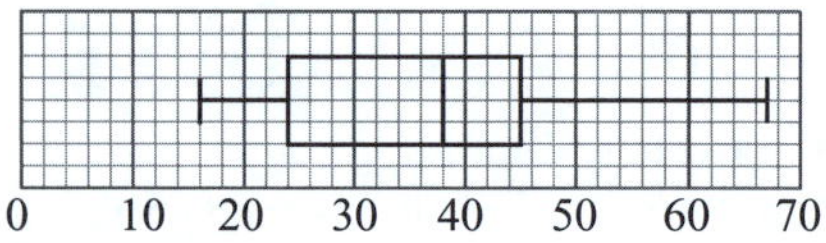

Page 182 (Exam Questions)

2 E.g. Comparing the box plots, the IQR of Rachel's distances is much smaller than the IQR of Harry's distances and Rachel's range is also smaller, so I agree that her distances were more consistent.
[2 marks available — 1 mark for correctly comparing the values of the range or IQR, 1 mark for a correct conclusion (supported by a correct comparison)]

3 a)

Exam mark (%)	≤ 20	≤ 30	≤ 40	≤ 50	≤ 60	≤ 70	≤ 80	≤ 100
Cumulative Frequency	3	13	25	49	91	107	116	120

[1 mark]

b)

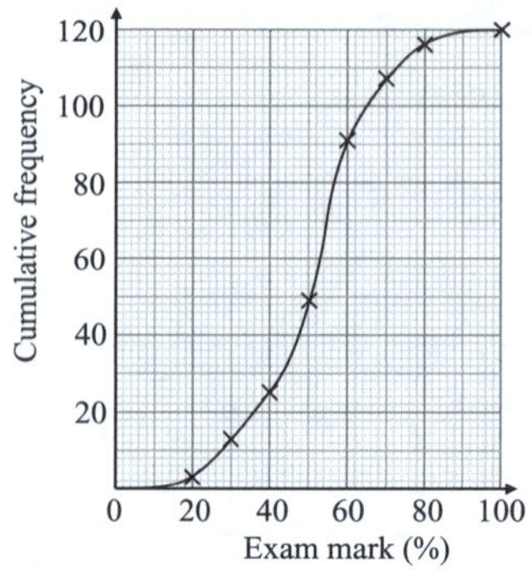

[3 marks available — 1 mark for all points plotted at correct class boundaries, 1 mark for all points plotted at correct heights, 1 mark for joining them with a smooth curve or straight lines]
A common mistake in exams is not plotting the points at the top end of the interval. But you wouldn't make that mistake, would you?

c) Median plotted at 60 gives a value of 53%
[1 mark, accept answers ± 1%]

d) Lower quartile at 30 gives a value of 43%
Upper quartile at 90 gives a value of 60%
Interquartile range = 60 – 43 = 17%
[2 marks available — 1 mark for correct method, 1 mark for correct answer, accept answers ± 2%]

e) $\frac{1}{5}$ of pupils got lower than grade 5,
$\frac{1}{5}$ of 120 = 24 pupils
So the 25th pupil was the first to get a grade 5. Reading from the graph at a cumulative frequency of 25 gives 40%, so the mark needed to get a grade 5 was about 40%.
[3 marks available — 1 mark for finding the number of pupils who got lower than grade 5, 1 mark for drawing a line across from 25 on the cumulative frequency axis, 1 mark for an answer in the range 38-42%]

Page 188 (Warm-Up Questions)

1 12 ÷ 3 = 4

2 a) Graph 1 — positive correlation.
b) Graph 2 — negative correlation.

3 3-point moving average

Page 189 (Exam Questions)

2 a)

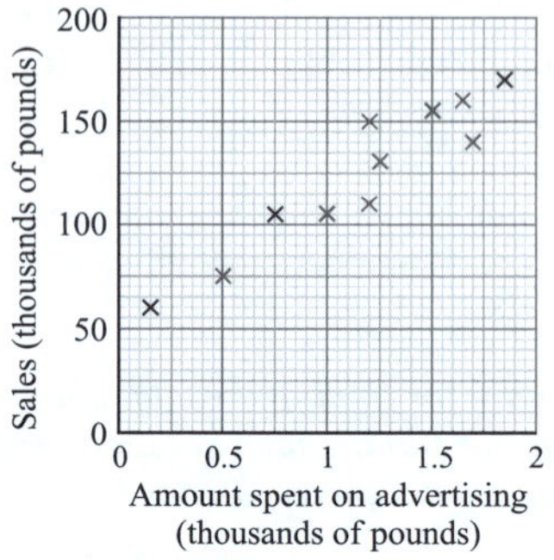

[1 mark if all three points are plotted correctly]

b) As the amount spent on advertising increases, so does the value of sales. ***[1 mark]***
Or you could say there's a positive correlation between the amount spent and the value of sales.

c)

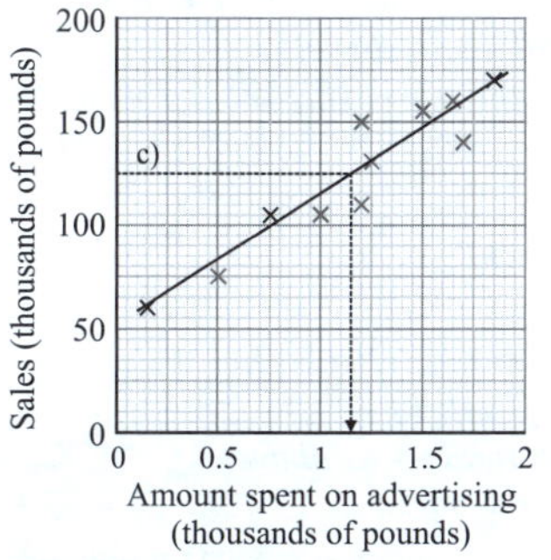

See graph — £1150
[2 marks available — 1 mark for drawing a line of best fit, 1 mark for reading off the correct answer, allow answers ± £100]

d) E.g. using the trend to predict sales for values over £3000 might be unreliable because those values are outside their range of data and they don't know whether the same pattern would continue. However, the data shows strong positive correlation, so the trend will probably continue.
[2 marks available — 1 mark for each sensible comment]

3 E.g. the median time taken by the boys is the same as the median time taken by the girls, so on average the boys and girls took the same time. The interquartile range for the boys is smaller than the interquartile range for the girls, so the times taken by the boys were more consistent than the times taken by the girls.
[2 marks available — 1 mark for a correct comparison of the median, 1 mark for a correct comparison of the interquartile range OR range (for both marks, at least one comparison must be given in the context of the data)]
'In the context of the data' means you need to explain what your comparison shows about the times taken by the boys and girls.

Page 190 (Revision Questions)

1 $\frac{8}{50} = \frac{4}{25}$

2 $\frac{1}{y^2}$

3 Divide the frequency of each result by the number of times the experiment was tried.

4 a)

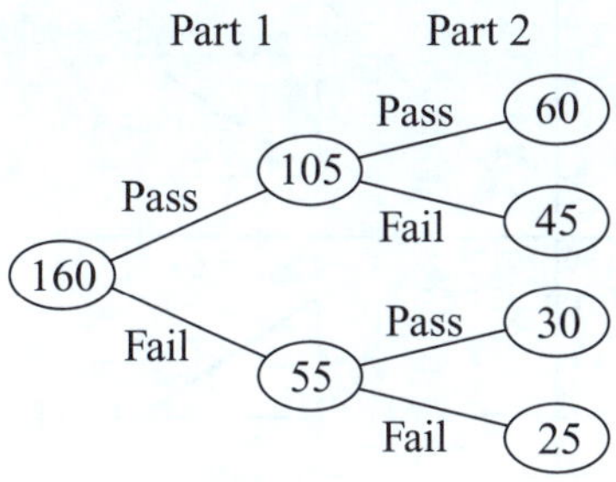

b) Relative frequency of:

pass, pass = $\frac{60}{160} = \frac{3}{8}$ or 0.375

pass, fail = $\frac{45}{160} = \frac{9}{32}$ or 0.28125

fail, pass = $\frac{30}{160} = \frac{3}{16}$ or 0.1875

fail, fail = $\frac{25}{160} = \frac{5}{32}$ or 0.15625

c) 113 people

5 $\frac{4}{81}$

6 $\frac{12}{20} = \frac{3}{5}$

7 a) $\frac{16}{2704} = \frac{1}{169}$

b) $\frac{24}{132600} = \frac{1}{5525}$

8 a)

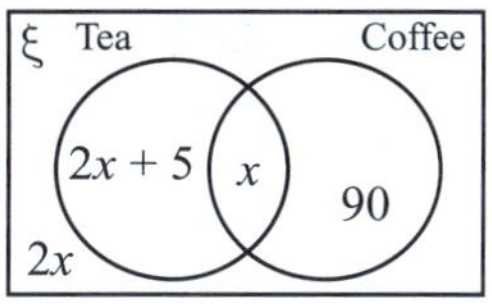

b) $x = 17$

c) $\frac{56}{180} = \frac{14}{45}$

9 A sample is part of a population. Samples need to be representative so that conclusions drawn from sample data can be applied to the whole population.

10 Qualitative data

11 Mode = 31, Median = 24
Mean = 22, Range = 39

12 a) Modal class is: $1.5 \le y < 1.6$.
b) Class containing median is: $1.5 \le y < 1.6$
c) Estimated mean = 1.58 m (to 2 d.p.)

13

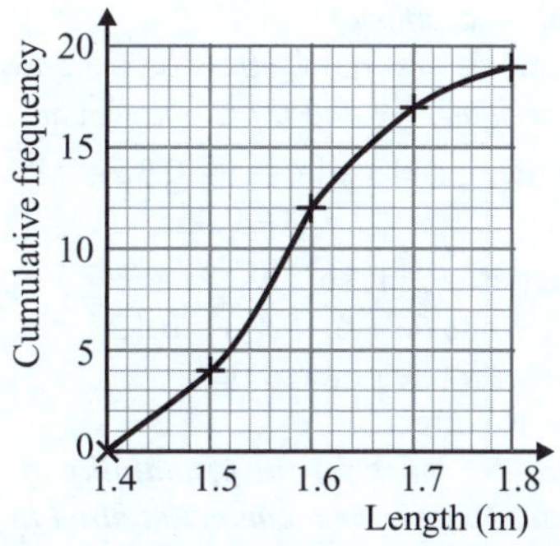

14 Calculate the bar's area or use the formula:
frequency = frequency density × class width.

15 a)

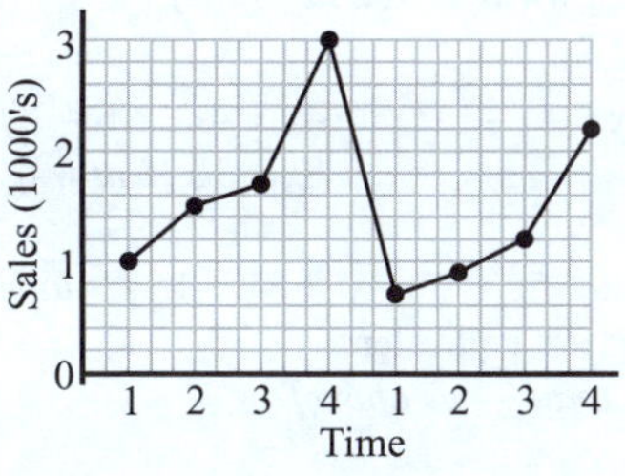

b) The data shows a downward trend.

16 a)

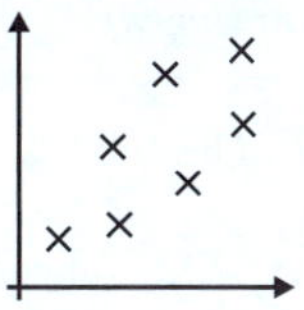

b)

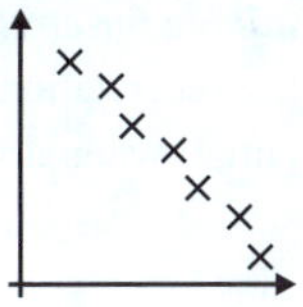

c)

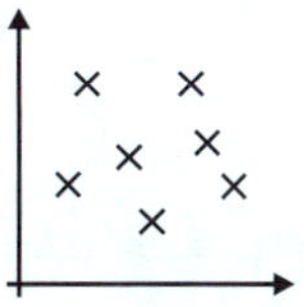

17 The median time in winter is lower than the median time in summer, so it generally took longer to get to work in the summer.
The range and the IQR for the summer are smaller than those for the winter, so there is less variation in journey times in the summer.

18 The runner's mean time after increasing her training hours has decreased from 147 seconds to 138 seconds, so this suggests that her running times have improved.

Practice Paper 1

1 The lower bound is $11 - 0.5 = 10.5$ cm ***[1 mark]***
The upper bound is $11 + 0.5 = 11.5$ cm ***[1 mark]***
[2 marks available in total — as above]

2 Use Pythagoras' theorem: $a^2 = 25^2 - 7^2 = 576$ and so $a = 24$
[2 marks available — 2 marks for the correct answer, otherwise 1 mark for showing a correct method]

3 E.g.

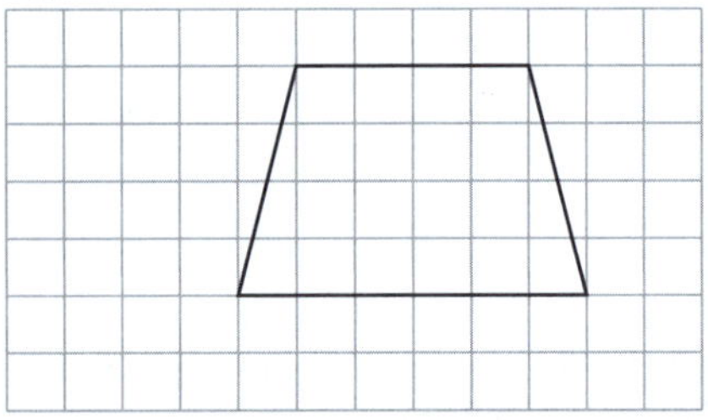

[2 marks available — 2 marks for drawing an isosceles trapezium, otherwise 1 mark for drawing a shape that has two of the listed properties]

4 Type the calculation directly into a calculator ensuring the whole of the denominator is square rooted.
$\frac{18.4 \times 2.56}{\sqrt{21.6 - 4 \times 1.55}} = 12.0032...$ ***[1 mark]***
$= 12.0$ ***[1 mark]*** (3 s.f.)
[2 marks available in total — as above]
The 0 at the end of the answer is important, otherwise the answer would only have 2 significant figures — and that's not enough.

5 Probability of getting a white or a black tie $= 1 - 0.35 - 0.2$
$= 0.45$

The ratio of black to white ties is 2 : 1, so
the probability of getting a white tie $= 0.45 \div 3 = 0.15$
So the probability of getting a white or a red tie
$= 0.35 + 0.15 = 0.5$, so she is correct
[4 marks available in total — 1 mark for the probability of picking a white or black tie, 1 mark for a correct method to find the probability of picking a white tie, 1 mark for the correct probability of picking a white tie, 1 mark for showing that the probability of picking a white or red tie is 0.5]

6 Amir's expression:
$(x + 4)^2 - 1 = (x + 4)(x + 4) - 1 = x^2 + 4x + 4x + 16 - 1$ ***[1 mark]***
$= x^2 + 8x + 15$ ***[1 mark]***

Orla's expression:
$(x + 5)(x + 3) = x^2 + 3x + 5x + 15 = x^2 + 8x + 15$ ***[1 mark]***
So the two expressions are equivalent.
[3 marks available in total — as above]

7 a) 125 kW ***[1 mark]***
b) See diagram in part d).
[1 mark available for circling the point shown]
c) Strong positive correlation ***[1 mark]***
d) Ignore the outlier when drawing a line of best fit.

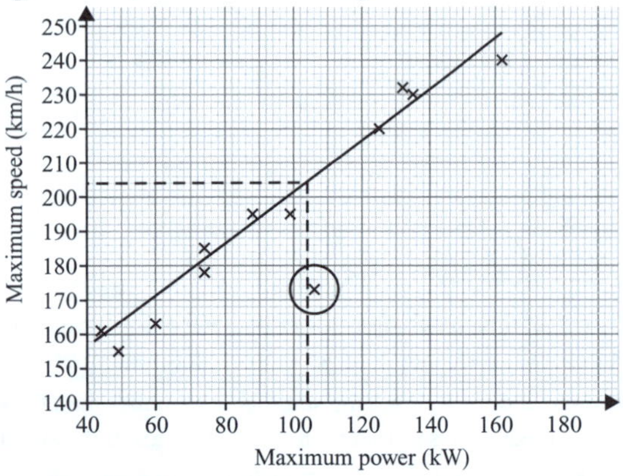

Maximum speed = 204 km/h (allow ± 2).
[2 marks available — 1 mark for drawing a line of best fit (ignoring the outlier), 1 mark for accurately reading from your graph the speed corresponding to a power of 104 kW]

e) 190 kW lies outside of the range of data plotted on the scatter graph. ***[1 mark]***

8 Exterior angle of a regular nine-sided shape $= \frac{360°}{9} = 40°$ ***[1 mark]***
So angle $ABC = 2 \times 40° = 80°$ ***[1 mark]***
Angle $ADC = 360° - 90° - 90° - 80° = 100°$ ***[1 mark]***
[3 marks available in total — as above]
Alternatively you could find the size of each interior angle (140°) of the regular nine-sided shape and use this to find angle ABC.

9 a) $5(3 - x) < x - 9$
$15 - 5x < x - 9$ ***[1 mark]***
$15 < 6x - 9$
$24 < 6x$ ***[1 mark]***
$4 < x$ or $x > 4$ ***[1 mark]***
[3 marks available in total — as above]
If you had rearranged the inequality to give $-6x < -24$, remember to reverse the inequality sign when dividing by -6.

b) $y = \frac{x^3 - 2}{5}$
$5y = x^3 - 2$ ***[1 mark]***
$5y + 2 = x^3$
$\sqrt[3]{5y + 2} = x$ ***[1 mark]***
[2 marks available in total — as above]

10 $2 + 3 + 4 + 6 = 15$ parts
1 part: £1200 ÷ 15 = £80 ***[1 mark]***
Eve gets 6 × £80 = £480 ***[1 mark]*** and so the sale price for the tyres is £480. The usual price for four tyres = 4 × £150 = £600 ***[1 mark]***
Percentage discount $= \frac{600 - 480}{600} \times 100$ ***[1 mark]***
$= 20\%$ ***[1 mark]***
[5 marks available in total — as above]

11 Volume of cylinder $= \pi \times 5.5^2 \times 12.4 = 1178.411...$ cm^3 ***[1 mark]***
Mass of cylinder = volume × density
$= 1178.411... \times 2.7 = 3181.710...$ g ***[1 mark]***
Mass of cube $= 3181.710...$ g so,
volume of cube = mass ÷ density
$= 3181.710... \div 10.5 = 303.020...$ cm^3 ***[1 mark]***
Side length $= \sqrt[3]{303.020...}$ ***[1 mark]***
$= 6.716...$ cm $= 6.7$ cm (to 2 s.f.) ***[1 mark]***
[5 marks available in total — as above]

12 a)

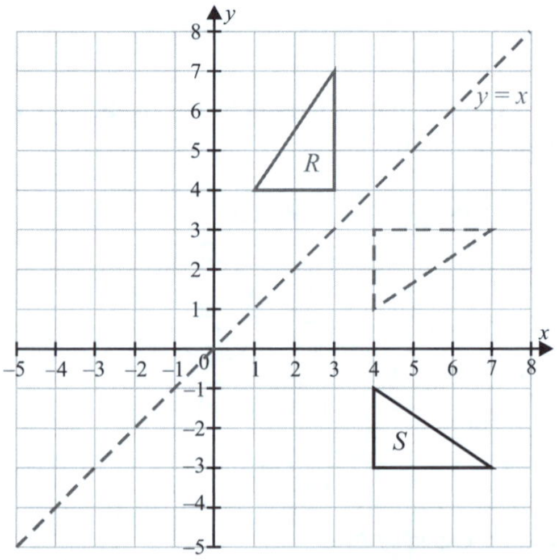

[2 marks available — 2 marks for correctly drawing and labelling shape S, otherwise 1 mark for a correct reflection in the line y = x]

b) The single transformation is a rotation 90° clockwise, about the origin.
[3 marks available — 1 mark for a rotation, 1 mark for the correct angle and direction, 1 mark for the correct centre]

13 Scale factor for heights $= 16 \div 10 = 1.6$ ***[1 mark]***
so scale factor for volumes $= 1.6^3 = 4.096$ ***[1 mark]***
Capacity of larger plant pot $= 4.096 \times 250 = 1024$ ml so the plant pot will hold 1 litre (= 1000 ml) of compost. ***[1 mark]***
[3 marks available in total — as above]

14 Angle $ABC = 90°$ ***[1 mark]*** (angle in a semicircle).

$\cos 52° = \frac{14.2}{AC}$ ***[1 mark]***

So $AC = \frac{14.2}{\cos 52°} = 23.0646...$ cm ***[1 mark]***

Circumference $= \pi \times 23.0646...$ ***[1 mark]***

$= 72.4596...$ cm $= 72.5$ cm (3 s.f.) ***[1 mark]***

[5 marks available in total — as above]

15 a) Total takings = 200 × £0.50 = £100 ***[1 mark]***

The possible scores for two spins of the spinner are

		First Spin				
	+	1	1	2	2	3
Second Spin	1	2	2	3	3	4
	1	2	2	3	3	4
	2	3	3	4	4	5
	2	3	3	4	4	5
	3	4	4	5	5	6

So there are 5 ways to get a 5 or more. ***[1 mark]***

The probability of getting a total score of 5 or more is $\frac{5}{25} = \frac{1}{5}$. ***[1 mark]***

$\frac{1}{5} \times 200 = 40$ winners expected ***[1 mark]***

Prize money = 40 × £2 = £80

So £100 – £80 = £20 profit ***[1 mark]***

[5 marks available in total — as above]

There are other ways that you could have found the probability of scoring a total of 5 or more, including systematic listing of the possible scores or a tree diagram.

b) The spinner is equally likely to land in each of the 5 sections.
[1 mark for any valid assumption linked to either the spinner being fair or the two spins being independent of one another]

16 a) Account 1:

The multiplier for an increase of 2.5% is 1.025.

Balance after 3 years = £6000 × 1.025^3

= £6461.34375

= £6461.34 to the nearest penny

Account 2:

Balance after 3 years = £6000 × 1.01 × 1.015 × 1.05

= £6458.445

= £6458.45 to the nearest penny

So Dev should invest in Account 1.

[4 marks available — 1 mark for correct method for working out the balance for Account 1, 1 mark for correct method for working out the balance for Account 2, 1 mark for finding correct final balances for both accounts, 1 mark for a correct conclusion following on from calculated balances]

Alternatively, you could compare $1.025^3 \approx 1.077$ with $1.01 \times 1.015 \times 1.05 \approx 1.076$ to see that the overall percentage change is greater for Account 1.

b) The multiplier for an increase of 2% is 1.02.

So for two years investment the multiplier is $1.02^2 = 1.0404$ which corresponds to a 4.04% increase. ***[1 mark]***

So £606 = 4.04% ***[1 mark]***

1% = £606 ÷ 4.04 = £150

So 100% = £150 × 100 = £15 000 ***[1 mark]***

So Nina invested £15 000.

[3 marks available in total — as above]

The question could be answered by forming and solving an equation: $x(1.02)^2 = x + 606$.

17 a) There are 8 + 23 + 40 + 21 + 18 = 110 nails in the jar.

40 + 21 + 18 = 79 of the nails are over 1.5 cm long.

$\frac{79}{110} \times 100 = 71.818... = 71.82\%$ (2 d.p.)

[2 marks available — 1 mark for a correct method, 1 mark for the correct answer]

b)

Length of nail, l cm	Frequency	Mid-point	Frequency × mid-point
$0.5 \le l < 1.0$	8	0.75	8 × 0.75 = 6
$1.0 \le l < 1.5$	23	1.25	23 × 1.25 = 28.75
$1.5 \le l < 2.0$	40	1.75	40 × 1.75 = 70
$2.0 \le l < 4.0$	21	3	21 × 3 = 63
$4.0 \le l < 6.0$	18	5	18 × 5 = 90
Total:	110	**Total:**	257.75

257.75 ÷ 110 = 2.343... cm = 2.34 cm (2 d.p.)

[4 marks available — 1 mark for all mid-interval values, 1 mark for calculation of frequency × mid-interval value, 1 mark for dividing sum of frequency × mid-interval values by sum of frequencies, 1 mark for the correct answer]

18 Factorise all the expressions.

$3x + 6 = 3(x + 2)$ ***[1 mark]***

$2x^2 - 32 = 2(x^2 - 16) = 2(x + 4)(x - 4)$ ***[1 mark]***

$x^2 - 2x - 8 = (x + 2)(x - 4)$ ***[1 mark]***

So $\frac{ax + b}{2x^2 - 32} \times (x^2 - 2x - 8) = 3x + 6$ becomes

$\frac{ax + b}{2(x+4)(x-4)} \times (x + 2)(x - 4) = 3(x + 2)$

Cancel common factors:

$\frac{ax + b}{2(x+4)\cancel{(x-4)}} \times \cancel{(x+2)}\cancel{(x-4)} = 3\cancel{(x+2)}$

$\frac{ax + b}{2(x+4)} = 3$ ***[1 mark]***

$ax + b = 3 \times 2(x + 4)$

$ax + b = 6x + 24$

$a = 6$ and $b = 24$ ***[1 mark]***

[5 marks available in total — as above]

19 a) Total length of rectangle $= 7 + 2x$

So area of rectangle $= x(7 + 2x)$ ***[1 mark]***

Area of semicircle $= \frac{\pi x^2}{2}$ ***[1 mark]***

So, $x(7 + 2x) + \frac{\pi x^2}{2} = 40$

$7x + 2x^2 + \frac{\pi x^2}{2} = 40$ ***[1 mark]***

$14x + 4x^2 + \pi x^2 = 80$

$(\pi + 4)x^2 + 14x - 80 = 0$ ***[1 mark]***

[4 marks available in total — as above]

b) Use the quadratic formula:

$\frac{-14 \pm \sqrt{14^2 - (4 \times (\pi + 4) \times -80)}}{2(\pi + 4)}$ ***[1 mark]***

$= 2.507...$ (or $-4.467...$)

So $x = 2.51$ (to 2 d.p.) ***[1 mark]***

[2 marks available in total — as above]

20 a) $y = \cos x$ has been translated 1 unit down.

The curve is $y = \cos x - 1$ ***[1 mark]***

b) $y = \cos x$ has been translated 30° to the right.

The curve is $y = \cos(x - 30)$ ***[1 mark]***

c) $y = \cos x$ has been reflected in the x-axis.

The curve is $y = -\cos x$ ***[1 mark]***

The curve could also have equation $y = \cos(x + 180°)$.

21 a) Go through the function machine backwards and reverse the signs: 154 × 3 = 462, 462 – 7 = 455, 455 ÷ 5 = 91

[2 marks available — 1 mark for a correct method, 1 mark for the correct answer]

b) $b = \frac{5a + 7}{3}$ ***[1 mark]***

$c = \frac{5\left(\frac{5a+7}{3}\right) + 7}{3}$ ***[1 mark]***

$= \frac{\frac{25a + 35}{3} + 7}{3} = \frac{\frac{25a + 35}{3} + \frac{21}{3}}{3}$ ***[1 mark]***

$\frac{\frac{25a + 56}{3}}{3} = \frac{25a + 56}{9}$ ***[1 mark]***

[4 marks available in total — as above]

22 a) $\overrightarrow{AC} = \overrightarrow{AB} + \overrightarrow{BC} = 6\mathbf{b} - 6\mathbf{a}$ ***[1 mark]***

b) $\overrightarrow{MN} = \overrightarrow{MA} + \overrightarrow{AB} + \overrightarrow{BN}$

$\overrightarrow{MA} = \overrightarrow{OA} - \overrightarrow{OM} = 6\mathbf{a} - 4\mathbf{a} = 2\mathbf{a}$ ***[1 mark]***

$\overrightarrow{BN} = \frac{1}{3}\overrightarrow{BD} = \frac{1}{3}(\overrightarrow{BC} + \overrightarrow{CD}) = \frac{1}{3}(-6\mathbf{a} + 6\mathbf{b})$

$= 2\mathbf{b} - 2\mathbf{a}$ ***[1 mark]***

So $\overrightarrow{MN} = 2\mathbf{a} + 6\mathbf{b} + (2\mathbf{b} - 2\mathbf{a}) = 8\mathbf{b}$ ***[1 mark]***

So $\overrightarrow{AB}$ and $\overrightarrow{MN}$ are parallel because $\overrightarrow{MN} = 8\mathbf{b}$ is a multiple of $\overrightarrow{AB} = 6\mathbf{b}$. ***[1 mark]***

[4 marks available in total — as above]

*6**b** and 8**b** have a common factor of **b** so the two vectors must each be a scalar multiple of the other.*

23 $x^2 + y^2 = 10$ is the equation of a circle with centre (0, 0).
A tangent is perpendicular to the radius at the point of contact.
Gradient of radius from (0, 0) to (3, 1) is $\frac{1-0}{3-0} = \frac{1}{3}$ ***[1 mark]***
The product of the gradients of perpendicular lines is −1, so gradient of tangent = −3 ***[1 mark]*** and so the equation of the tangent is $y = -3x + c$ ***[1 mark]***.
The tangent passes through (3, 1): $1 = -3 \times 3 + c$ so $c = 10$
The equation of the tangent is $y = -3x + 10$ ***[1 mark]***
[4 marks available in total — as above]

Practice Paper 2

1 a) m represents the gradient and c represents the y-intercept
The only graph with a positive gradient and a negative y-intercept is graph B ***[1 mark]***

b) y is inversely proportional to x has the equation $y = \frac{k}{x}$ which is shown by graph C ***[1 mark]***

2 a) $A = 60\%$ of $B = \frac{3}{5}B$ $B = 30\%$ of $C = \frac{3}{10}C$

$A = \frac{3}{5} \times \frac{3}{10}C = \frac{9}{50}C$

[3 marks available — 1 mark for giving A as a fraction of B, 1 mark for giving B as a fraction of C, 1 mark for the correct answer]

You could also have worked out A as a percentage of C and then converted it to a fraction.

b) $C = \frac{50}{9}A = \frac{50}{9} \times 9 \times 10^8$ ***[1 mark]***

$= 50 \times 10^8 = 5 \times 10^9$ ***[1 mark]***

[2 marks available — 1 mark for a correct method, 1 mark for the correct answer]

3 a) Area of $ABCD = 9 \times 12 = 108$ cm^2 ***[1 mark]***
Ratio of area of smaller and larger piece is 1 : 3.
So 1 + 3 = 4 parts = 108 cm^2
Area of DPQ = 1 part = 108 ÷ 4 = 27 cm^2 ***[1 mark]***
[2 marks available in total — as above]

b) $DP = \frac{2}{3}$ of 9 cm = 9 × 2 ÷ 3 = 6 cm ***[1 mark]***

Area of triangle $DPQ = \frac{1}{2} \times DQ \times DP$

So $27 = \frac{1}{2} \times DQ \times 6$ ***[1 mark]***

$DQ = 27 \div 3 = 9$ cm and $x = 12 - 9 = 3$ ***[1 mark]***

[3 marks available in total — as above]

4 Round all numbers to 1 significant figure.

$\sqrt[3]{\frac{785.3 \times 2.156}{0.1972}} = \sqrt[3]{\frac{800 \times 2}{0.2}}$ ***[1 mark]***

$= \sqrt[3]{\frac{1600}{0.2}} = \sqrt[3]{\frac{16000}{2}} = \sqrt[3]{8000}$ ***[1 mark]***

$= 20$ ***[1 mark]***

[3 marks available in total — as above]

5
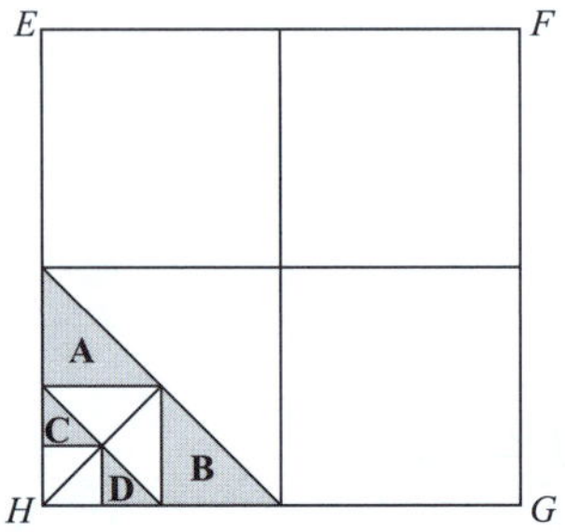

Work out the fraction of the square made up by each shaded region:

Region A = region B $= \frac{1}{4} \times \frac{1}{2} \times \frac{1}{2} \times \frac{1}{2} = \frac{1}{32}$

Region C = region D $= \frac{1}{4} \times \frac{1}{4} \times \frac{1}{4} \times \frac{1}{2} = \frac{1}{128}$

Overall fraction shaded $= \left(2 \times \frac{1}{32}\right) + \left(2 \times \frac{1}{128}\right)$

$= \frac{2}{32} + \frac{2}{128} = \frac{4}{64} + \frac{1}{64} = \frac{5}{64}$

[3 marks available — 1 mark for identifying the fraction of the square regions A and B make up, 1 mark for identifying the fraction of the square regions C and D make up, 1 mark for the correct answer]

6 a) 4th term $= 4^2 + (2 \times 4) + 5$
$= 16 + 8 + 5 = 29$ so Charlie is correct
[2 marks available — 2 marks for indicating that Charlie is correct and finding that the 4th term is 29, otherwise 1 mark for correct method to find the 4th term]

b) A term in the sequence is the sum of the two previous terms.
The next two terms are 12 + 19 = 31 ***[1 mark]***
and 19 + 31 = 50 ***[1 mark]***
[2 marks available in total — as above]

7 a) Elements of A are 3, 4, 5, 6
Elements of B are 1, 2, 3, 4, 6 (12 is not in the universal set ξ)

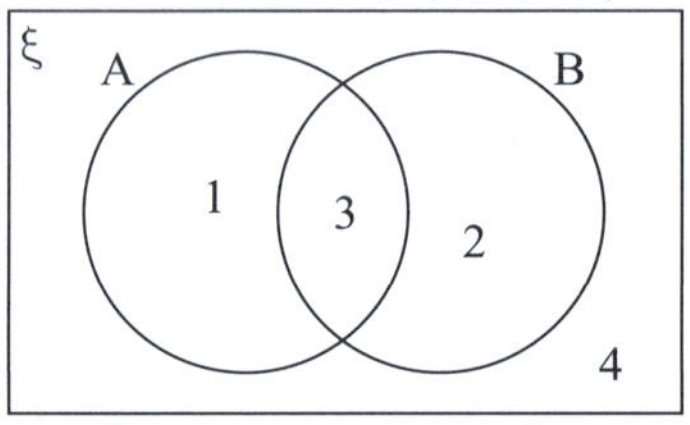

[3 marks available — 1 mark for identifying the elements of sets A and B, 1 mark for identifying the elements in the intersection, 1 mark for a completely correct Venn diagram]

b) There are 1 + 4 = 5 elements that are not in set B.
Only one of them is in set A so,
P(in A given not in B) $= \frac{1}{5}$ or 20% or 0.2 ***[1 mark]***

8 a) $2a + b = 10 \xrightarrow{\times 2} 4a + 2b = 20$ ***[1 mark]***

$4a + 2b = 20$ $4(3) + 2b = 20$
$\underline{3a + 2b = 17\,-}$ $2b = 20 - 12$
$a = 3$ ***[1 mark]*** $2b = 8$
$b = 4$ ***[1 mark]***

[3 marks available in total — as above]

b) $a\begin{pmatrix}2\\1\end{pmatrix} - b\begin{pmatrix}3\\2\end{pmatrix} = 3\begin{pmatrix}2\\1\end{pmatrix} - 4\begin{pmatrix}3\\2\end{pmatrix} = \begin{pmatrix}6\\3\end{pmatrix} - \begin{pmatrix}12\\8\end{pmatrix}$ ***[1 mark]***

$= \begin{pmatrix}-6\\-5\end{pmatrix}$ ***[1 mark]***

[2 marks available in total — as above]

9
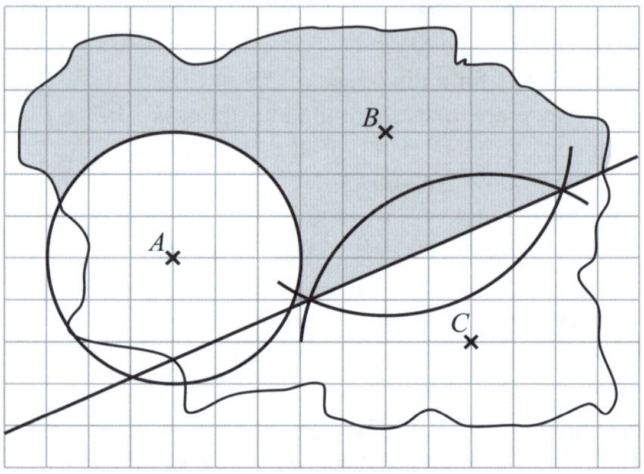

[3 marks available — 1 mark for drawing an arc centre A with radius 3 cm, 1 mark for drawing the perpendicular bisector of the line BC, 1 mark for indicating the correct region]

10 a) $p = 2^3 \times 3^2 \times 5 \times 7$

$q = \frac{6}{7} \times 2^3 \times 3^2 \times 5 \times 7 = \frac{2 \times 3}{7} \times 2^3 \times 3^2 \times 5 \times 7 = 2^4 \times 3^3 \times 5$

The lowest common multiple of p and q is $2^4 \times 3^3 \times 5 \times 7$

[3 marks available — 1 mark for finding the prime factorisation of q, 1 mark for a correct method to find the LCM, 1 mark for the correct answer]

b) $q = 2^4 \times 3^3 \times 5$

$r = \frac{4}{15} \times 2^3 \times 3^2 \times 5 \times 7 = \frac{2 \times 2}{3 \times 5} \times 2^3 \times 3^2 \times 5 \times 7 = 2^5 \times 3 \times 7$

The highest common factor of q and r is $2^4 \times 3 = 16 \times 3 = 48$

[3 marks available — 1 mark for finding the prime factorisation of r, 1 mark for a correct method to find the HCF, 1 mark for the correct answer]

11 a) $4.2 \times 10^5 = 420\,000$ ***[1 mark]***

$10\% = 420\,000 \div 10 = 42\,000$

$5\% = 42\,000 \div 2 = 21\,000$

$2.5\% = 21\,000 \div 2 = 10\,500$

So, $7.5\% = 21\,000 + 10\,500 = 31\,500$ ***[1 mark]***

Let x = number of left-handed people aged 75 or above, then number of right-handed people aged 75 or above $= x + 20\,500$

So, $x + x + 20\,500 = 31\,500$

$2x = 31\,500 - 20\,500$ ***[1 mark]*** $= 11\,000$

So $x = 5500$ ***[1 mark]***

[4 marks available in total — as above]

Alternatively, you could leave your answer in standard form through the calculation and converted it at the end.

b) Population of Deepland = $27 \times 4.2 \times 10^5$

First find 27×42:

$$\begin{array}{r} 2\ 7 \\ \times\ 4\ 2 \\ \hline 5\ {}_1 4 \\ 1\ 0\ {}_2 8\ 0 \\ \hline 1\ 1\ 3\ 4 \\ {}_1 \end{array}$$

So, $27 \times 4.2 = 113.4$ ***[1 mark]***

So population of Deepland $= 113.4 \times 10^5$

$= 1.1 \times 10^7$ (2 s.f) ***[1 mark]***

[2 marks available in total — as above]

12 $\frac{3x^2 + 14x + 8}{9x^2 - 4} = \frac{(3x+2)(x+4)}{(3x+2)(3x-2)} = \frac{x+4}{3x-2}$

[3 marks available — 1 mark for correctly factorising the numerator, 1 mark for correctly factorising the denominator, 1 mark for the correct answer]

13 Shape A: Area $= \pi \times 4^2 = 16\pi$

Shape B: Area $= \frac{80}{360} \times \pi \times 6^2 = 8\pi$

$16\pi = 2 \times 8\pi$ so the area of A is twice the area of B.

[3 marks available — 1 mark for the area of shape A, 1 mark for the correct method to find the area of shape B, 1 mark for the area of shape B and showing that the area of shape A is twice the area of shape B]

14 *Negative powers correspond to reciprocals and fractional powers correspond to roots.*

$\left(2\frac{1}{4}\right)^{-\frac{1}{2}} = \left(\frac{9}{4}\right)^{-\frac{1}{2}} = \left(\frac{4}{9}\right)^{\frac{1}{2}}$ ***[1 mark]*** $= \sqrt{\frac{4}{9}} = \frac{2}{3}$ ***[1 mark]***

$\left(2\frac{1}{4}\right)^{-\frac{1}{2}} \div \frac{2}{9} = \frac{2}{3} \div \frac{2}{9} = \frac{2}{3} \times \frac{9}{2}$ ***[1 mark]*** $= 3$ ***[1 mark]***

[4 marks available in total — as above]

15 a) The cumulative frequency diagram starts at 40 minutes so that is the quickest possible winning time in 2022.

Difference: $40 - 37 = 3$ minutes

[2 marks available — 1 mark for finding the quickest possible winning time in 2022, 1 mark for the correct answer]

b) To find the median, read across at 30 and then read down: Median for 2022 = 71.5 minutes. ***[1 mark]***

The teams were faster on average in 2022 than in 2021 as the median was lower. ***[1 mark]***

[2 marks available in total — as above]

c) To find the upper quartile read across and down at 45, to find the lower quartile read across and down at 15:

Upper quartile = 78 and lower quartile = 63

Interquartile range = 78 – 63 = 15 minutes ***[1 mark]***

The times of the teams were less spread out in 2022 than in 2021 as the interquartile range was smaller, so times were more consistent in 2022. ***[1 mark]***

[2 marks available in total — as above]

16 a) Read across and down at 36 °C.

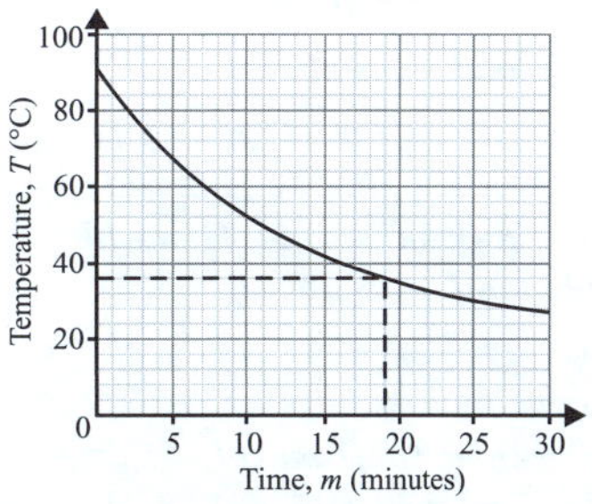

To reach 36 °C it takes 19 minutes ***[1 mark]***

b) At 16 minutes the tea is 40 °C, at 28 minutes the tea is 28 °C.

Percentage decrease $= \frac{40 - 28}{40} = \frac{12}{40} = \frac{3}{10} = 30\%$

[3 marks available — 1 mark for finding the temperatures at 16 and 28 minutes, 1 mark for a correct method to find the percentage decrease, 1 mark for the correct answer]

c) Draw a tangent at $m = 10$ (see graph) ***[1 mark]***

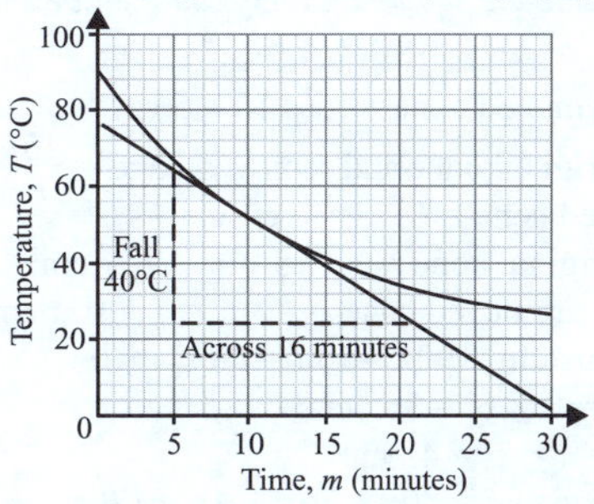

Gradient of tangent $= -40 \div 16$ ***[1 mark]***

$= -2.5$ °C/minute

So 10 minutes after the tea is made it is decreasing at a rate of 2.5 °C/minute. ***[1 mark for an answer in the range 2.3-2.7]***

[3 marks available in total — as above]

17 Rationalise the denominator of the first term:

$\frac{6}{\sqrt{3}} = \frac{6\sqrt{3}}{\sqrt{3} \times \sqrt{3}} = \frac{6\sqrt{3}}{3} = 2\sqrt{3}$ ***[1 mark]***

And simplify the second term:

$\sqrt{27} = \sqrt{9}\sqrt{3} = 3\sqrt{3}$ ***[1 mark]***

So $\frac{6}{\sqrt{3}} + \sqrt{27} = 2\sqrt{3} + 3\sqrt{3} = 5\sqrt{3}$ ***[1 mark]***

[3 marks available in total — as above]

18 Angle $DBA = 35°$ ***[1 mark]*** (alternate segment theorem)

Angle $OAD = 90° - 35° = 55°$ ***[1 mark]***

(a tangent and a radius are perpendicular)

Obtuse angle $COA = 2 \times$ angle CBA

$= 2(39° + 35°) = 2 \times 74° = 148°$ ***[1 mark]***

(angle at centre is twice angle at circumference)

Angle $CDA = 180° - 74° = 106°$ ***[1 mark]***

(opposite angles in a cyclic quadrilateral add up to 180°)

Angle $DCO = 360° - 106° - 148° - 55° = 51°$ ***[1 mark]***

(angles in a quadrilateral add up to 360°)

[5 marks available in total — as above]

There may be alternative ways of getting to the same answer.

19 a) $x = \frac{18(s + 2t^2)}{5t}$

$5xt = 18(s + 2t^2)$

$5xt = 18s + 36t^2$ ***[1 mark]***

$18s = 5xt - 36t^2$

$s = \frac{5xt - 36t^2}{18}$ ***[1 mark]*** or $s = \frac{5xt}{18} - 2t^2$

[2 marks available in total — as above]

b) $s = kt^2$ ***[1 mark]***
$s = 160$ and $t = 8$ so $160 = k \times 8^2$
So $k = 160 \div 64 = 2.5$ ***[1 mark]***
When $t = 6$, $s = 2.5 \times 6^2 = 90$ m ***[1 mark]***
Therefore from part a):
$90 = \frac{5x \times 6}{18} - 2 \times 6^2$ ***[1 mark]***
$90 + 72 = \frac{5x}{3}$
$x = \frac{162 \times 3}{5} = \frac{486}{5} = 97.2$ ***[1 mark]***
[5 marks available in total — as above]

20 a) $(n + 1)(n - 1)(n + 4) = (n^2 - n + n - 1)(n + 4)$ ***[1 mark]***
$= (n^2 - 1)(n + 4)$ ***[1 mark]***
$= n^3 + 4n^2 - n - 4$ ***[1 mark]***
[3 marks available in total — as above]

b) $n(n + 3)(n + 1) = n(n^2 + n + 3n + 3)$
$= n(n^2 + 4n + 3)$
$= n^3 + 4n^2 + 3n$ ***[1 mark]***
So $n(n + 3)(n + 1) - (n + 1)(n - 1)(n + 4)$
$= n^3 + 4n^2 + 3n - (n^3 + 4n^2 - n - 4)$
$= 4n + 4 = 4(n + 1)$ ***[1 mark]***
n is an integer so $4(n + 1)$ must be a multiple of 4. ***[1 mark]***
[3 marks available in total — as above]
An alternative method would be to notice that (n + 1) appears as a factor of both parts and so can be taken out as a factor of the whole expression, i.e. (n + 1)[n(n + 3) − (n − 1)(n + 4)]. The expression inside the square brackets can then be shown to simplify to give 4.

21 Slant height of the removed cone = 15 – 10 = 5 cm
Scale factor of the original cone and removed cone = $\frac{15}{5} = 3$
Radius of the removed cone = 6 ÷ 3 = 2 cm
Curved surface of original cone = $6 \times 15 \times \pi = 90\pi$ cm^2
Curved surface of the removed cone = $2 \times 5 \times \pi = 10\pi$ cm^2
Curved surface of frustum = $90\pi - 10\pi = 80\pi$ cm^2
Base of the frustum = $\pi \times 6^2 = 36\pi$ cm^2
Top of the frustum = $\pi \times 2^2 = 4\pi$ cm^2
Surface area of the frustum = $80\pi + 36\pi + 4\pi = 120\pi$ cm^2
[6 marks available — 1 mark for finding the scale factor between the original and removed cones, 1 mark for finding the radius of the removed cone, 1 mark for the curved surface of the original cone, 1 mark for the curved surface of the frustum, 1 mark for the surface area of the top or bottom of the frustum, 1 mark for the correct answer]

22 The turning point of a graph written in the form $y = (x - a)^2 + b$ is at the point (a, b). The turning point is at (2, 7) so the curve must have the equation $y = (x - 2)^2 + 7$ ***[1 mark]***
Substituting $y = 128$ and $x = k$ gives $128 = (k - 2)^2 + 7$ ***[1 mark]***
$121 = (k - 2)^2$
$\pm\sqrt{121} = k - 2$ ***[1 mark]***
$\pm 11 = k - 2$
So $k = 13$ or $k = -9$ ***[1 mark]***
[4 marks available in total — as above]
The equation could be solved by factorising or using the quadratic formula, but using the completed square form is simplest here.

Practice Paper 3

1 Probability of getting a number greater than 4 = $\frac{2}{6} = \frac{1}{3}$
Probability that they both get a number greater than 4
$= \frac{1}{3} \times \frac{1}{3} = \frac{1}{9}$
[2 marks available — 2 marks for the correct answer, otherwise 1 mark for finding the probability of getting a number greater than a 4]

2 When Darcie squares their number the final digit is 1, so their number must end in 1 or 9. ***[1 mark]***
It must be either 11, 19, 21 or 29, but Darcie's number is not prime so it must be 21. ***[1 mark]***
[2 marks available in total — as above]

3 The median is 12, so one of the unknown cards must be another 12. ***[1 mark]***
The mean is 13 so the sum of the cards is 13 × 4 = 52.
The last card must be 52 – 7 – 12 – 12 = 21 ***[1 mark]***
The range is 21 – 7 = 14 ***[1 mark]***
[3 marks available in total — as above]

4 When the input is 15 the output is (15 × 2) – 5 = 25
When the input is 25 the output is (25 × 2) – 5 = 45
So the ratio is 25 : 45 = 1 : 1.8 or 1 : $\frac{9}{5}$.
[3 marks available — 1 mark for finding the correct output for 15 or 25, 1 mark for finding the correct output for the other number and forming a ratio, 1 mark for the correct answer]

5 Plan view — Front elevation

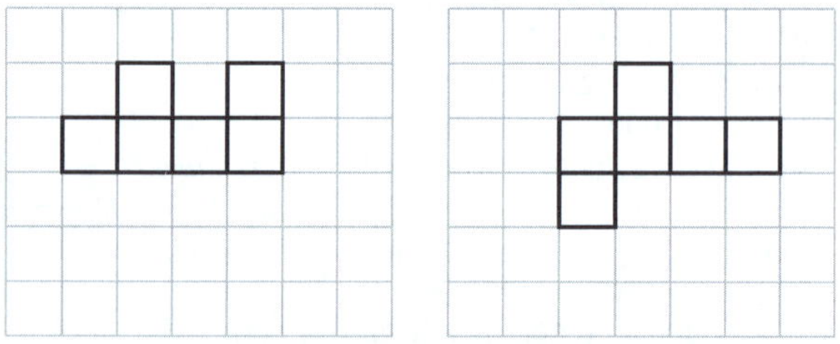

[3 marks available — 3 marks for both correct drawings, otherwise 2 mark for one correct drawing or 1 mark for a drawing made from 6 squares]

6 a) Work back to find the first term in Amaya's sequence:
17 – 3 – 3 – 3 = 8 ***[1 mark]***
So the first term of Leo's sequence is 8 ÷ 2 = 4 ***[1 mark]***
and the first five terms of Leo's sequence are 4, 10, 16, 22, 28.
Leo's 5th term is 28. ***[1 mark]***
[3 marks available in total — as above]

b) First find an expression for the nth term of Leo's sequence:
The sequence is "add 6" so the expression must contain $6n$
The first term is 4 so Leo's sequence is $6n - 2$ ***[1 mark]***
So the sum of the nth and $(n + 1)$th terms is:
$(6n - 2) + (6(n + 1) - 2)$ ***[1 mark]***
$= 12n + 2$ or $2(6n + 1)$ ***[1 mark]***
[3 marks available in total — as above]

7 a)

	Gradient equal to 2	Gradient not equal to 2
Passes though the point (2, 1)	$y - 2x = -3$	$2y = x$
Does not pass though the point (2, 1)	$y = 2x + 1$	$y = 2$ $3y = 2x + 2$

[2 marks available — 2 marks if all equations are in the correct places, otherwise 1 mark if 2 or 3 equations are in the correct place]

b) $2y = x$ ***[1 mark]***

8 Box A contains c chocolates.
Box B contains $c + 4$ chocolates.
Box C contains $2(c + 4) = 2c + 8$ chocolates.
Box D contains $c + 4 + 44 = c + 48$ chocolates. ***[1 mark]***
Total number of chocolates in Boxes A, B and C
$= c + c + 4 + 2c + 8 = 4c + 12$ ***[1 mark]***
This gives the equation:
$4c + 12 = c + 48$ ***[1 mark]***
$3c + 12 = 48$
$3c = 36$ ***[1 mark]***
So $c = 12$. The number of chocolates in Box A is 12. ***[1 mark]***
[5 marks available in total — as above]
An equivalent equation could be formed based on the total number of chocolates in Boxes A and C being 44 (i.e. 3c + 8 = 44).

9 a) Let I be the point directly below G on the line EH.
$IH = 15.5 - 10 = 5.5$ cm
$GH^2 = GI^2 + IH^2$ ***[1 mark]*** $= 4.8^2 + 5.5^2 = 53.29$ ***[1 mark]***
$GH = \sqrt{53.29} = 7.3$ cm ***[1 mark]***
[3 marks available in total — as above]

b) Call the missing angle x. Then $\tan x = \frac{4.8}{5.5} = 0.872...$ ***[1 mark]***
$x = \tan^{-1} 0.872...$ ***[1 mark]*** $= 41.112... = 41.1°$ (1 d.p.) ***[1 mark]***
[3 marks available in total — as above]
You know all three sides of the triangle, so you could have used sin or cos instead of tan and got the same answer — just make sure you put the right numbers in.

10 From 8:30 am to 11:30 am is 3 hours ***[1 mark]***
So Taylor's speed is $180 \div 3 = 60$ km/h ***[1 mark]***
Sasha's speed is $60 + 12 = 72$ km/h ***[1 mark]***
So it takes Sasha $180 \div 72 = 2.5$ hours ***[1 mark]***
2 hours 30 minutes after 9.00 am is 11.30 am
so Sasha arrives at the same time as Taylor. ***[1 mark]***
[5 marks available in total — as above]

11 Percentage of mortgage customers that are satisfied:
$\frac{153}{180} \times 100 = 85\%$ ***[1 mark]***
Percentage of savings customers that are satisfied:
$\frac{252}{320} \times 100 = 78.75\%$ ***[1 mark]***
So a greater percentage of mortgage customers are satisfied than savings customers. ***[1 mark]***
[3 marks available in total — as above]
Alternatively, you could compare decimals or consider the proportions of each type of customer that are not satisfied.

12 a) $a^0 = 1$ ***[1 mark]***

b) $3a^2b \times 2a^3b^2 = (2 \times 3)a^{2+3}b^{1+2} = 6a^5b^3$
[2 marks available — 2 marks for the correct answer, otherwise 1 mark if 2 out of 6, a^5 and b^3 are correct]

c) $(3a^3)^2 = 9a^6$ ***[1 mark]***
$\frac{9a^3b^{\frac{1}{2}}}{4b^{\frac{3}{2}}a^2} = \frac{9}{4} \times \frac{a^3}{a^2} \times \frac{b^{\frac{1}{2}}}{b^{\frac{3}{2}}} = \frac{9}{4} \times a \times b^{-1} = \frac{9a}{4b}$ ***[1 mark]***
so $(3a^3)^2 \div \frac{9a^3b^{\frac{1}{2}}}{4b^{\frac{3}{2}}a^2} = 9a^6 \div \frac{9a}{4b} = 9a^6 \times \frac{4b}{9a} = 4a^5b$ ***[1 mark]***
[3 marks available in total — as above]

13 For 840 m², amount of paint needed is $\frac{840}{30} \times 2 = 56$ litres
$56 \times \frac{3}{10} = 16.8$ litres of yellow (Kareem has this)
$56 \times \frac{5}{10} = 28$ litres of blue (Kareem only has 27.5 litres)
$56 \times \frac{2}{10} = 11.2$ litres of white (Kareem has this)
Kareem does not have enough paint to finish the hall as they do not have enough blue paint.
[4 marks available — 1 mark for dividing 840 by 30, 1 mark for 56 litres, 1 mark for correct method of scaling the mixture to find the amount of any colour of paint needed, 1 mark for finding 28 litres of blue are needed so they don't have enough]

14 a)

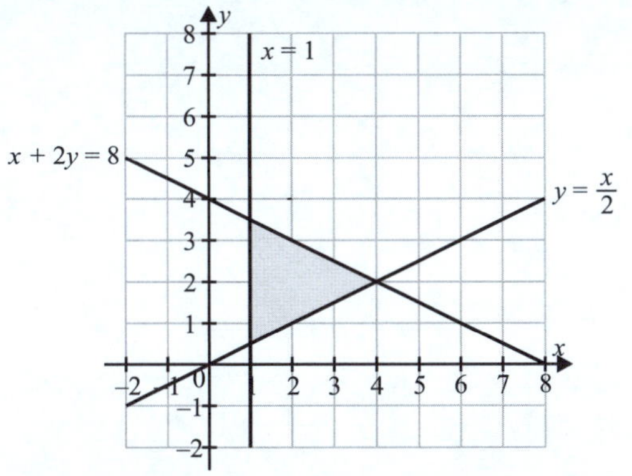

[4 marks available — 1 mark for drawing each line correctly, 1 mark for shading the correct region]

b) There are 8 pairs of integer coordinates within the region, so there are 8 pairs of integer solutions.
[2 marks available — 1 mark for an answer of 8, 1 mark for a correct explanation or for listing the coordinate pairs]
The integer solutions are (1, 1), (1, 2), (1, 3), (2, 1), (2, 2), (2, 3), (3, 2) and (4, 2).

15 a) Angle AEB = angle BDC = 90°.
Angle CBD + angle ABE = 90°
Angle EAB + angle ABE = 90°
So angle CBD = angle EAB.
Two angles in the triangles match up, so the third angles must also match up, so triangles AEB and BDC are similar.
[2 marks available — 1 mark for showing that 2 angles in the triangles match up, 1 mark for a correct explanation]

b) $AE = 12$ and $BD = 18$
$AE : BD = 12 : 18 = 2 : 3$
[2 marks available — 1 mark for finding the lengths of the two sides, 1 mark for the correct answer]

c) $DC = 27 - 12 = 15$ ***[1 mark]***
From b) you know that corresponding sides of the two triangles are in the ratio 2 : 3.
$2 : 3 = BE : CD = BE : 15$ so $BE = 10$ ***[1 mark]***
The coordinates of E are (12, 18 – 10) = (12, 8). ***[1 mark]***
[3 marks available in total — as above]

16 a) $p \propto \frac{1}{d^3}$ so $p = \frac{k}{d^3}$ ***[1 mark]***. When $p = 60$, $d = 2$, so $60 = \frac{k}{2^3}$.
$k = 60 \times 2^3 = 480$ ***[1 mark]***, so $p = \frac{480}{d^3}$ ***[1 mark]***
[3 marks available in total — as above]

b) $p = 100$, so $100 = \frac{480}{d^3}$. $d^3 = \frac{480}{100} = 4.8$ ***[1 mark]***
So $d = \sqrt[3]{4.8} = 1.6868... = 1.69$ cm (2 d.p.) ***[1 mark]***
[2 marks available in total — as above]

17 a) The class widths are not all the same, so the manager should have plotted the frequency densities on the vertical axis.
[1 mark for an explanation implying that the class widths are not all equal or mentioning that frequency densities should have been used]

b) Calculate frequency densities using frequency ÷ class width

Age, A years	Frequency	Frequency Density
$18 < A \le 20$	18	$18 \div 2 = 9$
$20 < A \le 25$	35	$35 \div 5 = 7$
$25 < A \le 30$	40	$40 \div 5 = 8$
$30 < A \le 40$	45	$45 \div 10 = 4.5$
$40 < A \le 60$	50	$50 \div 20 = 2.5$
$60 < A \le 70$	75	$75 \div 10 = 7.5$
$70 < A \le 90$	40	$40 \div 20 = 2$

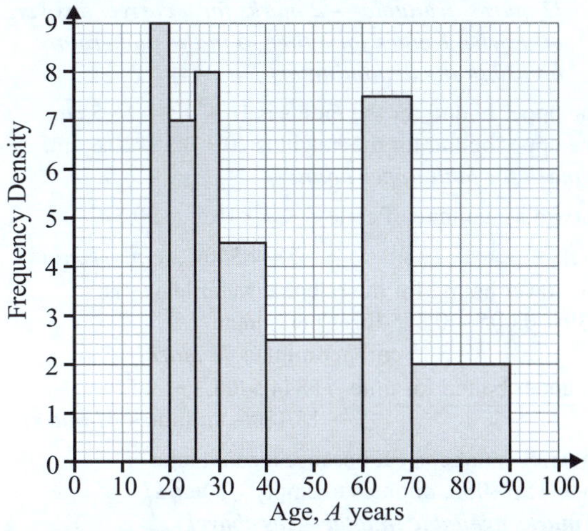

[3 marks available — 1 mark for finding all frequency densities correctly, 1 mark for bars correctly drawn, 1 mark for vertical axis scaled and labelled]

c) The ages of the members have a large range. There are a lot of people with ages much lower than the mean and a lot of people with ages much higher than the mean, so the mean is not a very typical age of the members.
[1 mark for a comment like these or that refers to the mean age being in a group which has a low frequency density]

18

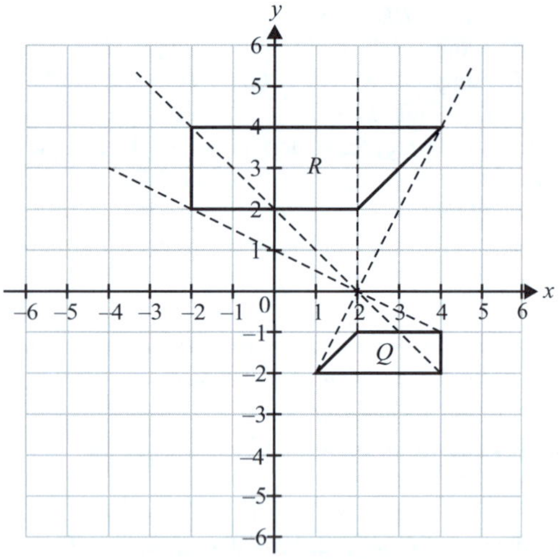

[3 marks available — 3 marks if R is in the correct position as shown above, otherwise 2 marks if 2 vertices of R are found correctly or 1 mark if an image of the correct size is drawn but positioned elsewhere on the grid]

19 a)

	Made in Europe	Made outside Europe	Total
Antique	75	50	125
Not antique	45	30	75
Total	120	80	200

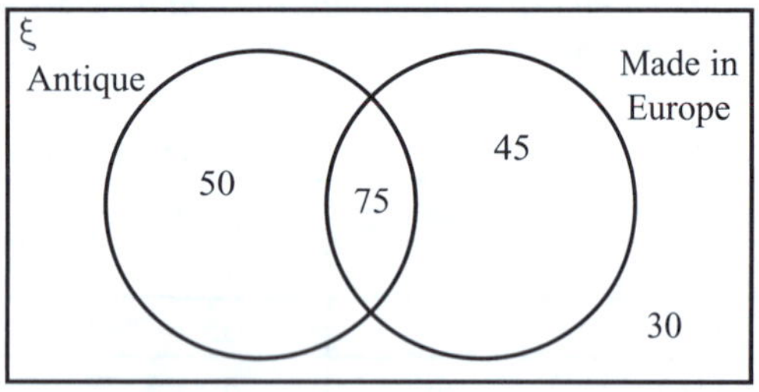

[3 marks available — 1 mark for putting both 30 and 75 correctly into the table, 1 mark for all remaining entries in table found correctly, and 1 mark for completing the Venn diagram using 45 and 50]

b) Probability = $\frac{75}{120} = \frac{5}{8}$

[2 marks available — 2 marks for a correct answer, otherwise 1 mark for either 75 in the numerator or 120 in the denominator]

20 The upper bound for the radius is 16 + 0.5 = 16.5 cm.
The upper bound for the height is 20 + 0.5 = 20.5 cm.
[1 mark for both upper bounds]

Maximum volume of cone = $\frac{1}{3}\pi \times 16.5^2 \times 20.5$
= 5844.5404... cm^3 ***[1 mark]***

The lower bound for the rate that water leaves is:
0.39 – 0.005 = 0.385 litres per minute
= 385 cm^3 per minute ***[1 mark]***

So upper bound for time = 5844.5404... ÷ 385
= 15.1806... minutes ***[1 mark]***

So Ariel is not correct because it could take up to 15.1806... minutes to empty. ***[1 mark]***
[5 marks available in total — as above]

21 Distance travelled in first 10 seconds = 0.5 × 10 × 16
= 80 m ***[1 mark]***

Distance between 10 and 60 seconds = 16 × (60 – 10)
= 800 m ***[1 mark]***

Distance travelled in final 60 seconds = 0.5 × (16 + 12) × 60
= 840 m ***[1 mark]***

Total distance = 80 + 800 + 840 = 1720 m ***[1 mark]***
[4 marks available in total — as above]

22 Let angle $DAB = x$.
Start by working out the size of angle x using the cosine rule.

$\cos x = \frac{AB^2 + AD^2 - BD^2}{2 \times AB \times AD}$

$\cos x = \frac{9^2 + 5^2 - 7^2}{2 \times 5 \times 9}$ ***[1 mark]*** = 0.633...

$x = \cos^{-1}(0.633...) = 50.703...°$ ***[1 mark]***

BC can then be found using the cosine rule in triangle ABC.

$BC^2 = 9^2 + 9^2 - (2 \times 9 \times 9 \times \cos 50.703...°)$ ***[1 mark]***
$= 81 + 81 - 162 \cos 50.703...°$
$= 59.4$

$BC = \sqrt{59.4} = 7.707... = 7.7$ cm (1 d.p.) ***[1 mark]***
[4 marks available in total — as above]

There are alternative methods you could use to get the correct answer — as long as you show your working and get the right answer you'll get full marks.

23 Let Emma's number be n, then Felix's number is $n + 1$

Form the equation: $\frac{1}{n} + \frac{1}{n+1} = \frac{5}{6}$ ***[1 mark]***

Therefore, $\frac{n+1}{n(n+1)} + \frac{n}{n(n+1)} = \frac{5}{6}$

$\frac{2n+1}{n(n+1)} = \frac{5}{6}$

$6(2n + 1) = 5n(n + 1)$ ***[1 mark]***

$12n + 6 = 5n^2 + 5n$

$5n^2 - 7n - 6 = 0$ ***[1 mark]***

$(5n + 3)(n - 2) = 0$ ***[1 mark]***

So $n = -\frac{3}{5}$ or $n = 2$

Emma's number is negative, so it was $-\frac{3}{5}$. ***[1 mark]***

[5 marks available in total — as above]

The quadratic formula could be used to solve the equation.

Index

Formulas in the Exams

GCSE Maths uses a lot of formulas — that's no lie. You'll be scuppered if you start trying to answer a question without the proper formula to start you off. Thankfully, CGP is here to explain all things formula-related.

These formulas will be given to you in the exam on a **formula sheet:**

Volume of Prism:

area of cross-section × length

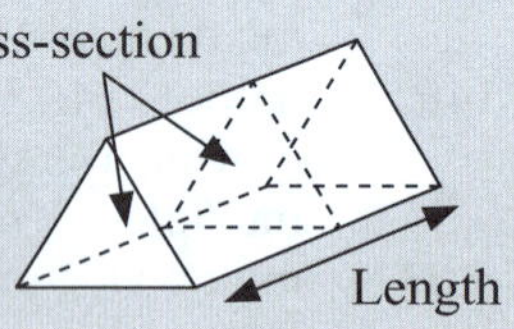

Circles:

Circumference = $2\pi r$

Area = πr^2

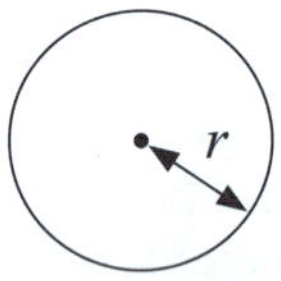

Trigonometric ratios:

$\sin x = \frac{a}{h}$

$\cos x = \frac{b}{h}$

$\tan x = \frac{a}{b}$

where h is the hypotenuse.

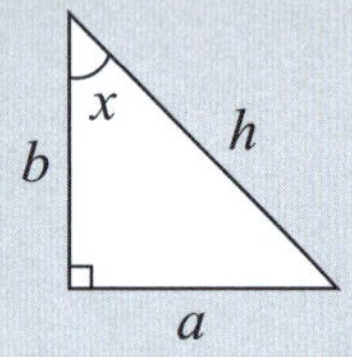

Pythagoras' theorem:

$a^2 + b^2 = h^2$

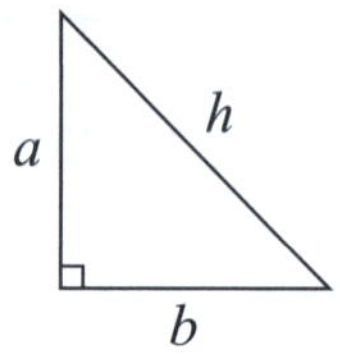

Area of trapezium:

$\frac{1}{2}(a + b)h$

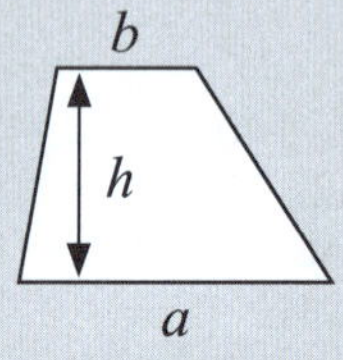

Compound Interest:

The total accrued, P_n, after n periods is:

$$P_n = P_0 \times \left(1 + \frac{r}{100}\right)^n$$

where P_0 is the initial amount and r is the percentage rate of change.

Probability:

P(A) and P(B) are the probabilities of events A and B respectively.

P(A or B) = P(A) + P(B) – P(A and B)

P(A and B) = P(A) × P(B given A)

The Quadratic Formula:

The solutions of $ax^2 + bx + c = 0$, where $a \neq 0$, are:

$$x = \frac{-b \pm \sqrt{b^2 - 4ac}}{2a}$$

For any triangle ABC:

Area of triangle = $\frac{1}{2}ab \sin C$

Sine rule:

$$\frac{a}{\sin A} = \frac{b}{\sin B} = \frac{c}{\sin C}$$

Cosine rule:

$$a^2 = b^2 + c^2 - 2bc \cos A$$

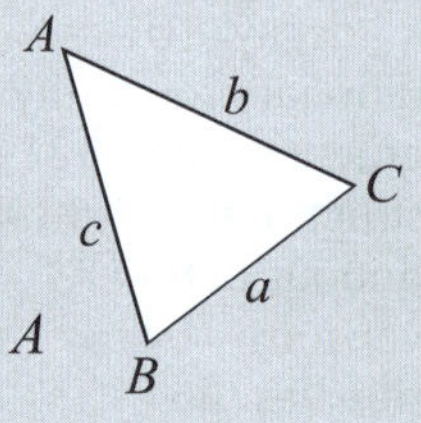

Some formulas will only be given to you if you need them for a **specific exam question:**

Spheres:

For a sphere with radius r:

Volume = $\frac{4}{3}\pi r^3$ Surface area = $4\pi r^2$

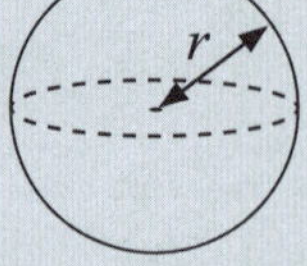

Pressure:

Pressure = $\frac{\text{Force}}{\text{Area}}$

Cones:

For a cone with base radius r, slant height l and perpendicular height h:

Volume = $\frac{1}{3}\pi r^2 h$ Curved surface area = πrl

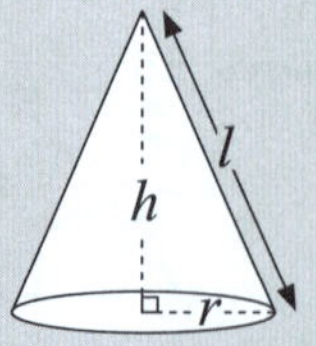

It's worth getting familiar with these last formulas as they may **not** be given to you in the exam:

Compound Measures: Speed = $\frac{\text{Distance}}{\text{Time}}$ Density = $\frac{\text{Mass}}{\text{Volume}}$

MCHS41